CAMBRIDGE LIBRARY COLLECTION

Books of enduring scholarly value

Technology

The focus of this series is engineering, broadly construed. It covers technological innovation from a range of periods and cultures, but centres on the technological achievements of the industrial era in the West, particularly in the nineteenth century, as understood by their contemporaries. Infrastructure is one major focus, covering the building of railways and canals, bridges and tunnels, land drainage, the laying of submarine cables, and the construction of docks and lighthouses. Other key topics include developments in industrial and manufacturing fields such as mining technology, the production of iron and steel, the use of steam power, and chemical processes such as photography and textile dyes.

History of the Manchester Ship Canal from its Inception to its Completion

By the late nineteenth century, charges imposed on Manchester companies for the use of Liverpool's docks and the connecting railway had created an atmosphere of resentment within the business community. The Manchester Ship Canal was to play a major part in the city's regeneration following the depression of the 1870s, but it took a lengthy battle for the scheme to gain the backing of Parliament and for construction to begin in 1887. In this two-volume work of 1907, Sir Bosdin Leech (1836–1912) traces the canal's conception, planning and construction. Volume 1 discusses the historical and economic factors that led to the creation of the waterway, as well as the bitter political fight to make it a reality. The work includes a large amount of illustrative content, enhancing the light shed on the landscape and notable personalities of Manchester at that time.

Cambridge University Press has long been a pioneer in the reissuing of out-of-print titles from its own backlist, producing digital reprints of books that are still sought after by scholars and students but could not be reprinted economically using traditional technology. The Cambridge Library Collection extends this activity to a wider range of books which are still of importance to researchers and professionals, either for the source material they contain, or as landmarks in the history of their academic discipline.

Drawing from the world-renowned collections in the Cambridge University Library and other partner libraries, and guided by the advice of experts in each subject area, Cambridge University Press is using state-of-the-art scanning machines in its own Printing House to capture the content of each book selected for inclusion. The files are processed to give a consistently clear, crisp image, and the books finished to the high quality standard for which the Press is recognised around the world. The latest print-on-demand technology ensures that the books will remain available indefinitely, and that orders for single or multiple copies can quickly be supplied.

The Cambridge Library Collection brings back to life books of enduring scholarly value (including out-of-copyright works originally issued by other publishers) across a wide range of disciplines in the humanities and social sciences and in science and technology.

History of the Manchester Ship Canal

from its Inception to its Completion

With Personal Reminiscences

VOLUME 1

BOSDIN LEECH

CAMBRIDGE
UNIVERSITY PRESS

CAMBRIDGE
UNIVERSITY PRESS

University Printing House, Cambridge, CB2 8BS, United Kingdom

Cambridge University Press is part of the University of Cambridge.

It furthers the University's mission by disseminating knowledge in the pursuit of
education, learning and research at the highest international levels of excellence.

www.cambridge.org
Information on this title: www.cambridge.org/9781108071192

© in this compilation Cambridge University Press 2014

This edition first published 1907
This digitally printed version 2014

ISBN 978-1-108-07119-2 Paperback

HISTORY OF

THE MANCHESTER SHIP CANAL

" Floreat Semper Mancunium "

DANIEL ADAMSON, FIRST CHAIRMAN OF THE MANCHESTER SHIP
CANAL COMPANY.

Elliott & Fry.   *Frontispiece.*

HISTORY

OF THE

MANCHESTER SHIP CANAL

FROM ITS INCEPTION TO ITS COMPLETION

WITH PERSONAL REMINISCENCES

BY

Sir BOSDIN LEECH

NUMEROUS PLANS, PORTRAITS AND ILLUSTRATIONS

IN TWO VOLUMES

VOL I.

MANCHESTER AND LONDON:

SHERRATT & HUGHES

1907

THE ABERDEEN UNIVERSITY PRESS LIMITED
THE ABERDEEN UNIVERSITY PRESS LIMITED

THESE VOLUMES ARE DEDICATED

TO THE

LORD MAYOR AND CORPORATION OF THE

CITY OF MANCHESTER

IN COMMEMORATION OF THE PUBLIC SPIRIT DISPLAYED BY THAT CITY IN COMING TO

THE ASSISTANCE OF THE MANCHESTER SHIP CANAL AT A CRITICAL

STATE OF ITS AFFAIRS, AND IN THE HOPE THAT THEIR EXAMPLE MAY STIMULATE

FUTURE GENERATIONS TO SIMILAR LOCAL PATRIOTISM

PREFACE.

THE early struggles and ultimate triumph of the Manchester Ship Canal constitute a subject of absorbing interest. In the history of Manchester, and indeed of South Lancashire as a whole, no other event or enterprise can compare with it in its far-reaching effects. The story, too, in many respects contains all the elements of a romance. It is the relation of a desperate and almost hopeless fight against opposition of the most powerful and uncompromising character, and it is meet that the names and qualities of the men engaged in the strife, and the nature of the difficulties which they encountered and overcame, should find a permanent record. To rescue both individuals and incidents from oblivion, and to give a connected narrative of the course of events from the conception to the completion of the canal, is the object of the present work. The task has not been an easy one, nor has it been lightly undertaken. If, however, a personal association with the enterprise from the outset, a close and intimate knowledge of all its engineering and constructive phases, and the possession of a unique collection of materials, correspondence, reports and records of every kind, may be regarded as qualifications necessary, if not even indispensable, I can at least lay claim to these.

My first chapters are devoted to the early history of water communication, both at home and abroad, and to the rise and development of the port of Liverpool. The conditions which called the Manchester Ship Canal into existence are next dealt with, and it will be seen that the actuating motive was in no sense one of hostility to Liverpool or to other existing interests, but was the mere instinct of self-preservation. The trade of Manchester, and in fact of the whole district of which Manchester is the centre, was being strangled by the heavy charges of various kinds imposed by Liverpool, aggravated as they were by the high rates of railway carriage between the two cities. As a remedy for a state of affairs beyond endurance, the Ship Canal has been a signal success. The average freight charges to Manchester on imported goods are now about one-third what they were twenty-five years ago. The pity is that some portion of the great saving thus effected has not gone into the pockets of the non-trading canal shareholders, by whose exertions and sacrifices these striking economies have been brought about.

It is interesting to note that the conception of a navigable waterway between Manchester and Liverpool dates back for at least two centuries, and I have given a brief account, accompanied by maps and plans, of the various schemes propounded since 1712. With reference to the Ship Canal itself, plans are inserted, showing the changes and modifications, both in the course of the waterway and in the position of the docks, which were made from time to time in the original design.

Two chapters in each of the years 1883-4-5 are devoted entirely to the Parliamentary fight. These necessarily give a repetition of evidence, and may be passed over by the casual reader ; they, however, contain extracts of speeches by eminent counsel, and are full of interest to any one wishful to become master of the questions in dispute.

My intention, at the outset, was to bring the story of the Ship Canal to the present time. My labours thus far, however, have been sufficiently heavy, and I content myself by expressing the hope that some abler hand will chronicle the subsequent developments and successes of the undertaking. Manchester people recognise with regret and even humiliation that those who have benefited most by the canal have helped it least. It is within the power of the manufacturing and trading interests of the district to place the enterprise within a very short space of time upon a paying level. A determined effort ought to be made to break down the pernicious rebate system of the shipping rings, and new lines should be established in Manchester itself if shipowners persist in boycotting the canal. If the above-named interests will rise to the occasion and be animated by the lofty patriotism displayed by the humble shareholders, they have it in their own power to make the canal a commercial success of which Manchester would indeed be proud.

Original shareholders, who have not been fortunate enough to recoup themselves by indirect advantages, have the satisfaction of knowing that the noble self-sacrifices they have made have not been thrown away, but will some day be acknowledged with gratitude by those who come after them.

It must be understood that this work has no pretentions to literary merit, but has been written with the sole object that future generations may be aware of the great struggle their forefathers had in order to convert Manchester into a port, and thus add to her commercial prosperity.

BOSDIN T. LEECH.

MANCHESTER, *25th December*, 1906.

MEMORANDUM.

I DESIRE to acknowledge my indebtedness to the many friends who have rendered me valuable assistance whilst compiling this history, especially to Sir Leader Williams and the Ship Canal officials, also to Mr. Marshall Stevens for his statistical information.

My gratitude is also due to the gentlemen who have assisted me in revising facts and figures, especially to Councillor Plummer, who has been most kind in rendering valuable counsel and advice. Also to Earl Egerton and to Messrs. Thos. Agnew & Sons for their ready consent to the reproduction of Mr. B. W. Leader's " Mount Manisty," and to Mr. Bythell and my co-directors for the help they have given me.

I wish particularly to thank Mr. Thos. Birtles, of Warrington, and Messrs. H. Birch-Killon and James Barningham, of Manchester, for the use of their excellent photographs of the Ship Canal works, taken during construction, and also the following firms who have permitted me to use their photographs and engravings, *viz.:* Messrs. Elliott & Fry, Baker Street, London ; Messrs. Thomson, New Bond Street, London ; Messrs. Valentine & Sons, Dundee ; Messrs. Lewis's, Ranelagh Street, Liverpool ; Messrs. Bradbury, Agnew & Co., *Punch ;* Messrs. George Newnes, Limited, *Tit-Bits ;* Messrs. Lafayette, Limited, Manchester ; Messrs. Wilkinson Brothers, Manchester ; Messrs. Brown, Barnes, & Bell, Manchester ; Mr. Franz Baum, Manchester ; Mr. Warwick Brooks, Manchester ; Mr. J. Ambler, Manchester ; Mr. W. P. Gray, Cable Street, Liverpool ; Mr. A. Coupe, Withington ; Mr. J. White, Dumfries, and Messrs. Macmillan & Co., Ltd., London.

My thanks are also due to Messrs. Chorlton & Knowles and Messrs. Wilson & Hudson, of Manchester, for their care and attention in reproducing the various plans and photographs.

B. T. L.

CONTENTS.

CONTENTS

LIST OF ILLUSTRATIONS.

CARTOONS, AUTOGRAPH LETTERS, ETC.

PLANS.

PLANS IN POCKET.

PORTRAITS.

LIST OF ILLUSTRATIONS XV

PHOTOGRAPHS.

CHAPTER I.

NAVIGABLE RIVERS OF ENGLAND—THE MERSEY AND IRWELL—THE MERSEY BAR.

There be three things which make a nation great and prosperous, a fertile soil, busy workshops, and easy conveyance for men and commodities from one place to another.— Francis Bacon, Lord St. Albans.

OUR townsman, Dr. Joule, in a communication to the Manchester Literary and Philosophical Society, on the question of navigation, suggested the idea that the monarchs of the deep were the first models for shipbuilders. Propulsion by the side fins was copied in the paddle-wheel, and by the tail fin in the screw-propeller. The closer we follow Nature in its proportions, the more certain are we to obtain speed and safety. He instanced the porpoise, with its bluff figurehead, attaining a velocity of over thirteen miles an hour, whilst voracious fishes are so constructed that they can obtain a much greater velocity. He advocated a study of natural proportions to those who wish to be successful shipbuilders.

Ever since the days of Noah (the first known boat-builder) waterways have been valuable agencies, not only of civilising the world, but of providing for the wants of its people. Prior to the seventeenth century, pack-horses and navigable rivers were the chief means of carrying on the trade of the country. Till then little had been done by artificial means to widen and deepen rivers and make them navigable. Weirs had been placed here and there, but mainly for the purpose of obtaining water-power for grinding corn.

The Thames and the Severn, fed as they are by huge watersheds, have from the earliest times been easy of navigation, and have required but a small expenditure to keep them fit for traffic. The Dee and the Lune on the West Coast, once maritime thoroughfares, are samples of neglected rivers that have become to a large extent silted up. The first Act for the improvement of the Ribble was that of 1806, in which it is recited: "Whereas the Port of the town of Preston is of great antiquity, and whereas the Ribble, from the uncertainty and changeable course of the Channel and the beds of gravel, sand, and other obstructions, is become very difficult and dangerous for ships," etc.

Amongst the projectors were Sir H. P. Hoghton, Sir T. D. Hesketh, Lawrence Rawsthorne, Esq., and others. The Ribble navigation, however, became worse until a few years ago, when the Preston Corporation in a plucky way took it in hand. After making commodious docks, they found dealing with the Ribble estuary a serious trouble, but the cost, together with the fact that Preston possesses no markets of its own, and only supplies a limited area, makes it doubtful if the waterway will be anything more than a moderate success.

The Clyde, the Tyne and the Tees are examples of naturally bad navigations being changed by artificial means into good and useful rivers, a credit to the corporations who took them in hand, and a source of prosperity to the districts in which they are situated.

We next come to consider the history of the Mersey, with its tributary the Irwell. The earliest mention of the river Mersey is in a deed in the reign of Ethelred (A.D. 1094). The origin of the name is not easy to determine, but it seems only reasonable to conjecture that it has some connection with the name of the kingdom of Mercia, of which it formed the northern boundary. In old Acts of Parliament the name is generally spelt " Mercy," and it is sometimes referred to as " The Water of Mercy".

About two hundred years ago there was an awakening in Lancashire. Manchester had become celebrated for her productions, and she needed larger supplies of food as well as raw material for her manufactures. Liverpool, at the time, was fast becoming an important mercantile port, and the avenue by which foreign productions were introduced into England.

It is a question who first conceived the idea of improving the communication between these rising towns, or who introduced canals. Baines in his *History of Lancashire* writes: " In this county the canal system of modern times originated with the Sankey Canal".

That the need for improved waterways was being felt is evident from a curious letter sent by Mr. Thomas Patten, of Bank Hall, Warrington, to Mr. Richard Norres, of Speke, dated "Ye 8th day of January, 1697. What a vast advantage it would be to Liverpool if the river were made navigable to Manchester and Stockport. Since I made it navigable to Warrington, there have been sent to Liverpool 2,000 tons of goods a year, and I believe as much by land, which, if the river were cleared of the weirs (fish weirs), would all go by water, for the river to Manchester is very capable of being made navigable at a very small charge." From this we may infer that this ancestor of the late Lord Winmarleigh had shown his local patriotism by improving the river to Warrington.

In 1714 a number of gentlemen, called in those days "the undertakers," formed a company and applied for an Act " For making the rivers Mersey and Irwell navigable from Liverpool to Manchester". They defined their object to be " Keeping the rivers Mersey and Irwell in the counties Palatine of Lancaster and Chester navigable and passable for boats, barges, lighters and other vessels from Liverpool to Hunt's Bank, in Manchester, which will be very beneficial to trade, advantageous to the poor, and convenient for the carriage of coals, cannel, stone, timber and other goods, wares and merchandise, to and from the towns and parts adjacent, and will very much tend to the employing and increase of watermen and seamen, and be a means to preserve the Highways". Among the undertakers are found many names perpetuated in the nomenclature of Manchester streets—Oswald Mosley of "Ancotes," Joseph Byrom, James Marsden, James Bradshaw, Thomas Garside, and an ancestor of my own, John Leech. Another of the family, Edward Leech, was one of the commissioners nominated in the Act to arbitrate disputes.

A special clause was inserted to prevent injury being done "to the working, going or grinding of any of the corn mills which now are, or may be erected or built, at which the inhabitants of Manchester are bound by any custom within the said town to grind any of their corn, malt or other grain". The maximum toll to be authorised was " 3s. 4d. for the carriage between Bank Quay, Warrington, and Manchester, of every ton of coal, cannel, stone, slate, timber or other goods, wares, merchandises and commodities".

The power for haulage was to be supplied "by winches and other engines, in convenient places, and by and with the strength of men, horses and beasts". The jurisdiction of the company extended from Liverpool to Hunt's Bank, Manchester; the distance between the termini was fifty-seven miles, and this was reduced to fifty miles by cuts. Inasmuch as the river had always been navigable up to Bank Quay, Warrington, no charge was to be made for the use of the river up to that point. Owners of land within five miles of the navigation could carry manures, marl, etc., free from toll.

Manchester stands at an elevation of 93 feet 3 inches above the sea-level. The vertical range of the tides at the Port of Liverpool, on the average of the springs (exclusive of the equinoctial tides), is 27 feet 6 inches, the level of low water 8 feet 8 inches below the datum level of the old dock sill.   The rise of the river from the west side of the town may be taken as nearly 70 feet.   The river is tidal up to Woolston, about fourteen and a half miles from Manchester.

An Act was passed in 1719 to make the river Douglas navigable from the river Ribble to a place called Miry Lane End, Wigan ; the undertakers being "Wm. Squire and Thomas Steeres, gentlemen, both of Liverpool," and the conditions being very similar to those imposed on the undertakers of the Mersey and Irwell navigation.   Though the undertakers first applied in 1714 for their Act to make the Mersey navigable, it was 1720 before they got clear away from Parliament. In the same year the first Liverpool Dock Act was obtained.

In 1766, about the time the Bridgewater Canal was completed, there were fourteen flats trading on the river from Liverpool to the "Old Key" in Manchester, and six flats to the "Salford Key".   It should be stated that for the first fifty years the Mersey and Irwell navigation did not pay.

The river navigation was offered in 1776 to the Duke of Bridgewater for £5,000 which he declined, and in 1779 it was sold to a new proprietary for £10,000.   They got an Act in 1794 empowering them to improve the river, but, instead of doing this, they applied their money in paying large dividends. Under this Act the company who had the river Mersey in hand was incorporated under the title of the Company of the Proprietors of the Mersey and Irwell Navigation.   There were then thirty-nine undertakers, of whom the Duke of Bridgewater was one.   They were restrained from building warehouses within a mile of Bank Quay, Warrington, or from demanding toll between Liverpool and Bank Quay— the river being already navigable to that point.

Though at first prosperous, the Mersey and Irwell Navigation Company had a hard struggle for existence.   To protect their trade they had to purchase the plant of many of the by-carriers (who were cutting one another's throats by ruinous competition) and to become carriers as well as toll takers, and as such assumed the name of the "Old Quay Company".   Opposition on their own waterway had been started by the New Quay Company, and it had to be bought off. But the chief difficulty was the Bridgewater Canal, which as a competitor had great advantages.   Its maximum toll was 2s. 6d. per ton ; it was linked at the Manchester end with the Rochdale, Leeds and Liverpool Canals, and the Worsley Collieries, and at the Runcorn end with the Trent, Mersey and Weaver Canals.

The distance by canal to Runcorn was thirty miles, whilst, in consequence of a tortuous waterway, the distance by river was much greater. To remedy such drawbacks, several short bends of the river (as at Sandywarps, Stickens and Butchersfields) were cut off, and the ends joined by cuts and locks. By these means, and the new Runcorn and Latchford Canal, made in 1804 at a cost of £40,000, the distance from Manchester to Runcorn by river was reduced to twenty-eight miles. Boats were enabled to leave the river at Latchford and to re-enter it by the Runcorn Locks "by which means they entirely avoided the neap tides, and conveyed goods in two days from Liverpool regularly".

It is worth recording that the Runcorn and Latchford Canal was made under the powers of the Mersey and Irwell Act, and without plans being submitted. A superior court held that the owners had powers to do this. When eighty years later (in 1883) the Ship Canal promoters sought to do work without submitting plans, they were blocked and had to apply for the suspension of standing orders.

In 1827 the Corporation of Liverpool brought an action against the Mersey and Irwell Navigation Company and their workmen, which was tried at Lancaster. They maintained that, by taking water from Woolston in troughs to supply the Runcorn and Latchford Canal, the defendants were causing the main channel of the river to silt up. Though the plaintiffs' case was ably advocated by Mr. (afterwards Lord) Brougham, the verdict was given in favour of the defendants.

In 1807 packet boats for passengers were first run to Runcorn daily, and there was another connection with packets to Liverpool. In 1840 the through fare was first class, 3s.; second class, 2s.; and in 1850 it was reduced to first class, 1s. 3d.; second class, 10d. In 1829 the through journey (41 miles) was performed in seven hours.

In 1821 Mr. Sandars, a Liverpool merchant, published a pamphlet, and got a declaration drawn up and signed, which set forth that—

The present establishment for the transport of goods is quite inadequate, and that a new line of communication has become absolutely necessary to conduct the increasing trade of the County with speed, certainty and economy.

Manchester joined in this movement, and sent a deputation which waited on the Mersey and Irwell representatives and on Mr. Robert H. Bradshaw, on behalf of the Bridgewater Trustees, but without effect. Mr. Bradshaw declined to make any concessions, and ridiculed the idea of railway competition.

In 1830 the Mersey and Irwell Navigation Company purchased the Junction Canal, and afterwards made the Manchester and Salford Canal, to connect the

Rochdale Canal with the Irwell. This passed under the site of the present Central Station, and a portion of it is now blocked up. In 1827 Messrs. Rennie and Giles reported on the navigation of the Mersey, and Sir John Rennie made a further and very interesting report in 1838.

In 1810, after many years of strenuous opposition, the Mersey and Irwell Navigation formed a conference with their younger rivals, the Bridgewater Trustees, in order to put an end to the existing ruinous competition, and identical rates were published in the *Manchester Mercury* on 15th June. This arrangement held good with one exception (1842) till the purchase by Lord Ellesmere of the Mersey and Irwell Navigation.

Originally there were 500 Mersey and Irwell shares of the nominal value of £100 each. The company after 1794 prospered exceedingly; the shares became very valuable, at one time paying a dividend of £35 per annum on each £70 paid-up share, and before the advent of the Manchester and Liverpool Railway, the £70 share realised £1,250. Afterwards they depreciated to £800, at which price, in 1845, Lord Ellesmere bought them, taking upon himself also a bonded debt of £149,000 owing by the company. Altogether he paid £550,800.

When the Bridgewater Trustees got an Act of Parliament allowing them to buy, this purchase was transferred to them.

In view of the advent of railways, every effort was made to improve the navigation, and a costly work was carried out in 1840. By widening and deepening, the river was made navigable for vessels of 300 tons burden. A system, too, was instituted of placing goods in bond at the warehouses belonging to the company. The journey from Manchester to Warrington took at that time three hours. Under the novel title of "The Port of Manchester," the *Manchester Guardian* (in 1841) announced that a vessel had arrived in Manchester direct from Dublin, with a cargo of potatoes, and was discharging at the Old Quay Company's Wharf.

We have now to add (says the editor in a subsequent number of this paper) that this vessel, the *Mary*, Captain John Hill, having discharged her cargo, will, it is expected, "clear," and sail this day direct for Dublin, with a cargo of coals from the colliery of Mr. Fitzgerald at Pendleton. Such are probably the small beginnings of Manchester's future greatness as one of the Ports of the United Kingdom.

THE SHIP CANAL SONG.

THE *MARY*: A YARN.

Air: The Ram of Darby.

[The following song, purporting to be written by "Poor Jack," was published in 1840, on a small handbill, by Wilmot Henry Jones, of Market Street, Manchester, whose name is well known in literary circles owing to his having been the first printer of Philip James Bailey's poem *Festus*.]

The Union flag is flying,
  By the Company's wharf, Old Quay,
And *Mary* of Dublin lying
  Unloading her Murphies to-day.

In the Irish Sea I hail'd her,
  As I stood in the packet boat ;
With equal pride I never saw
  A merchant sloop afloat.

" Your name ? "   " The *Mary*, Captain Hill ! "
  " Your cargo ? "   " Pratees, sir ! "
" Where from ? "   " Dublin."   " Whither bound ? "
  " The Port of Manchester ! "

Eighteen hundred and forty,
  October the twentieth day,
At half-past four in the evening
  She anchored by the quay.

It always does my heart good
  To see the Union Jack,
So here's success to *Mary*,
  And soon may she come back.

And soon may scores of others
  Perform the trip with her,
And trade and commerce double
  In noble Manchester.

At one time the Irwell was largely used both for passengers and heavy goods traffic. Country people coming to Manchester travelled by swift packet, and the river was alive with boats, many of them carrying produce to, and manure from Manchester for the farmers on its banks. In 1882 the river was hopelessly choked with silt and filth. The following figures will show the result of neglect by the proprietors. Out of 311 working days the river was navigable for 50-ton boats—

In 1878 . . . 201 days.
In 1880 . . . 110 „
In 1881 . . . 50 „
In 1882 . . . 47 „
In 1883 at the rate of 32 „

In October, 1883, it was not navigable *for a single day*.   From the time of the purchase of the Mersey and Irwell Navigation in 1845 by the Bridgewater Trustees, its future history may be considered bound up in the annals of the purchasing company.   The Bridgewater Trustees worked both waterways with varying success, till they were taken over by Sir Edward Watkin and his friends in 1872, and formed into the "Bridgewater Navigation Company".   The history of the transfer, and the circumstances connected with it, will be found in a subsequent chapter.

In 1842 it was felt expedient that further provision should be made for preserving the navigation of the Mersey estuary, and an Act was passed appointing a Board of Mersey Commissioners for the conservancy of the river, to consist of the First Lord of the Admiralty, the Chancellor of the Duchy of Lancaster, and the Chief Commissioner of the Woods and Forests (since changed to the President of the Board of Trade); this Board was to have authority over the Mersey and Weaver from Warrington and Frodsham Bridges to the sea, with power to employ an Acting Conservator.

In 1876 Upper Mersey Commissioners were appointed who were to look after the lighting and buoying of the river below Warrington, and in 1879 further powers were granted.   Thus, when the Parliamentary fight of 1883-84 took place, there were, at least, three distinct authorities on the Mersey.   There can be no doubt the Mersey and Irwell Act, passed in the time of George I., gave powers which, to some extent, came into conflict with subsequent Acts.   The Ship Canal opponents argued they were obsolete, because they did not sanction the expenditure of money in the estuary of the Upper Mersey, and because it would be an outrage against the public to carry out the power of changing fixed into swing bridges.   On the other hand, the promoters maintained that all rights had been kept alive, and could be put in force by the owners of the navigation.   They were willing, however, to submit their proposed alterations to the Acting Conservator for his sanction.

At present several channels give entrance to the mouth of the Mersey, but there is a reasonable supposition that in time past the river had two mouths, the present one, and a second, passing through the Wallasey Pool, crossing the Wirral

To face page 8.

Calamanco Lock.   On the Old Irwell Navigation.

Killon.

Peninsula, and having its exit through the present Leasowe embankment. Probably the sea was not able to maintain the latter, for it silted up, and to prevent another invasion the Leasowe embankment was erected.

In 1693 Captain Granville Collins, Hydrographer to the Government, gave as an instruction :—

Keep close along Hyle Sand, and so into High Lake, and anchor. Here, the great ships that belong to Liverpool put out one part of their lading, till the ships are light enough to sail over the flats to Liverpool, where is 3 fathoms (18 feet) at low water on the Bar, but this place is not buoy'd or bekon'd, and so not known.

Ships lye aground before the town of Liverpool—'tis bad riding afloat before the town by reason of the strong tydes that run here; therefore ships that ride afloat ride up at the Sloyne where is less tyde.

This shows that so far back there were 18 feet on the bar; when Admiral Denham next surveyed in 1833 there were 12 feet at low water spring tides at the shallowest part in the new channel. The Queen's Channel did not then exist; it appeared for the first time the following year. In 1863 there were 10 to 13 feet on the Queen's Channel bar, but in 1873 there were only 7 to 8 feet. As the trade of Liverpool increased, and the ships entering the port became larger, the accommodation decreased, and large vessels had to wait outside the bar for a sufficiency of water. Whilst in 1693 there were 18 feet on the bar, in 1873 it had decreased to 7 or 8 feet. All this was very graphically stated by Mr. John Laird, M.P., when he addressed a meeting of the Dock Board in January, 1874. He then related the marvellous work that had been done on the Clyde and the Tyne, where shallow rivers had been dealt with, and good ports practically made. He reminded them that in spite of the protestations of Admiral Evans, the Acting Conservator, who for the last thirty years had been urging them to improve the channel, they had done nothing; on the contrary they had been tipping dock dredgings, filth, and mud into the fairway of the channel. This, in 1872, amounted to 600,000 tons. He was aware the Conservancy Board had power to deal with the question, but they had no money, and he suggested the lighting and buoying charges should be raised to provide some. After reproaching the Board for their inaction, he said :—

Every manufacturing interest in the Kingdom is deeply concerned in this matter, and I think I may safely say, without fear of contradiction, that they will not only justify but applaud the Board for taking steps, and spending any necessary money to place the port in an efficient condition.

Mr. Laird's resolution was modified to one instructing the Marine Committee to communicate with Admiral Evans, and to report if it was desirable that any, and if so, what measures should be adopted with the view of obtaining an increased depth of water on the bar.   Also to make inquiries and to report as to the rubbish being tipped into the Mersey.   This was a quiet way of shelving the matter for another ten years.   The depth of the bar slightly improved, and beyond a few indignant letters, things slept till the Ship Canal agitation brought the question again to the front.

After the river navigation in the upper reaches had been allowed to become blocked for some years, the price to be paid for it by arbitration could not have been a serious one; but when the Ship Canal Company wanted to come to terms for the Mersey and Irwell Navigation, the Bridgewater Company pleaded it would become an opponent to their canal, and if they sold one they must sell both.   Arrangements were therefore come to in the sessions 1883-84.   When in the latter year the Ship Canal was rejected, the Bridgewater Navigation changed their tactics, and bringing forward a scheme in 1885 to make the river applicable for barges, became strenuous opponents of the Ship Canal Company.   They proposed to make the waterway 10 feet deep, to straighten it, to shorten the distance by two miles, to reduce the number of locks from eleven to five, and to make them 154 feet long and 32 feet wide.   They would make the waterway suitable for vessels of 300 to 400 tons at a cost of £324,000.   After a prolonged and costly struggle, arrangements were come to, and £1,710,000 were given by the Ship Canal Company for both navigations.   Thereafter, the upper reaches of the Mersey became part of the Ship Canal.

CHAPTER II.

ENGLISH CANALS—HISTORY OF THE BRIDGEWATER CANAL.

The history of Francis Duke of Bridgewater is engraved in intaglio on the face of the country he helped to civilise and enrich.—Quarterly Review.

THERE are numerous canals and canalised rivers, communicating with the whole of England, that are linked with the Manchester Ship Canal, by the Bolton and Bury Canal, the Runcorn and Weaver Canal, the Rochdale and the Bridgewater Canals. Unfortunately, since the advent of railways many of these waterways have been allowed to fall into disuse and disrepair. England for a long time was indifferent to the welfare of her canals, and the railway companies were allowed to purchase certain sections, and by placing prohibitive tolls on them to demoralise the whole system. They now possess 1,139 miles out of a total length of 3,907 miles throughout the country. In consequence they make no improvements to assist traffic, and utterly neglect the deepening of waterways and the enlargement of locks necessary to keep pace with the increased size of boats needed for cheap carriage. The Board of Trade returns of canals and navigations for the year 1898 showed that while the traffic on the independent canals in Great Britain had increased 5,000,000 tons since 1888, the traffic on the railway-owned canals had decreased 2,000,000 tons in the same period. Whilst it is true that in some agricultural districts canals were made that have not justified their construction, it is equally true that water communication is and must remain the cheapest form of carriage for minerals and heavy goods where speed is not required. On the independent and up-to-date canals great changes are taking place : the narrow 20-ton boat drawn by one horse has given place to the steam tugs drawing four boats of at least 50 tons each, and the time does not seem far distant when the canals of this country will all be worked by electricity. If this should come to pass the Ship Canal could distribute to, and collect from the whole of England, for the canal system extends from Manchester to London, Bristol, Birmingham and all parts of the Kingdom.   It was given in evidence before the 1883 Committee that the average cost of English canals per mile was £3,350, whilst that of railways was £46,000, also that the cost of carriage on canals was one-third of that by railway.   A few brief facts about the most important canals may therefore be interesting.

The principal canals affecting Lancashire are :—

        1. The Grand Junction Canal.
        2. The Leeds and Liverpool Canal.
        3. The Rochdale Canal.
        4. The Ashton Canal and the Bolton and Bury Canal.
        5. The Bridgewater Canal.

THE GRAND JUNCTION CANAL.

The Grand Junction Canal, as its name implies, connects the northern and southern English canals, and has suffered through the London and North-Western Railway Company getting possession of a portion of the route.   The North Staffordshire Canal has also suffered from the same cause.   At the Parliamentary Commission in 1883 evidence was given that the railway commands $7\frac{1}{2}$ per cent. of the Grand Junction through route, and for this they demand and get 33 per cent. of the freightage earned.   The same company stopped the Warwick Canal eleven days out of a month for repairs, and in one case it was stopped for twenty-five days, the traffic meantime going to the railway.   Mr. Morton, before the same Commission, said : "The canal system was ruined by want of unity.   Between Wolverhampton and London, carriers had to pass over six separately owned canals, each with a costly staff of managers and engineers.   Manufacturers declare they are beaten by foreigners because of the cost of transit in England.   In his opinion the best way to obtain cheap and efficient carriage, and prevent combination and monopoly, was to have good waterways."   On the Midland canals are many long tunnels—the Sapperton 3,808 yards, the Lappal 3,795 yards, and the Dudley tunnel 3,172 yards long.   On these and many other tunnels the boats are "legged," *i.e.*, pushed through by men who lie on their backs and use their legs against the sides.   If an inhabitant of Japan were to see them, he would say our tunnels were very antiquated compared with theirs, and he would be quite correct.   As an example of the difference between a railway-owned and a free waterway, corn carried in 1883 from the sea to

Birmingham, *via* Sharpness, cost only half as much as from the sea to Manchester, although in the former case the distance was nearly double.

THE LEEDS AND LIVERPOOL CANAL.

This canal had the misfortune to get into railway hands who doubled the rates, and charged 16s. from Leeds to Liverpool, when by the quicker railway it was only 15s. The lease, however, expired in 1874, and though in the interval traffic was diverted and much damage done to the interests of the waterway, the canal revived when it came back into private hands, and has paid some good dividends, though like all carrying companies it is not now quite so prosperous.

THE ROCHDALE CANAL.

This canal is most intimately connected with the Ship Canal, inasmuch as it is linked to it by means of the Bridgewater Canal. It also supplies the Bridgewater undertaking with a large portion of the fresh water required to work it. A small boat of 50 tons can pass right acoss England from the Ship Canal, *via* the Rochdale Canal, and emerge *via* the Aire and Calder Navigation at Goole.

The first Bill for the Rochdale Canal was obtained in 1794. The navigation extends from its junction with the Bridgewater at Knott Mill to the Calder Navigation at Sowerby Bridge, and forms a link in the through communication between the North and Irish Seas. Power is given to the manufacturers on the banks to use the water free of charge for heating their mills and condensing steam, provided, after using it, the water is returned into the canal. For a great many years the manufacturers had an unstinted free use, but about 1844 some official (sharper than the rest) discovered there was no power to take water for raising steam. This gave rise to much litigation, because the water was useless for condensing unless it could be used for raising steam. Ultimately it was decided that those who had used the water for twenty-one years for the dual purpose might continue to do so without charge, but that new-comers must pay. Like all other canals the Rochdale felt the competition of railways, and they naturally harassed each other by cutting rates. To stop this, and for other purposes, the Lancashire and Yorkshire and other railway companies in 1854 took a lease of the Rochdale Canal, with the usual result that they equalised the cost of carriage, and by offering a more regular service and a quicker route, soon began to undermine the vitality of the canal by letting it fall into a state of disrepair. The result was that when the lease expired the canal was in a very impoverished condition. Since then there has been a revival, but it is a question if the canal will ever recover from the mistaken policy then adopted. Unfortunately, difficulties arose between the company and the Ship Canal, both as regards towage by steam on the canal and the use of water, but I propose to deal with these disputes later on.   Happily they have now been settled.

THE ASHTON AND BOLTON AND BURY CANALS.

These are both in railway hands, and it was given in evidence in 1883 as regards the latter, that manufacturers on its banks conveyed their goods by railway because, though it is well known that water is the cheapest possible means of carriage, it was practically denied to them, being made dearer than by railway.

THE BRIDGEWATER CANAL.

This canal has always a special interest for Lancashire men, and its history will bear repeating. It brought forth native talent to a remarkable extent, and whilst other canals were the work of company promoters and undertakers, the Bridgewater Canal was the conception of one man, who was prepared to face herculean difficulties and to spend his fortune in order that " he might found canals and navigations for the public good ". These remarkable words are to be found in the Duke of Bridgewater's will, proved at Doctors' Commons in 1803.

It would appear that Scroope, the first Duke of Bridgewater, and father of the great canal maker, obtained in 1737 an Act of Parliament to make the Worsley Brook navigable, with the intention of forming a waterway from the Worsley Collieries to the Irwell, but the scheme was never carried out. The Duke of Bridgewater was one of the Commissioners, the undertakers being Jonathan Lees, Samuel Beardman, Josiah Byrom, Jeremiah Bradshaw, Isaac Clegg and others.

How Francis, the second Duke, came to embark his money and devote all his energies to carrying out his great work is explained by the rumour that in his youth he enjoyed the pleasures of London Society and applied himself assiduously to win the hand of a fair lady (Miss Gunning), who, however, rejected his addresses. Stung by the rebuff he determined to leave the gaieties of the capital, retire to his country estate, and try to forget the disappointment by leading a country life and devoting his energies to the improvement of his property in Lancashire. Coal of good quality had been found there, and he wished to get it into the Manchester market ; therefore, carrying out his father's idea, in 1758, he applied for, and obtained an Act " to make and maintain a navigable river, cut or canal from any part of the croft or meadow known by the name of Master Cooke's Tenter Croft in the town-

Manchester Febᵗᵃʳʸ 22 1758

Madam

The inclosed Scheme having met with Incouragement from the Gentlemen and Traders in Manchester and Telford I take this Method of communicating it to you as a proprietor of Lands thro which the Cut is intended to be made, and if it shall receive your Approbation, and Concurrence it will much Oblige Your Most Obedient Humble Servant

Bridgewater

FACSIMILE OF ORIGINAL LETTER FROM THE DUKE OF BRIDGEWATER.

ship of Salford to or near Worsley Mill and Middlewood in the Manor of Worsley and from thence to or near Hollin Ferry". The latter place was close to the mouth of the Glaze and almost a dozen miles from Salford. This scheme was, however, given up as impracticable. Father and son up to this time had intended to reach Manchester by the Irwell, deeming it impossible to take the canal over the river. Whilst the Duke was modifying his plans he had introduced to him by his agent (Mr. Gilbert) a shrewd but illiterate countryman, James Brindley. He soon found out that his new acquaintance was a genius, and would be an able adviser. Brindley, after making what he called an "ochilor (ocular) servey or a ricconitoring," conceived the idea of making a bridge or aqueduct over the Irwell. His scheme was laughed at, and the Duke consulted an eminent engineer, who concluded his report with— "I have often heard of castles being built in the air, but never before saw where any of them were to be erected". Having matured his plans, the Duke of Bridgewater obtained his second Act on the 24th March, 1760. It recites that "whereas the said Duke of Bridgewater hath begun the said intended navigation and made a considerable part thereof," the canal so begun "may be continued and taken over the river Irwell at or near a certain bridge called Barton Bridge". The Duke himself was of a scientific turn, with a taste for engineering, but his previous idea had been to cross the river by a series of locks on either side. When before Parliament, Brindley puzzled both Counsel and Committee by his dialect and the local technical terms he used ; and when he talked of stopping water by "puddling" they were nonplussed. Next day he took with him some clay, which, by the use of water he converted into puddle, and moulded into coffer dams. Again, when Brindley wished the Committee to realise his intended bridge, he produced a model of it carved out of a cheese. These ocular demonstrations made the matter clear and convinced the Committee. The Honble. Francis Egerton, writing from Paris, says he had it from the old Duke's own mouth that when the late Duke made Barton Bridge, "Mr. Brindley was so timid during the filling that he ran away from it to Stretford, and never appeared again until the bridge had proved secure. Afterwards heavy rain came and Mr. Brindley feared it would fall. 'He did very ill,' for he weighted the sides ; but the bridge was saved by Mr. John Gilbert, who took just the contrary method. He weighted the arch in danger, and lightened the sides ; taking the material from the sides and placing it regularly on the top of the arch. He then put a layer or covering of straw on the arch ; then he clayed it again. He let the whole remain to settle till late in the following spring. The structure now stands but it is not a regular arch."

The late Sir Thomas Bazley confirms this: "He (Brindley) sought refuge in Stretford, and in his absence the Duke and Gilbert saw the waters softly and slowly run to their new abode of laborious duty. No flinching of the aqueduct, and no accident betokened present or future disappointment in the triumph which was then achieved. Brindley was brought, if not excavated, from his temporary retirement, and the bloodless victory of skill and labour was complete. All honour to the courageous Duke, to the valiant-minded Brindley, to the faithful Gilbert, and to the honest workmen whose labours raised this monument to utility." People came from far and near to see what they termed "the canal in the air"; and there were plenty of prophets who said "that man Brindley were ruinin' th' Duke," but by skill and perseverance he carried the canal to the mill at Worsley, and connected it with the pit workings.

The aqueduct was opened on 17th July, 1761, by a boat-load of coal passing over it. It was 250 feet long, 36 feet wide and about 39 feet above the Irwell, which there is 50 yards wide. The value of the canal was speedily felt, as it reduced the price of coal by one-half in Manchester, namely, from 7d. to 3½d. per cwt.

This aqueduct at first was one of the wonders of the district. In 1768 it was viewed by the King of Denmark, and our present King and Queen visited it in 1869. It was a specimen of admirable masonry, and very hard to dismember. When it was pulled down in 1891, I secured some stones with the mason's marks on them just as fresh as when they were inscribed.

Before the Duke obtained his first Act, the owners of the Mersey and Irwell Navigation refused to carry his coals, etc., for less than 3s. 4d. per ton from Barton to Manchester, and he had further to pay the cost of conveyance to the former place. Now they voluntarily offered his Grace the use of their navigation at 6d. per ton, to induce him to connect his canal with their river. We largely owe the Bridgewater Canal to the unreasonableness of a monopoly.

His success with the Worsley to Manchester Canal induced the Duke to enter upon a new task, *viz.*, linking his collieries with Liverpool. This he proposed to do by bifurcating at Stretford and taking a new branch to Liverpool, the idea being to make a shorter and cheaper communication between Manchester and Liverpool, *via* Runcorn. Not many engineering difficulties were in the way, as it was decided to take the canal on a dead level to Runcorn, and there descend to the river by a series of locks, having a fall of 90 feet. To avoid crossing valleys, the canal was to be carried by a detour along hill sides, thus lengthening its course, but still leaving it shorter than the winding river navigation. In those days people were not very particular, and when the Ship Canal bought the undertaking, it was found the Duke had in some cases simply paid for the land and taken possession, regardless of title deeds, believing that when once the canal was made he would never be disturbed. The maximum toll on the canal was to be 2s. 6d. per ton, and the maximum freight between Manchester and Liverpool 6s. per ton.

When the Duke asked Parliament to pass his Bill, he was met with a vigorous opposition. His attempt to cheapen and facilitate the passage of goods was bitterly opposed by the proprietors of the River Navigation, notwithstanding they themselves had obtained exclusive rights on the river "only in view of the advantage to the public," which could not mean to give a monopoly to any mode of water carriage if a more advantageous one could be found.

The Duke in his application said: "The River Navigation is very precarious, imperfect and expensive, and there has been no attempt to improve the navigation to Warrington Bridge, or between that place and Manchester, where are many shallows, which it is very difficult to pass with loaded vessels". Again, the new canal did not depend on the tide, and offered to reduce carriage from 12s. to 6s. per ton, and was nine to ten miles shorter in distance.

The owners of the River Navigation contended that Parliament had given them a monopoly, and that, as it was proposed the owners of the canal should become carriers, they could monopolise the freightage of the canal by consolidating tonnage, freightage and wharfage into one charge, and so drive off all by-carriers. Further, that 2s. 6d. per ton was to be charged, regardless of distance. Parliament could not be persuaded to throw out the Bill, but passed it in 1762.

In purchasing land the Duke was by no means particular as to title, feeling sure that if once the waterway were made it would be difficult to dispossess him. Often there was a letter of acceptance, and a receipt for the money was all that passed between the buyer and the seller. On page 15 is an original letter from the Duke to a lady from whom he wished to purchase land. It will be noticed that the Duke writes of "the cut intended to be made". Even to the present day country people often speak of the "cut" instead of the canal.

Before the completion the Duke's fortunes were brought to a very low ebb. So hard up was he that he had to pledge his estates, borrow money from his friends, and adopt at times all kinds of expedients to raise the weekly wages. To be on the spot he built himself Bridgewater House, Runcorn, a modest mansion now in

The Most Noble Francis Egerton, Duke of Bridgewater,
Constructor of the Bridgewater Canal.

To face page 18.

the possession of the Ship Canal Company. Here at one time were many relics of old days, and there is still to be found an excellent likeness of the old Duke. Fortunately, he lived to surmount all his difficulties, and to see the canal a complete success.

During the war with France the Duke showed his generosity and patriotism by subscribing £100,000 to the Loyalty Funds.

The following specimen of the accounts passing between the Duke and Brindley (his right-hand man) is interesting :—

		£	S.	D.
1761.				
Nov. 18. Mesuring a cross from Dunham to Warburton, Morely and Thalwal. Dunham for 2 diners 1s. 3d. for the man 1s. at Thalwal 1s. 2d. all night at Worington 3s. 11d.		0	7	4
Nov. 19. Set out from Chestar for London and returned back—going to London, and at London, then back to Worsley. Charges Hors and myself .		4	8	0
Dec. 9. Coming back from Hamston. Charges at Wilderspool all night .		0	8	0
At Worington to meet Mr. Ashley dining		0	4	2
Dec. 10. Chaind the Turnpike Rode 2s. 6d. and again on ye 12 ye rode 3s. 6d. .		0	6	0
Dec. 21. To inspect ye flux and reflux at Hamston 2 dayes charges . . .		0	6	0
		£6	0	0

26th Dec., 1761. Received the contents of the above bill by the hand of John Gilbert Esq.

JAMES BRINDLEY.

This bill speaks well as regards economy. The trio, *viz.* the Duke, Gilbert and Brindley, when a difficult problem had to be solved, used often to meet either at Worsley Old Hall or at the Village Inn, and sit till they had done so.

Reilly, in his *History of Manchester*, says that the Runcorn Locks were opened on 10th January, 1773, and that the undertaking was completed at a cost of £220,000. The length from the water meeting to Runcorn (twenty-four miles) was opened in 1766. The Worsley branch to Leigh, for which powers were obtained in 1795, increased the total length of canal to forty miles.

Litigation followed the making of the canal, for in carrying it across the Mersey at Stretford by a bridge, so narrow an aperture was left that when the Mersey was in flood the water could not get through fast enough. In consequence it backed up and flooded the Stretford and Chorlton meadows : indeed on some occasions it broke its banks and destroyed much valuable property. The Trafford Estate held the Bridgewater Trustees responsible for obstructing the river by the canal bridge and sued for damages. After the great flood of 1828 the Trustees turned the tables and sued Mr. Trafford, because that he and his tenants by raising the banks, had prevented the water having a free course.   Hence the floods. When the case, Trustees of the late Duke of Bridgewater *v.* Thomas Joseph de Trafford, was tried in 1829, judgment was given in favour of the plaintiffs on the ground that the defendant had raised the river banks, and by preventing the water flooding the meadows caused damage to the canal arch over the river.   But when the case by writ of error was taken before the Court of Exchequer in 1832, the decision was reversed and a new trial ordered.   In the end an arrangement was come to whereby a new weir was to be erected nearer Chorlton, with a lip low enough to allow water to pass away before it reached the bridge, and so relieve it. The surplus water then flowed down to the Urmston Ees, below Urmston Hall, where it re-entered the river at a lower level.   By agreement dated 10th June, 1838, the necessary work to prevent damage was to be paid for in the following proportions :—

Bridgewater Trustees and the Earl of Ellesmere . . .	£1,500
Commissioners of the Turnpike Road	500
Thomas Joseph Trafford, Esq.	1,000
Wilbraham Egerton of Tatton, Esq.	1,000
	£4,000

The Bridgewater Trustees to pay Mr. Trafford £500 for the wayleave.   Future maintenance to be borne in the following proportions :—

Bridgewater Trustees	one-half.
Thomas Joseph Trafford, Esq.	one-fourth.
Wilbraham Egerton, Esq.	one-fourth.

The first overflow embankment was washed away in 1840, but was shortly afterwards replaced by one which has an inscription on it, and is still doing its work.

In 1766 an anonymous Manchester writer published a book entitled *The History of Inland Navigations, particularly those of the Duke of Bridgewater in Lancashire and Cheshire.*   Printed for T. Lowndes in Fleet Street, London, MDCCLXVI.   It was written whilst the Bridgewater Canal was in construction, and contains some pertinent and quaint passages.   In the preface it says :—

It has been customary to erect monuments in honour of men who have during their lives distinguished themselves by patriotic services; but I wish to see your Grace represented in the prime of life by an elegant statue, or a distinguished pillar, fixed in the centre of St. Ann's Square.   This I propose as one grateful record of your fame which the history

VIEW OF BARTON AQUEDUCT.

From Aitken's *History of Manchester*, 1795.

To face page 20.

of these times will spread through Europe, and I hope, my Lord Duke, to see your navigation finished, and to bring you annual treasures such as few peers can boast of.

It does seem singular that nothing exists in Manchester to commemorate the name of a man who perhaps helped more than any other man to lay the foundations of the prosperity of the city. Rumour says that at one time Mr. Algernon Egerton, then M.P. for the county, offered a handsome contribution by the Bridgewater Trustees towards a statue, but that the proposition fell through. It is to be hoped that some day Manchester will, in an appropriate way, show her gratitude to the memory of a man who devoted his life to opening out the trade of the district and securing for her cheap carriage and cheap coal.

The before-mentioned author claimed that China was the pioneer of canals and understood their advantages long before more civilised nations adopted them.

But the great Canal called the Royal Canal, which is 300 leagues in length, is without comparison; which, at infinite expense, and with amazing industry, is carried on through many provinces upon which all the riches of the North and South are conveyed, and by its communications with other canals and rivers, the Chinese can travel or convey goods very commodiously from Pekin, the capital, to the farthest part of the Empire, being about 600 leagues, by water: they commonly have a fathom and a half of water in their canals.

The writer states that whilst of old the carriage from Manchester to Liverpool by road was 40s. and by river 12s. per ton, the legal maximum by canal will be 6s., and that timber will be carried at half the previous rate. Further that coals are then being sold at the Cornbrook Wharf for 3½d. for seven score pounds—an enormous reduction.

History reproduces itself, and it would seem the Bridgewater Canal had to encounter very similar opposition to that offered 125 years later to the Ship Canal. The author wrote :—

To have the means of conveyance so greatly facilitated ; the price of carriage so much diminished ; old manufactures encouraged ; new ones established ; estates greatly improved ; plenty widely diffused ; and the country, in general, rendered still more affluent, populous and secure, are considerations of such weight as cannot fail to interest all benevolent and public-spirited persons in the success of this important undertaking.

To conclude, it would be happy for this country if private interest, prejudice, ignorance or obstinacy were not employed to discredit such patriotic undertakings, as must redound so greatly to the honour and welfare of the kingdom : but such is the tax ever laid upon attempts for the public emolument ; let them be proved ever so salutary by the most convictive and forcible reasons, some opposition will be made if it only flow from the natural vanity or malevolence of mankind. But it is time, in the present critical circumstances of the nation, when rivals in trade and manufactures are taking every advantage over us, when enormous taxes, and the advanced price of the necessaries of life oppress our manufactures and our poor, that we unite as one man in promoting those designs which will contribute to raise our drooping commerce, to find employment for our labourers, and enable us to bear the burden of our numerous taxes with some degree of chearfulness (*sic*) and patience ; by which we may once more raise up our heads and recover what we have lost.   Let us say, at least, in the language of the poet :—

> ' 'Tis not in mortals to command success ;
> But we'll do more, we'll deserve it.' "

From a description of the Bridgewater Canal by Arthur Young, published in 1769, it would appear that the Duke was puzzled how to get lime for building walls, but that he met with a chalky kind of substance which he tempered, like clay, and then burnt.   This was called "lime marle".   By the use of this for the sides of the canal he saved many thousand pounds.

Speaking of crossing Sale Moor, he writes :—

This part of the navigation, from the lowness of the Moor below the level of the canal, was pronounced by many to be impracticable, and Mr. Brindley's *ne plus ultra ;* but this difficulty was removed by perseverance and spirit ; a complete bed was made for the canal, raised at bottom as well as the sides, sufficient for conducting the water on a level.   This was effected by making a vast case of timber for the whole work.   Great piles of deal were fixed as a mound to keep the earth in a proper position to form the banks ; and when they were raised, the piles removed on for answering the same work again, and the water brought forward by degrees, to the astonishment of those who pronounced the work impracticable.

At one time the Duke contemplated carrying a branch canal from Sale Moor to Stockport.   He also, with Mr. Brindley, viewed the river at Runcorn and expressed the opinion that it was practicable to bridge it over and by means of an aqueduct to carry his canal into Liverpool.   Young says :—

The number of foreigners who have viewed the Duke of Bridgewater's present navigation is surprising ; what would it be if his Grace was to extend it over a boisterous arm of the sea.   To exhibit a navigation afloat in the air, with ships of a hundred tons sailing full masted beneath it.   What a splendid idea !

But this courageous nobleman was not permitted to carry out his ambitious designs.

When the Duke died he left an eccentric will.   He made the second son of the Duke of Sutherland, Lord Francis Egerton, his heir, but he so tied up his property, that for one hundred years neither the heir nor his successors had anything to do with the management of the estate.   This was vested in Trustees, one of whom was to be salaried, and to be the manager of the property. Mr. Bradshaw was the first superintendent, and he was succeeded in turn by Mr. Loch and Mr. Algernon Egerton. The Trustee was all powerful, and Lord Francis Egerton and his successors have had simply to draw the income. The will was typical of the man. He was determined his pet work (the Bridgewater Canal) should not suffer through neglect in after days, so he created a trust that should exist during the lifetime of all his contemporaries, and for twenty years after. The last survivor of the great number of persons named or indicated in the will died in October, 1883, and it was not till 1904 that the trust ended, and the vast Bridgewater Estate passed into the sole possession of the present Earl of Ellesmere.

As has been recited in the history of the Mersey and Irwell Navigation, after a severe competition the two companies entered into a working arrangement in 1810, and finally, in 1842, became merged, and the Bridgewater Trustees became owners of both navigations. So fierce at one time was the struggle with the railways that passengers were carried from Manchester to Liverpool at the low figure of 3d. each. Goods came down in 1849 from 9s. per ton to 2s. 6d. per ton. Cotton was carried at this rate from Liverpool to Manchester. At length the rivals came to terms and the canal entered the railway conference.

In time, to reduce expenses, the Trustees decided mainly to work the canal, and keep the river navigation as a kind of stand-by in case of need. They continued however, to use the Old Quay warehouses, bringing goods by canal and then taking them through the Hulme Locks to the river for warehousing purposes. Copying from a newspaper correspondent ("B. L.") in 1882, I give the relative cost of articles of same weight by the Mersey and Irwell (1795), by the combined water carriers in 1810 and by the same bodies when railway competition was threatened (1822):—

	1795.	1810.	1822.
1 chest starch . . .	1s. 3d.	2s. 10d.	2s. 4d.
2 hhds. sugar . . .	11s. 0d.	£1 14s. 3d.	£1 5s. 10d.
1 hhd. tobacco . . .	2s. 6d.	5s. 4d.	4s. 3d.
6 boxes soap . . .	4s. 5d.	9s. 0d.	7s. 6d.

Though from 1845 onwards competition by water ceased, there was a stiff struggle with the railways for traffic. The Bridgewater Canal was a thorn in their side, because they could carry more cheaply by steam haulage, and they delivered goods almost as quickly as the railways. Cotton sent from Liverpool one day could be delivered in Manchester the next. Railways then began to buy or lease canals, and by throttling them to create a monopoly, just as the Lancashire and Yorkshire had taken a lease for twenty-one years of the Rochdale Canal.   But when in 1871 the railway companies wanted to buy the Bridgewater Canal, the commercial bodies of Manchester and elsewhere were in arms.   The Manchester Chamber of Commerce opposed it, and Parliament refused permission for a Bill to enable the leading railway and canal companies to take over the Bridgewater Trustees' property on the ground that it was not public policy to allow a monopoly of the kind.   More than this, in 1872 an Amalgamation Bill was passed forbidding railways to buy up canal property.   But what could not be done openly was done by a side wind.   A number of railway magnates, captained by Sir Edward Watkin, banded themselves together and bought the Bridgewater Trustees' interest in their waterways, and then created shares which were apportioned amongst those railway shareholders who desired to take them up.   Henceforth the canal virtually became railway property, and matters were so arranged that there was no active competition, heavy material and goods not requiring speed being sent by canal.   In order to disarm public anxiety, Sir Edward Watkin and Mr. Fereday Smith, principal agent, waited on the Chamber of Commerce and assured them that the transfer would be found for the advantage of the public, that they were opposed to monopolies or restrictions on trade, that they were against any attempt to treat a canal as anything but an open highway, that it was intended to develop the traffic on the canal and not make it over to railway control, and that it was proposed to limit the dividends to 5 per cent. for a certain time,   to apply the surplus to improvements.

In order to appease the corporation Sir Edward Watkin, in a very diplomatic way, gave Sir Joseph Heron, the Town Clerk of Manchester, an honorary seat on the Board of the Bridgewater Navigation Company, but I am not aware he ever took any part in their proceedings.   To do justice to Sir Edward Watkin, he always showed a friendly disposition towards canals, but there is no doubt railway interests were his first consideration.   When the Ship Canal Bill was introduced in 1883, the Bridgewater Navigation Company, consisting as it did of railway magnates, was in a position to offer a powerful opposition, and as narrated in the preceding chapter, the Ship Canal Company thought it prudent to buy their whole estate rather than take only the river navigation.   The following figures will be interesting :—

The cost of making the Bridgewater Canal.	£240,000
The Bridgewater Trust in 1845 paid for the Mersey and Irwell Navigation . .	550,800
	£790,800

The Bridgewater Navigation in 1872 paid the Bridgewater Trustees for their Canal
and carrying properties £1,120,000

In addition to this sum £10,000 was paid to Sir Edward Watkin and Mr. Price, the vendors, and very heavy legal expenses were incurred. The sum of £600,000 was the price fixed on the canal portion of the property.

The Ship Canal in 1885 paid the Bridgewater Navigation Co. . . . £1,710,000

That the purchase has not been a bad one is evident from the fact that in 1900 the Ship Canal sold the Duke's Dock, or thirteen acres of land in Liverpool, to the Liverpool Dock Board for £522,000. This was a valuable yet a very small portion of the estate purchased.

The £1,710,000 paid to the Bridgewater Navigation caused much litigation. The capital of the company was—

100,000 Ordinary Shares, £10 each, £3 10s. paid up	£350,000
Preference Shares, 5 per cent.	300,000
	£650,000

After all capital, liabilities and expenses were paid, there was the substantial sum of £550,000 clear profit left for division.

The ordinary shareholders claimed that the preference shareholders were not entitled to a share ; they had had 5 per cent.—all they bargained for. The preference shareholders claimed a share, as their capital had carried on the concern. The matter went into court, and it was decided by Justice North and the Court of Appeal that the profit must be divided *pro rata* on the paid-up capital among all the shareholders. Against this the ordinary shareholders appealed to the House of Lords, who reversed all previous decisions, and said the ordinary shareholders, though they had only paid up £350,000, had taken all the risks of £1,000,000, and as they would have had to pay that amount in case of failure, they must participate to the same extent in case of profit, so that their share was calculated on the £1,000,000 instead of on £350,000. The litigation lasted a long time, and the lawyers must have had a good harvest.

When Mr. J. F. Bateman, C.E., who at one time was engineer for the Bridgewater Trust, was examined before a Parliamentary Committee as to the filthy condition of the canals in Manchester, he replied : " The canals in Manchester are certainly black and disagreeable looking, but, like the Devil, not so black as they are painted ".

CHAPTER III.

CANALISED RIVERS AND SHIP CANALS: ENGLISH AND FOREIGN—THE MANCHESTER SHIP CANAL.

Navigable waterways play an important part in the production of the wealth of a country. It has been found that navigable waterways and railways are not destined to supplant but to support one another. Each has its particular attributes. Railways take the least cumbrous traffic—that which requires speed and regularity and bears most easily the cost of carriage. Waterways take heavy goods of low value, and their mere existence checks and moderates the rates on goods which are sent by railway.—M. FREYCINET.

FOR centuries our forefathers had to depend mainly on the roads of the country. The coach sufficed for passengers, and the pack-horse and waggon for the collection and distribution of goods. Rivers were to some extent utilised by the construction of locks which conserved the water and made them navigable. Many of these in time became linked with canals and thus provided an arterial system of water navigation throughout the country.

The Thames Navigation.—The first Act for the improvement of a river was for the Thames in the year 1423. The earliest locks and weirs were made by landowners rather with the intention of securing water-power than of improving the navigation. They were of a rude type, and heavy tolls were levied by the landowners. In 1730 the river was placed in the hands of Commissioners, who had ample powers, and created a good barge navigation. In 1783 an Act was passed to unite the Thames and the Severn by a canal forty miles long. This worked successfully, but on the introduction of railways it was gradually allowed to fall into disrepair. Lately a public trust has been formed to resuscitate what ought to be a most important link in the English waterways.

The Severn Navigation.—Various Acts were passed from 1503 to 1811 to secure the maintenance and navigation of the river, but it was not till 1842 that Commissioners, representing the various interests in the river, were appointed to take tolls and improve the navigation. In 1869 the canalisation of the river was completed, and there is now a minimum depth of 10 feet of water from Gloucester to Worcester, enabling vessels of 300 to 400 tons to reach the latter city. The Severn Navigation is an important link between the Midland towns and the Bristol Channel ports, and it connects with the Gloucester and Berkeley Ship Canal at Gloucester. The Midland Railway Company tried to purchase it, but Parliament would not allow them to do so.

The Aire and Calder Navigation was authorised in 1698, and is now one of the best of English waterways. It is 85 miles long and has thirty-one locks. Total cost, £2,761,807. It starts at Goole and ends at Leeds by a junction with the Leeds and Liverpool Canal, thus forming a continuous navigation from the east to the west coast. In 1860 great improvements were made : the depth, once 3 feet 6 inches, was increased to 9 feet on the lock sills ; the locks were made 215 feet long, and now vessels of 170 tons can use the canal. By a novel system designed by the manager, Mr. W. H. Bartholomew, trains of boats propelled by a steamer convey 700 to 900 tons of cargo for shipment at Goole at a minimum of cost.

The Weaver Navigation was created under the powers of an Act in 1721, which enabled the Trustees to use any profits in repairing bridges and other public charges in the county of Chester. Later a canal 4 miles long was made to avoid the lower tidal portion of the river and obtain a better entrance to the Mersey at Weston Point.

An Act obtained in 1866 authorised further improvements, including the Anderton Lift, an ingenious piece of mechanism constructed by Sir Leader Williams. Since their completion the Weaver has been the best navigation in England for small coasters and large barges. The locks are in pairs, the largest being 220 feet long by 42 feet wide with 15 feet depth, sill depth about 10 feet 6 inches. They have intermediate gates so as to pass small craft without wasting water. The whole of the salt trade outwards passes down this navigation, and salt can thus be carried far cheaper than by the railways.

Abroad canalised rivers have received much more Government assistance than they have in England. France paid £2,500,000 to open the Seine for navigation. It was made toll free, and now over 4,000,000 tons pass over it yearly.

Germany between 1880 and 1893 spent £17,875,350 in creating and improving waterways, and in 1895 there were 6,214 miles open. On the Rhine the size of vessels has been increased from 800 to 1,300 tons, and vessels of 500 to 600 tons can now go direct from Breslau to Hamburg. In order to foster water carriage the tolls levied are only about one-third of the cost of maintenance, the State paying the rest.

SHIP CANALS IN ENGLAND AND IRELAND.

The earliest English ship canals were the Gloucester and Berkeley Canal and the Exeter Canal. The Newry Canal dates from the eighteenth century.

The Gloucester and Berkeley Canal was first visited by me in 1882 when searching for information, and I well remember the strange sensation felt on viewing its course one fine day from the top of Gloucester Cathedral. Far as the eye could reach was a luxuriant country, through which meandered the canal like a silver streak. This connected the sea with the port of Gloucester, where, in the midst of an agricultural county, were to be seen forests of masts and huge grain warehouses with vast stacks of timber on every side.

The Act authorising the construction of the canal was passed in 1793. It was completed in 1827. The first engineer was Robert Milne, but the celebrated Thomas Telford revised the plans and completed the work. The navigation of the Severn from Sharpness to Gloucester, being both dangerous and circuitous, it was decided to make an artificial channel 18 feet deep between these points and thus reduce the distance from 28 miles to $16\frac{1}{2}$ miles. The level is uniform with basins at both ends for shipping, and with locks down to the Severn. The water supply is obtained from the river Frome, and by means of the Stroudwater Navigation, which crosses it, there is a connection with the Thames and Severn Canal, and other inland navigations. Foreign grain and timber ships drawing too much water lighten their cargoes at Sharpness and the surplus is barged up. Since 1869 steam tugs have been used to bring up sailing ships.

The Exeter Canal.—This was the first canal made in Great Britain to enable sea-going ships to reach an inland port. It was made in 1566 by the Corporation of Exeter and lengthened in 1675. Previous to 1820 only small ships of 9 feet draught could pass up the Exe to Exeter. By the construction of a canal 3 miles long, from Exeter to the river, and the subsequent extension down to the tidal estuary, vessels drawing 14 feet of water can now pass up to the basin and wharves at Exeter. The change was effected by raising the banks and making new locks. The works were carried out by Mr. Green, under the advice of Mr. Telford. Coasting vessels still use the canal, but it is too shallow for much business.

The Newry Canal.—Ireland has a great number of arms of the sea penetrating far into the interior, and making it impossible for any one to live more than 50 miles from salt water. Carlingford Lough is one of these arms. At its head is the port of Warren Point, and here the river Newry enters the sea. The town of Newry is a few miles up the river. To enable ships to reach it, a canal was constructed about 150 years ago. First a canal 2 miles long was made so as to avoid the shallows on the upper reach of the river, and afterwards, in 1829, a private company got powers to continue it 1½ miles seaward to deeper water, and improve the river to Carlingford Lough. Sir John Rennie was the engineer.

FOREIGN SHIP CANALS.

The chief continental ship canals are the two connecting Amsterdam with the North Sea, the Baltic, the Ghent, the Brussels and the Bruges Canals. Farther afield are the Suez Canal and the Corinth Canal. In America and Canada the chief canals are the Erie, the Sault Sainte Marie and the Welland Canals.

On determining to take up the question of a ship canal for Manchester, I made up my mind to visit all the existing ship canals, also those in the process of construction. I propose now to give a brief description of them.

The Dutch Canals.—Holland is largely dependent on artificial waterways for her wealth and importance : they form the commercial roads of the country.

In 1825 was completed the North Holland Ship Canal. It connected Lake Y, on which Amsterdam is situated, with Niewediep on the North Sea; formerly, ships used to pass through the shallows of the Zuider Zee, and then through the Texel Roads to the ocean. This passage was blocked with ice in the winter, and at no time could ships of any size reach Amsterdam, the available depth being only 11½ feet. The North Holland Canal is 52 miles long, 18 feet deep, has a bottom width of 33 feet and cost £916,000. There is a lock at either end with three intermediate locks.

In time this ship canal became insufficient for the growing trade of Amsterdam. The passage was circuitous and too shallow for large ships. Indeed one wonders why it was ever made, when by a cut of about 16 miles direct access to the North Sea could be obtained, especially as this would pass across Wyker Meer, and reduce the excavation considerably. It is presumed the varying water levels and the difficulty of protecting the entrance into the North Sea daunted the original projectors.

Greater proficiency in the construction of harbour work having been attained, it was decided to make an embankment across Lake Y and construct the North Sea Canal in a direct line to the sea.   This was started in 1866 and completed in 1872.   Originally there were two locks at Ymuiden, where it enters the North Sea, and where protecting breakwaters and extensive harbour works have been built.   Soon after the opening of the canal, it was recognised that the proportions were insufficient, it was therefore deepened, widened, and larger locks were constructed.   It is now 28 feet deep up to Amsterdam, with a bottom width of 82 feet and 105 feet at the sidings.   The top water level varies from 328 to 426 feet. New locks added at Ymuiden and Amsterdam will pass ships 722 feet long and 65½ feet wide at the former place, and 567 feet long and 51 feet wide at the latter. The canal now belongs to the State.   The cost was £4,853,000, but this has been reduced by sales of reclaimed land to the value of £1,166,000.   An idea of the increase of traffic may be gained from the fact that in fifteen years the business increased five fold.   Sir John Hawkshaw was one of the consulting engineers, and the work was carried out by English contractors.

The Baltic Canal.—This waterway gives access from the North Sea to the Baltic for the German Navy and Merchant Service.   It was started in 1887 and completed in 1895, at the cost of about £8,000,000.   The length is about 61½ miles, depth 29½ feet, width at bottom 72 feet.   Every 2½ miles there are wide places to enable large ships to pass one another.   The canal runs chiefly through the Schleswig-Holstein territory, which Prussia took from Denmark half a century ago. Previously, German ships could only reach the Baltic ports through the Sound, commanded by Denmark on the one side and Sweden on the other.   By means of the canal the approach to the Baltic is now entirely through German territory, and in time of war battle-ships can pass through in safety.

My first visit was by steamer from Kiel to the Baltic end at Holtenau.   Here we passed through spacious locks and under several magnificent fixed and swing bridges, the clear headway to the former being 138 feet.   Afterwards I went to the Kuden Lake length and thus to Brünsbuttel, where are the locks through which the canal is entered from the North Sea.   The levels of the connected seas vary very little, the locks at each end being used only to equalise any little variation there may be.   The excavated material, in all about 106,000,000 cubic yards, together with the dredgings have been largely put on the low land adjoining the banks, and it is wonderful to see how fertile the made land has become, growing most excellent crops.

Belgium, compared with most continental nations, is in an unfortunate position. She has only a small sea frontage, and on it Ostend is the only port of any importance.   Her access to the sea is mainly by the Scheldt, a free navigation running into the interior, and this is the means by which her imports and exports are chiefly conducted.   The growth of the population and importance of Antwerp is phenomenal, and it is now one of the great emporiums of the Continent.   But Belgium has also many other internal towns, hives of industry, and it became necessary they also should have cheap and good communication with the sea.

Brussels possesses a ship canal 17½ miles long to the Rupel which connects with the Scheldt.   By this means ships of 400 tons can reach the capital.   The present canal has three locks, and is only 19½ feet deep, but the State, the city, and the neighbouring communes propose, by a joint expenditure of £1,500,000 to increase the depth to 20 feet, and this will allow vessels of 2,000 tons to pass up to Brussels.

Bruges used to be a port connected with the sea by the Zwyn estuary. This, however, became choked with sand and the port was ruined.   A large canal to Ostend was of little use, and in 1896 the State and local bodies determined to try and restore the past commercial prosperity of Bruges by making a ship canal. It runs to Zeebrugge on the sea-coast and is 6¼ miles long, 26¼ feet deep, and has a bottom width of 72 feet with flat slopes, pitched with stone near the water level. The country through which the canal runs, being flat, presents few difficulties; the great cost is on the sea-coast, where embankments, jetties, and a breakwater 5,000 feet long are being constructed.   To resist the heavy seas, the outer part of the latter is built of concrete blocks weighing 250 to 300 tons each, the upper courses weighing 50 tons each.   The harbour formed by the breakwater will have an area of about 270 acres.

Ghent is, perhaps, the most enterprising of all the Belgian towns.   The present is her third effort to reach the sea by means of canals.   When an earlier canal silted up, one 20 feet deep, with a bottom width of 56 feet, was made to Terneusen on the Scheldt.   There are two locks 295 feet long and 39⅓ feet wide. As Holland owns both sides of the Scheldt at its mouth, a convention was entered into with that country in 1883 to secure the working of the canal.   Shortly after it was opened I travelled its whole length by one of the Grimsby steamboats, and I was exceedingly pleased with the neatness of its banks and the quaintness of the scenery.   The surrounding country being very flat our boat towered over it, affording us an excellent view. The port of Ghent has commodious docks and warehousing accommodation of a modern type.

The Suez Canal.—So much has been written about this canal that I do not propose to go into its history. Its complete success after years of failure did more than anything else to stimulate the promoters of the Manchester Ship Canal. Originally the canal was 26 feet deep, with a bottom width of 72 feet. The Committee appointed in 1884 recommended that the depth should be increased to 27 feet 6 inches, and the bottom width to 121 feet 4 inches. This was completed in 1898. Now another effort is being made, and dredgers are at work dredging to 31 feet. The width varies from 220 to 260 feet. Ships pass one another at certain fixed places, and can, with the aid of electric light, travel by night as well as by day. It has been my lot to pass through many times, and the quickest passage, when in a P. & O. mail steamer, was thirteen hours. The canal is 101 miles long.

Canada has a large number of ship canals, made chiefly to assist the navigation of the St. Lawrence. Of these the most important is the Welland Canal, connecting the waters of Lakes Erie and Ontario, across the Niagara Peninsula. The locks are 370 feet long and 45 feet wide. Depth of canal, 14 feet. It is 27½ miles long with twenty-six locks, and has a rise of 327 feet. In 1842 the Canadian Government bought the old canal, and at once commenced to improve the navigation. From Port Colborne, on Lake Erie, to near Thorold is a uniform level, and on this length the old course was widened and deepened. Then there is a long series of locks with an average fall of 12 to 14 feet. Before coming to the first lock there are guard gates, in case of accident. These series of locks take a ship down almost to the level of Lake Ontario, and from the foot of the bottom lock Government made a new channel, running by the side of the old one, for nearly 12 miles, to the lock at Port Dalhousie, which has to be passed on entering Lake Ontario. On the old portion are wooden locks, 180 feet by 24 feet 6 inches wide. It takes about eight hours to pass through the twenty-five locks, and two to three minutes to get through each of twenty-six road swing bridges. There are towpaths all along the canals, and sailing vessels are sometimes hauled by horses. I paid my visit of inspection to this canal in 1884, and I was much struck with the magnificent stone-work on the canal, and the conveniences provided. The whole length is lighted by gas, and can be worked the full twenty-four hours. There are telegraphic connections the entire distance. From the top of the locks the view across the country to Lake Ontario, 350 feet below, is simply magnificent.

Sault Sainte Marie Canal.—Another important Canadian waterway is the Sault Sainte Marie Canal, connecting Lakes Huron and Michigan with Lake Superior. The depth is 20½ feet, and the locks are 900 feet long and 60 feet wide. On the American side are other locks 800 feet long and 100 feet wide. One would suppose that with this ample accommodation there would be no delay, yet so huge is the business in corn, iron ore, coal and other merchandise, that there is often a string of huge steamers waiting to go through. If the Manchester Canal could do one-fourth of the weight that passes through this canal, her success would be assured. The business is done chiefly in huge long whaleback steamers, of moderate draught, and it was one of the impressions of my life when I viewed a string of these monsters passing through. In my travels I met with the President of the Company, who made a railway from Pittsburg to the shores of Lake Erie, and I repeat his story to illustrate the wonderful energy and enterprise that has made America so successful. In his early years he had distinguished himself in the struggle between the North and the South, and had become a colonel. When the war was over he joined other capitalists in floating and carrying out various enterprises. Hearing that Mr. Carnegie was anxious to bring Canadian ore from the shore of Lake Superior to his Pittsburg works, he interviewed that gentleman, and got the offer of 5,000,000 tons of ore per year if he would make a line 156 miles long and carry ore at 2s. per ton, or ·16 of 1d. per ton per mile. The colonel and his company accepted what was thought by most people a ridiculous offer. They made a straight and cheap line through the wilderness to the selected point on the Lake, and worked it with powerful engines and steel trucks each carrying 50 tons. By this means they carried the 5,000,000 tons at the least possible cost, and they secured return cargoes of Pennsylvanian coal, which was wanted in Canada. In the end their effort was a complete success.

America does not possess any deep ship canals. The *Erie Canal* is only about 9 feet deep, yet an immense traffic (to a large extent in grain) passes over it. The canal divides with the railways the 30,000,000 of tons of merchandise that in 1899 passed both ways through the Detroit River.

THE MANCHESTER SHIP CANAL.

I propose now to describe the constructed canal, the substance of the information and the figures being kindly supplied to me by Sir Leader Williams and other engineers engaged on the works.

The total length of the canal is 35½ miles. It runs for 12¾ miles alongside the Mersey estuary to Runcorn, thence inland for 8¼ miles to Latchford, near Warrington. Here is the first lock, and to this point it is tidal. From Latchford to Manchester (14½ miles) the canal follows the course of the Mersey and Irwell. Eastham, where the canal commences, is 6 miles above Liverpool. The entrance adjoins a good low-water channel, communicating with the Sloyne deep at Liverpool. It is close by the Eastham Hotel and Pleasure Gardens, and is quite picturesque in its character. There are three entrance locks parallel with each other, *viz.*, 600 × 80 feet wide, 300 × 50 feet and 150 × 30 feet. These maintain the water level in the canal at the height of a tide rising 14 feet 2 inches above the Liverpool datum, which is rather below mean high water level; when the tide rises above that height, the lock gates are opened, and it flows up to Latchford, giving on high water spring tides an additional depth of about 7 feet. On the ebb tide this water is returned to the Mersey through large sluices at Randles Creek, and at the junction of the river Weaver with the canal; the level of the canal is thus reduced to its normal height. The minimum depth of the canal is 26 feet; the lock sills are fixed 2 feet lower to enable the canal some day to be made 28 feet deep. The minimum bottom width is 120 feet. This allows large vessels to pass one another everywhere. The width is much increased at the locks and other parts. The slopes are as a rule 1½ to 1 but flatter in places and nearly vertical in rock cuttings. From Eastham to Runcorn the canal skirts the estuary: in some places clay embankments have been formed to exclude the tidal waters, and in others it runs inland and cuts off promontories. One of these has been used for the deposit of spoil, and it has assumed such huge dimensions as to look like a mountain in the distance. In honour of the departmental engineer, a son of Mr. Justice Manisty, it was christened " Mount Manisty," and it has been made famous on canvas by a picture painted by Mr. B. W. Leader, R.A., for Earl Egerton, to illustrate the works in progress. To protect the clay embankments they are faced with heavy coursed stone on each side. Where the foundation was bad, sheeting piles of timber had to be used. Facing Ellesmere Port there is an embankment 6,200 feet long, and to secure its safety 13,000 whole sheeting piles 35 feet long were driven, water jets under pressure, through 1½ inch wrought-iron pipes, being used to assist in sinking the piles, which were found most difficult to drive by ordinary means.

At the river Weaver ten Stoney's roller-sluices are built, each 30 feet span, with heavy stone and concrete piers and foundations. At Runcorn, where the Mersey is narrow, a concrete sea-wall 4,300 feet long was substituted for an embankment. At various points cast-iron syphon pipes are laid under the canal to carry into the estuary any land drainage which is at a lower level than the canal; the largest of these syphons was constructed to allow of the tidal and fresh water of the river Gowey to pass under the canal at Stanlow Point, between Runcorn and Ellesmere Port. Five 12-inch syphons are placed close together, built of cast-iron segments: they are 400 feet long, and were laid on concrete 4 feet below the canal. From Runcorn to Latchford the canal is nearly straight, the depth of cutting varying from 35 to 70 feet, partly in rock, but generally in alluvial deposit. The canal passes through the new red sandstone formation, with its overlying beds of gravel, clay, sand and silt. Retaining walls of stone and brick-work had to be built in these places to

EASTHAM CUTTING, WITH MOUNT MANISTY IN THE DISTANCE. SPECIALLY PAINTED FOR EARL EGERTON, BY B. W. LEADER, R.A.
From Engraving by Thos. Agnew & Sons.

To face page 34.

maintain the sides of the canal from slips and injury by the wash of steamers. Unfortunately, the red sandstone when depended upon as a wall to the canal proved treacherous in character. What seemed at first very good rock turned out to be laminated by veins of sand, and gave way under the action of water and weather. The brick-work necessary to remedy defects has been an unexpected addition to the cost of the canal. The canal from Latchford to Manchester is in a deep cutting through the valleys of the rivers Mersey and Irwell. Both these rivers abound in bends, and only small portions of the old course could be used; an almost straight line was therefore adopted, and this involved many crossings of the old river channels. These had to be kept open for the discharge of flood and land water, and in some places temporary cuts of great length had to be made for the same purpose. In November, 1890, and December, 1891, the workings were drowned out, many miles of the unfinished canal being filled, and great damage being done to the slopes. In all 23 miles of the canal had to be pumped out before the work could be completed. When the cutting between the lengths of the old river was finished, the end dams were removed, and the rivers Irwell and Mersey turned into the new channel now forming the upper portion of the Ship Canal.

The total rise to the level of the docks at Manchester, from the ordinary level of the water in the tidal portion of the canal below Latchford Locks, is 60 feet 6 inches; this is obtained by an average rise of about 15 feet at each of the locks at Latchford, Irlam, Barton and Mode Wheel. These are respectively 14½, 7, 5 and 1½ miles from Manchester. At the upper end of the canal the bottom width is 170 feet, to allow vessels to discharge cargo on wharves without interfering with the general traffic of the canal. The interior locks are in duplicate, the largest being 600 feet long by 65 feet wide, the other 350 feet by 45 wide, each with four Stoney's sluices adjacent. Both the locks have intermediate gates, in order to pass small vessels with the least possible waste of water. They are filled or emptied in five minutes by large culverts on each side, with side openings into the lock. Concrete with blue brick facing is largely used, and the copings, quoins and fender courses are of Cornish granite. The lock gates are made of greenheart timber from Demerara. The sluices are 30 feet span. Ordinarily, surplus water passes over the top of the sluices, which are kept closed; in flood times they are raised sufficiently to pass off floods with only a small rise in the canal. There are eight hydraulic installations, each having duplicate steam engines and boilers; the mains exceed 7 miles in length, the pressure being 700 lb. to the inch. They work the cranes, lifts and capstans at the docks, lock gates and culvert sluices, coal-tips, swing bridges and aqueduct.

At Barton, near Manchester, means had to be found to maintain the Bridgewater Canal, and yet permit large steamers to use the Ship Canal beneath it. Brindley's Canal is one level throughout its whole length; it draws its water from the Rochdale Canal, a few streams *en route*, and an auxiliary supply from the Medlock, but this is only sufficient for the docks at Runcorn. To lower down to the Ship Canal at Barton would have meant the waste of a lock of water, and caused serious delay to the traffic. Sir Leader Williams solved the problem by means of a swing aqueduct for the Bridgewater Canal, which, when closed, allows of the traffic passing as heretofore. The water in the swing portion of the aqueduct when opened is maintained by closing gates at each end, similar gates being shut at the same time across the fixed portion of the aqueduct. The swing portion is a large steel trough, carried by side girders 234 feet long and 33 feet high in the centre, tapering 4 feet to the ends: the waterway is 19 feet wide and 6 feet deep. The whole works on a central pier, and is turned like a swing bridge; it has two spans over the canal of 90 feet each. It is somewhat singular that the first fixed canal aqueduct in England should, after the lapse of 136 years, be replaced by the first swing aqueduct ever constructed. The structure is moved by hydraulic power, and has never given any trouble in working, even in times of severe frost. The weight of the movable portion, including the water, is 1,600 tons.

The manner of dealing with the existing five lines of railways crossing the proposed route of the canal was of importance, both in the interests of the travelling public and the canal traders. The main line of the London and North-Western to Scotland was, perhaps, the most important that had to be crossed. No doubt swing bridges to cross navigations are both dangerous and inconvenient. Tunnels, like the Severn Tunnel, are costly and difficult to make, besides often requiring permanent pumping power. Eventually high level deviation lines were adopted for each railway crossing the canal. Parliament had never hitherto sanctioned such extensive alterations, and it was only the fact that clauses for swing bridges existed and the proved necessity of a ship canal to Manchester, which secured the requisite power against the strong opposition of the combined railway companies. By means of embankments, made close to and parallel with the old lines, and started about a mile and a quarter from the canal on each side, a viaduct was thrown over the canal itself, to give a clear headway of 75 feet at ordinary water level. The provision of fidded, or telescopic masts, will enable vessels of any size to use the canal. The gradients on the railways rising up to the viaducts are 1 in 135. The span of the viaducts is so arranged as to maintain the full width of the canal for navigation; and as the railways generally cross the canal on the skew, this necessitates girders in some cases of 300 feet span. There are nine main roads across the canal, all requiring swing bridges; those below Barton have a clear waterway of 120 feet. The width of these bridges varies from 20 feet to 36 feet, and they are constructed of steel, their weight varying from 500 to 1,000 tons each. They work on a live ring of conical cast-iron rollers, and are moved by hydraulic power supplied by steam, gas, or oil engines. The Trafford Road Bridge at the Manchester Docks is the heaviest swing bridge on the canal; being of extra width it weighs 1,800 tons.

The canal being virtually one long dock, wharves can be placed anywhere for the erection of works. At Ellesmere Port coal-tips, sheds and a pontoon have been erected, and the canal is in direct communication with the docks there, as well as at Weston Point and Runcorn, where a large trade is carried on with the Potteries and the salt districts.

At Partington, branches from the railways connect the canal with the Lancashire and Yorkshire coal-fields, and the canal is widened out 65 feet on each side to accommodate six hydraulic coal-tips. At Mode Wheel there are extensive abattoirs and lairages erected by the Manchester Corporation; also large petroleum oil tanks, a graving dock and a pontoon, cold-air meat stores and other accommodation for traffic. At Manchester the area of the docks is 104 acres, with 152 acres of quay space, having over 5 miles of frontage to the docks, which are provided with a number of three-storey transit sheds, thirteen seven-storey, and seven four-storey warehouses and a large grain silo. These are constantly being added to as the trade increases. The London and North-Western, and the Lancashire and Yorkshire Railway Companies have made branch lines to the docks, the railway sidings at which are over thirty miles in length. Much traffic is also carted or dealt with by inland canals in direct communication with the docks. The deepening, widening and straightening of the Irwell and Medlock, and the use of fixed sluices in the place of weirs, has been of great advantage to the district passed through, by the prevention of floods. In some cases, land that used to be regularly flooded is now built upon.

The total amount of excavation in the canal, docks and subsidiary work was fifty-four million cubic yards, nearly one-fourth of which was sandstone rock; the excavated material was used in forming the railway deviation embankments, filling up the old beds of the rivers, and raising low lands near the canal. In the latter case a covenant was sometimes entered into to place 18 to 24 inches of good soil on the top, and this has produced excellent crops.

As many men were employed on the works as could be obtained, but the number never exceeded 17,000, and the greater part of the excavation was done by about eighty steam-navvies and land dredgers. For the conveyance of excavations and materials, 228 miles of temporary railway lines were laid, and 173 locomotives, 6,300 waggons and trucks, and 316 fixed and portable steam engines and cranes were employed, the total cost of the plant being about £1,000,000. The expenditure on the works, including plant and equipment, to 1st January, 1900, was £10,327,666. The purchase in all of the Mersey and Irwell and Bridge-water Canal Navigations, £1,786,651. Land and compensation, £1,223,809. Interest on capital during construction, £1,170,733. These items with parliamentary and general expenses bring up the total cost of the canal to £15,248,437.

Each year since the completion of the canal has been marked by extensive additions, such as the purchase of the race-course at Mode Wheel, and of 50 acres adjoining from Colonel Clowes, the construction on the added area of a huge dock 900 yards long with a co-extensive block of warehouses built of fine concrete on the Hennebique system ; also other extensions and additions too numerous to mention, which really belong to the history of a later period.

CHAPTER IV.

HISTORY OF THE LIVERPOOL DOCKS—ORIGIN OF THE DOCK AND TOWN DUES—THEIR OPPRESSIVE CHARACTER—THE LIVERPOOL TOLL BAR.

The fact that persons live in Liverpool is no reason why they only should take part in the government of the docks. On the other hand, we say that those whose interests are bound up in the successful rearrangement of the port of Liverpool ought not to be excluded from the government of it because they do not happen to live in Liverpool.—Speech of Mr. Calvert, Q.C.

ANY history of the Ship Canal movement would be incomplete that did not give a brief account of the controversy that raged for at least half a century in respect to the Liverpool dock and town dues and the management of the Dock Trust. The Dock Authority has been by turns in conflict with the Liverpool Corporation, the merchants of Liverpool, Birkenhead, Manchester and other Lancashire towns. Certain it is that the heavy shipping charges and dues exacted by the Dock Board at Liverpool mainly caused the advent of the Ship Canal.

In 1207 King John granted a charter to Liverpool, and made it a Royal borough. "Liverpool" was endowed with "all liberties and free customs which any free borough on the sea hath in our land".

In the reign of Henry III. another charter was granted creating freemen in Liverpool, who could buy and sell in their own port and every free borough in the kingdom without paying any dues. They were also exempt from market tolls and ferry rates.

Sir James Picton, in his *Memorials of Liverpool*, says that an Act passed in 1544 included Liverpool in the list of decayed towns: "There hath been in times past many beautiful houses which are now falling into ruin". Leland, in his *Itinerary*, commenced in 1533, says: "At Lyrpole is a small costome payid that causeth marchantes to resorte". Further, "Good Marchandis at Lyrpole, moch

(38)

Irisch yarn that Manchester men do by ther". On this Sir James Picton remarks: "The yarn here mentioned was doubtless linen yarn spun in Ireland and woven in Manchester and the neighbourhood. This is the first intimation we have of the textile manufactures of South Lancashire—the first feeble rill of that manufacturing industry and commerce which has swelled into such a mighty stream."

Ralph Sekerston in addressing Queen Elizabeth in a letter in 1566 calls himself "your poor subject, of your Grace's decayed town of Liverpool," and asks that the charter governing the town may remain in her hands, pleading, "Liverpool is your own town. Your Majesty hath a castle and two chauntries clear, the fee farms of the town, the ferry boat, two windmills, the custom of the duchy, the new custom of the tonnage and poundage, which was never paid in Liverpool before your time; and the commodity thereof is your majesty's. For your own sake suffer us not utterly to be cast away in your Grace's time, but relieve us like a mother."

In 1581 the merchants of Chester, being jealous, tried to repress the rising town of Liverpool, which had hitherto been considered a mere dependency of the Dee port. Liverpool thereupon invoked the help of their neighbour the Earl of Derby, who brought the matter before the Privy Council whence it was referred to the Master of the Rolls. Secretary Walsingham in making this communication describes the Earl of Derby as "the chief person in those parts, and Patron of *that poor town* of Lyverpoole".

In the whirligig of time what wonderful changes take place and how history repeats itself! We read that the flourishing port of Chester tried to retard the progress of "*that poor town* of Lyverpoole" just as three centuries later the wealthy and important port of Liverpool tried to prevent Manchester redeeming her fortunes by becoming a centre for shipping. Strange to say an Earl of Derby in both cases befriended the town struggling for commercial freedom. It was largely owing to the assistance rendered by the Earl of Derby that Lord Winmarleigh in 1883 was enabled to overcome the opposition of Lord Redesdale to the Bill in the House of Lords.

It remained a royal borough till 1628, when "in consideration of a past and certain future loans from the City of London," Charles I., after excepting a rent due to Sir Richard Molyneux of £14 6s. 8d., made over to trustees "his town and Lordship of Liverpool". This included the market tolls and perquisites of the courts, the ferry, all the Customs and anchorage and all rights appertaining to the property to *low-water mark*. In 1635 the City of London sold its bargain to

Lord Richard Molyneux for £450, and he again sold it for £700, subject to an annual rent of £14 6s. 8d. to be paid to the Crown. Afterwards Lord Molyneux redeemed this rent, and let his entire rights over the town on lease for 1,000 years at £30 per year. His successors sold their reversionary interest for £2,250. By that purchase the Liverpool Corporation obtained possession of land, which, prior to 1856, brought in over £50,000 per annum, irrespective of the town dues.

In 1691 the freemen of Lancaster complained of the tolls levied on their goods at Liverpool, and there was litigation. The Attorney-General, on behalf of Liverpool, said they owned the soil to low-water mark; they kept the haven in order by clearing away rocks and sand banks; they provided warehousing, and found scales and weights for weighing goods; and that for these services they had a prescriptive right to levy town's duty or town's custom. They made no mention of any claim by purchase from Lord Molyneux. Liverpool lost the case, and it was decided that freemen of other ports were free of Liverpool and not liable to any tolls. In 1708 the Liverpool Corporation asked Parliament to let them make a dock and increase the dues. On consideration of the services to be rendered to shippers and merchants they were allowed to do so, and all future rights to levy dues on goods specially named were to be on a scale of charges which were given with great minuteness. The Corporation retained the management of the docks till 1785, at which date they had given fifty-six acres of land for dock purposes, and had realised large sums for adjoining land, which they had sold on seventy-five years' leases, chiefly for storage purposes.

The next change was a transfer from the Corporation to a Dock Trust, and separate accounts were kept, but as the Trustees were all members of the Corporation they dominated it. It enabled the Corporation, however, to sell to the Trust hundreds of acres of land (some of it river bed, which cost them next to nothing) at a very substantial price; and this has largely been the means of making Liverpool one of the richest Corporations in England. In 1840 the Corporation sold to the Dock Trustees (virtually themselves) the site of the Albert Dock for £221,000, and in 1846 another plot for £250,000. These, and other lands sold, cost a mere trifle, and yielded a large sum to the Corporation; it will be readily seen how the Dock Board debt had mounted in 1851 to six millions. No wonder that at that time the merchants and shipowners turned restive because of the ever-increasing dues laid upon them. It is quite clear that, instead of making a charge simply for services rendered (the principal one being the improvement of the river and port), hundreds of thousands of pounds that ought to have been used for the reduction of the dues on the imports and exports, chiefly from Lancashire, have been applied to pay the taxes of the freemen of Liverpool, to build its churches, to pave its streets, and generally to reduce its local taxation. The Dock Trust seems to have made its own scale of charges, for whilst in 1674 there were only sixty-one articles subject to dues, in 1854 there were 695 articles, the increase being unauthorised.

The original grant defined the port of Liverpool as extending from the Dee to Warrington; so that not only Lancashire inland towns paid dues on goods passing through Liverpool, but traders on the river, who did not use the port at all, had to pay dues, and often were charged on articles not scheduled in the charter.

In 1852-53 the town dues levied principally on Manchester and other towns amounted to £115,000 clear of expenses. Of this, only £4,770 was spent on the purposes for which it was raised, *viz.*, the improvement of the port and river. The remainder was spent thus :—

Grant to Mayor of Liverpool	£2,000
Law expenses	3,000
Town Hall and Sessions' expenses	4,500
Judge's lodgings	551
Mayor's stables	164
Churches	3,400
Public schools	2,000
Library, Museum and Observatory	3,000
Rates, Sanitary Act	5,000
Cost of elections	1,500
Constabulary and Police	54,000
Lighting, etc.	5,000
Borough Gaol	7,000
Prosecutions and Clerk of Peace	2,500

Mr. Lloyd, of Birmingham, calculated that his town paid £4,500 in dues for the benefit of Liverpool. If a freeman were the agent he could draw the dues, and then they would be remitted, and he could pocket them.

Prior to 1855 Birkenhead had been pushing to the front, impelled largely by the Great Western Railway, who had made their terminus in that town. Liverpool looked with a jealous eye upon the progress of her neighbour, and eventually a costly Parliamentary fight took place, into which ultimately Manchester was drawn. The exposition was an eye-opener to the traders of Lancashire, who had been "as lambs led to the slaughter". Birkenhead and Manchester had been paying heavy and questionable dues which had been used to reduce the taxes of the

Liverpool ratepayers. Whilst Bristol had surrendered obsolete and unjustifiable privileges, gained under an old charter, and so freed her port; and whilst London had given up her octroi duty on provision sellers who were not freemen, Liverpool exacted all she could get from traders in the interior, and thereby became known as a "dear" port. At the same time, the Corporation was coining money by selling land for building dock walls down to low-water mark, thus crippling the river, and damaging the Birkenhead side. They were also buying land in Birkenhead, on the border of the Wallasey Pool, in order to prevent the construction of competitive docks, and they were allowing the Mersey to fall into a condition that would have been fatal to Liverpool itself if it had not speedily been taken in hand.

When the Ship Canal Company applied for their Bill they were fully aware the Mersey Docks and Harbour Board would be amongst their most strenuous opponents, and that in order to justify their attempt to emancipate Manchester and district from the exactions of the Dock Board, they had to prove that that body had by no means acted up to their duties, and that they unnecessarily placed heavy burdens on trade and commerce; further, that they could not give the amount of relief from existing burdens which was imperatively required by the trade in the district. The charges they made against the existing system were :—

A. The retention by Liverpool of town dues.
B. Misapplication of funds, which ought to have gone towards the reduction of port charges in goods.
C. Wasteful and extravagant expenditure of the funds of a national trust.
D. That local and personal interests had been fostered at the expense of trade and commerce.
E. That though ample funds had been raised by dock and town dues, the Liverpool bar had not been dealt with.

A. *Town dues* were claimed on the ground of a royal grant from Charles I. In 1854 Liverpool had a Bill to borrow £260,000 on the security of the town dues, and apply it to making a new street from the Town Hall to St. George's Hall, but Manchester and Birmingham opposed the appropriation of money, part of which they had paid in the shape of dues, and the Bill failed. In 1855 the Dock Trust was at a low ebb, and the Dock Trustees (the Corporation having preponderating power on the Board) proposed to borrow three and a half millions, principally for warehousing accommodation and new docks. As security they offered the dues of the port. The same year Birkenhead had a Bill to give power to lease their docks. Parliament recommended that the docks on both sides of the Mersey should be incorporated, and that Liverpool should purchase the Birkenhead Docks and complete them rather than extend in Liverpool. In 1856 Liverpool asked for power to buy the Birkenhead Docks, apparently with the intention of shutting them up. This was refused, and Parliament insisted that Liverpool should buy the Birkenhead Docks and complete them by 1st September, 1858, at a cost of £1,000,000. When Manchester found Liverpool proposed to borrow large sums on the security of the dues, she joined forces with Birkenhead and the Great Western Railway to obtain a Trust Act, whereby the docks should be vested in trustees, and made a national and not a local trust, which should exist solely for the benefit of the commerce of the nation, and not for the benefit of merely local interests. When the Bill went before Parliament it was urged by Sir Joseph Heron, on behalf of Manchester, that the town dues were being illegally raised and extravagantly and improperly expended. In the end a compromise was effected, and an Act passed in 1857, by which it was provided that the newly constituted Dock Board should pay to the Corporation of Liverpool £1,500,000 as compensation for giving up the town dues, and the question of the legality of the dues was left over. It was certainly understood that the new Dock Board would levy no more town dues than would suffice to discharge the money to be paid to the Liverpool Corporation and the interest thereon. Though Manchester and Sir Joseph Heron were largely responsible for the reconstruction of the Dock Trust and the passing of the 1857 Act, it ought to be made quite clear they had nothing to do with the Bill of the previous year, which obliged Liverpool to take over the Birkenhead Docks. The widened area from which the new Dock Board was elected gave hope that local feelings would disappear, and that the Trust would be managed as a national one, reducing, as far as possible, all burdens that were in restraint of the trade of the country. But these hopes were doomed to disappointment. The new Dock Board, through the peculiar way in which it was elected, represented a limited number of interests—sail and steam shipowners having a decided majority—and the intention to secure a public trust that would represent national trade was frustrated. It was to be expected the town dues would bear an equal incidence on goods and ships; instead of which the former were made to pay 62 per cent. of the dues levied and the latter 38 per cent. This caused the Liverpool Chamber of Commerce to make a vigorous protest in 1867. They asked for a rectification of the charges, and that, inasmuch as the dues collected had been sufficient to repay the Dock Board the £1,500,000 paid to the Liverpool Corporation, the town dues should cease.   They also invited the assistance of Manchester to make the Dock Board act equitably.

RESOLUTION PASSED BY THE LIVERPOOL CHAMBER OF COMMERCE (TOWN DUES), 2ND APRIL, 1867.

That the payment of £1,500,000 by the Mersey Docks and Harbour Board to the Corporation of Liverpool as compensation for the town dues and anchorage having now been repaid to them by the dues collected, or will be upon the 23rd of June, the mercantile community have a right to be relieved from those imposts.   That the revenue of the dock estate should be augmented, in case of need, by an increase in the dock dues on ships and goods, levied upon equitable principles, whereby tonnage on ships and dues on goods should produce an equal amount in accordance with the agreement of 1809.   That representations to this effect be made to the Mersey Docks and Harbour Board, the Corporation of Liverpool, the Chamber of Commerce of Manchester and the Railway and Canal Companies having termini in Liverpool; as also, at a time hereafter to be fixed by the Council, to the Board of Trade and Parliament.

In 1873 a highly respected merchant of Liverpool, Mr. John Patterson, gave evidence that the £1,500,000 due to the Corporation had long been repaid, and although this money had all come back into the Dock Board coffers, traders were still obliged to pay town dues on goods.   In his opinion this and other heavy charges were diverting trade from the port.   Other well-known merchants confirmed Mr. Patterson.   In 1879 a Special Committee of the Liverpool Chamber of Commerce reported that goods passing through Liverpool were unjustly taxed, and they called on the Dock Board to give adequate relief.   In 1880 Mr. John Patterson again pointed out in the Press that, of the £1,500,000 paid to the Corporation, the Upper Mersey Trust had found £105,000, and had abolished town dues on the Upper Mersey; and

This latter body, acting as faithful stewards, have long since repaid the high proportion they had paid, and as a consequence flour can be, and is, imported into Garston free of town dues chargeable in Liverpool; which flour is actually carried past our empty southern docks in Liverpool.

Further, that the Dock Board had collected more than sufficient to pay off the rest of the £1,500,000; yet they had not done so, and "the blistering and trade-diverting imposition still remained".

When accountants, on behalf of the Ship Canal, investigated the Dock Board accounts in 1882, they found that not only had the one and a half millions, and interest, been raised in dues and paid to the Corporation, but that a further sum of £3,312,407 had been extracted from the traders of the country in the shape of town dues. Whilst in 1857 the dues brought in £109,000, in 1882 they had increased to £245,797. In fact, the Board had not only increased the rate at which the dues were levied, but had increased them sufficiently to make them pay, in addition, the interest on the £1,500,000 for compensation. Strange to say, even the interest paid by the Mersey Commissioners was carried to capital account.

There have been no severer critics of Dock Board policy than leading Liverpool merchants; indeed, the Chamber of Commerce and the Dock Board have been constantly at variance, but in face of the Ship Canal (a common enemy) they closed up their ranks. In 1879 Mr. Coke, a leading member and for some time Chairman of the Liverpool Chamber of Commerce, compared the relative costs of the port then with those of 1846 and 1856. In the former year, on the authority of Mr. Bramley Moore, articles that in Hull were charged 1s. 3d., 1od., 1s. 9d. per ton were charged in Liverpool 1s., 9d. and 1s. 6d. respectively. Now the figures in Hull were 2s. 4d., 2s. 1od., 1s. 1od., whilst the relative costs in Liverpool were 4s. 11d., 5s. 1od., 5s. There had been a rapid rise, and it was quite evident that Liverpool was now the dearest port in the kingdom. Speaking of the comparison between 1856 and 1879, Mr. Coke said the dock and town dues on cotton had been raised 50 per cent. As a rule the increased charges had been largely caused by town dues: for instance, whilst the average advance had been 42 per cent. on flour in sacks, the town dues had gone up 1oo per cent. Mr. Coke went on to say that whilst Liverpool used to be considered the gate to England for trade from the West, in 1878 its trade from the United States had fallen from 75 per cent. to 46 per cent. Glasgow, Hull and other ports had meanwhile increased their business.

It was stated that the dues and tolls in Liverpool were higher than any other port except London, and in confirmation Mr. Harold Littledale[1] showed the comparative port charges to be—

Liverpool .	.	.	.	.	.	3s. 4½d. per ton.
Hull .	.	.	.	.	.	1s. 10¾d. „
The Clyde .	.	.	.	.	1s. 7¾d. „

The constant complaints inside and outside Liverpool in 1880 had some effect, and in that year a deduction of 1o per cent. on the Liverpool charges was made by the Dock Board. In 1883 the Chamber of Commerce again approached the Dock Board, and reminded them of their previous application for the extinction of town

---

[1] A member of the Liverpool Dock Board.

dues, and asked that justice should be done to importers and exporters, and suggested that a reduction of 25 per cent. should be made for the next four years, when the charge would die away.   They pleaded that the Dock Board could afford the reduction; that they had a large unappropriated revenue of £666,562; that a rectification of the terms for loans might effect a large saving; and they clearly showed that previous reductions had been quickly made up by increased trade.

Mr. C. B. Paris, Chairman of the Committee on the Charges of the Port, said: "If the Dock Board wisely resolved to remit, or entirely abolish, town dues, trade would be certain to prosper, but if they declined to touch the subject the pressure would be such that the Lancashire people would go to Parliament on the question, in which case they would get the dues remitted.   These were a constant source of irritation and discontent, and the sooner the matter was disposed of one way or another the better for all interests."

Mr. John Patterson, at a meeting of Dock Ratepayers, "thought that the old standing town dues ought to be reduced, for, by the existence of heavy dues, traffic was driven from Liverpool, and other ports benefited.   As an instance of that, he might point to Garston, where the town dues of the Upper Mersey had been abolished.   Since 1879 the dock tonnage of sailing vessels at Liverpool had been reduced 14,000 tons; and in the same period the steam tonnage had increased 38 per cent.   But what were the figures during the same period in regard to the Upper Mersey?   In sailing ships there had been an increase of 22¾ per cent. and in steam tonnage of 70 per cent.

"The Manchester Ship Canal had assumed very formidable proportions, and if the agitation should take the form of an attack on the Liverpool town dues, there would be nothing to do but surrender at discretion, and make the best possible terms with the enemy.

"If the town dues were abolished, goods and shipping would contribute equally to the support of the dock estate; and this, he thought, was an argument for the Board at once addressing itself to their abolition."

At the February meeting of the Dock Board in 1884, a letter was read from Mr. Ismay (White Star Line) saying, though he was a shipowner, he would be glad for the dues on produce to cease, and all revenue be collected from a tonnage on ships.   It was so on the Suez Canal.   He would sweep away dock and town dues, making the port absolutely free as regards produce.   There would be economy in collection—a saving of annoyance and of time, a great benefit to the prosperity of the port, and an increased employment of tonnage, which meant more work for the people.   He believed the costly, unremunerative and dangerous project of the Manchester Ship Canal owed its origin to a desire (which it would certainly fail to satisfy) for reducing the charges on the raw material and on manufactured goods.   To remove the dues would go far to remove existing complaints, and act beneficially on the trade of the port.

The *Liverpool Courier* wrote :—

Mr. Ismay's letter to the Dock Board, proposing the abolition of dock and town dues, and the collection of the whole revenue from ships, seems to have excited a more lively interest in business circles than at the Dock Board.   The General Brokers' Association were emphatic in their opinion that the proposed change would largely increase the importation of produce, bring Liverpool vessels from London and elsewhere, and conduce not only to the benefit of the trade of the port, but it would also increase the Dock Board's revenue.

"A Steamship Owner," writing to the *Liverpool Daily Post* in 1885, said :—

A good deal of the American trade has already been diverted.   Surely the Dock Board authorities are susceptible to the lessons which these new docks and facilities teach. Some other gigantic scheme may be hatched.   It may come from any quarter, as the high rates of Liverpool are a byeword in commercial centres.

B. *Misapplication of Funds.*—The grievance of Manchester, and of the trade generally, was that the £100,000 per year paid to the Sinking Fund, as per arrangement, instead of being applied to the wiping out of town dues, and in reducing the rates on food, was used in capital expenditure on new works ; and that a large fund kept in reserve as unappropriated revenue was not applied to the reduction of port charges.   This was objected to by Mr. Williams, the Dock Auditor, who actually resigned his post rather than vouch accounts which he considered were incorrect in principle.   He also considered that as borrowing powers were exercised for improvements, the large sums received for the sale of lands ought to go in the cancelment of borrowing powers.

A leading Liverpool paper, in a financial article, wrote : " How is it that the Dock Board throw away probably £100,000 per annum in extravagant interest ?   Why should they be paying 4½ per cent. when their immediate neighbours, the Corporation of Liverpool, only pay 3½ per cent. ?   It is not the doubtful credit of the estate, or the adequacy of the security : it is the financial incapacity of the Board."   The finances were not as well managed as in some other towns.   The *Times* on 24th October, 1883, thus commented on the Dock Board finance : " No less than four and a quarter millions of the debt stands at 4½ per cent. . . . It is an antiquated and wasteful system, when it is remembered that a reduction of ½ per cent. represents a saving of no less than £80,000 per year."

The Board had resorted to the novel expedient of issuing perpetual annuities ; a very costly system compared with the Metropolitan Board of Works, who found no difficulty in borrowing at 3 per cent.

C. *Wasteful and Extravagant Expenditure of Trust Funds.*—People who do not find or earn money are often careless about its expenditure, and it was felt that the extravagant and unnecessary expenditure of the Dock Board was the great first cause of the heavy charges on all Lancashire exports and imports. A huge debt had been created, and the interest had to be found very largely by outside traders. In 1857 the Dock Board debt was £6,099,657; in 1882 it had increased to £16,373,451, nearly trebling the interest that had to be paid. In Birkenhead the estimated expenditure to complete the docks was £1,000,000, and, though nearly £5,000,000 had been spent, they were to a great extent a failure. Quoting from a speech of Mr. Henry Coke in 1879 (the same gentleman who in 1885 gave evidence in favour of the Dock Board) to the Liverpool Chamber of Commerce :—

If in the time of Mr. Bramley Moore he thought Liverpool a dear port, " I represent a party who, in the present time, consider, and with justice, that we are not fairly treated by the Dock Board. We go to them and lay our complaints before them in the language of the people ruled by Rehoboam—our burdens are too heavy to be borne—our previous rulers gave us a light yoke, the present one is grievous. Practically our constitutional rulers of the Dock Board reply, our fathers chastised you with whips, but we will chastise you with scorpions."

Instances without end can be quoted of the wasteful and extravagant expenditure which had increased the debt of Liverpool and made it a dear port :—

1. When the Egerton Dock was made, the sides were so high that barges could not land their goods without cranes, and these had not been provided. The dock was made for certain railway companies at a cost of £120,000, but was so inconvenient they would not use it.

2. The Morpeth warehouses at Birkenhead cost £42,500, but, through fault of construction, they were practically uninsurable, and have been worked at a serious loss. Other Birkenhead warehouses, that cost £225,000, do not pay 2 per cent. on their outlay. A tenant took them on the estimate they would hold 42,000 tons, but as they only held 28,000 tons he repudiated his lease.

3. The Wallasey Dock, Birkenhead, through bad access, has been a perfect failure. Mr. Turner, Chairman of the Warehouse Committee of the Dock Board, in a speech delivered in May, 1878, twelve months after the dock was opened, graphically described its condition : " Vessels scarcely ever entered it, and not a package was in the warehouses. Day after day, week after week, the same state of things continued. It was melancholy to observe this. Not a hungry cat or a

starved rat was to be seen about the locality, and the very policeman shuddered at the sight of the place. The cost of the alterations of this dock was £322,000." A correspondent to the *Daily Post* wrote: "Those (in Liverpool) who think the Birkenhead Docks a mistake were justified in opposing the converting the low-water basin tooth and nail, but when Parliament decided against them, the sensible course in the interests of the port was to expend the money as judiciously as possible; make the best dock they could, and not spoil it for the sake of fulfilling their own prophecies of its failure".

4. The Langton Graving Docks were made in 1874-75 at a huge cost, but soon after completion it was found they were too shallow, and vessels like the *City of Berlin, Polynesian, Virginian* and many others were compelled to go into dock at Birkenhead.

5. The Cost of the Birkenhead Docks, through reckless management and extravagance, was £6,000,000, or twice the estimated cost. The Dock Board ought to have supplied deep water accommodation for the growing trade of Liverpool. It was because of their faulty construction, and the determination of Liverpool not to use them, that the north and south extensions of 1877, costing £4,000,000, became necessary. The Birkenhead Docks will be found to have yielded about 1 per cent. on the outlay; at the same time the Dock Board were paying 4 to 4½ per cent. on the money. The difference was a tax largely on the traders of Lancashire.

6. The rapid growth of cost in the management of the trust is shown by the fact that whilst in 1868 the cost per acre was £493, in 1878 it had risen to £852 per acre.

D. *The Local and Personal Interests that preyed on Trade and Commerce.*— When Liverpool through its proximity to the manufacturing districts became the most important port on the West Coast of England, there grew up numerous guilds or trades unions which monopolised certain processes and enabled a few favoured ones to obtain advantages or levy blackmail on all goods passing through the port. This caused the charges to exceed those of other ports except London.

1. *Master Porterage.*—In consequence of the careless way goods at one time were dumped down on the quays, Parliamentary powers were obtained for Master Porterage. There must be a Master Porter for each ship discharged, and he has many duties to perform, such as receiving, sorting, weighing, measuring, marking, watching, sending off, etc., etc. For these he receives a stipulated charge, and in

addition he may charge extra for other services. The office is a lucrative one; often the shipowner takes a licence and is his own Master Porter. This applies to open docks. Some are called closed docks, and here the Dock Board themselves do the work. It had been expected the system would work well, and so it has done in the closed docks, but in the open docks there has been chaos, full charges being made for goods to be tumbled out of ships on to the quay. Consignees could not get their goods, and then they were charged quay rent for not removing them. Messrs. Chambres, Holder & Co. in January, 1883, wrote a bitter letter of complaint :—

We are thus put to the expense of sending men and carts down, only to find the cargoes so mixed up that the latter have often to come away empty, or with only a portion of a load; and so our commission is eaten away with extra charges, besides *heavy mending expenses to put the bales in merchantable condition.*

After the rejection of the first Ship Canal Bill in 1883, a Liverpool correspondent, urging the Dock Board to take the opportunity of putting their house in order, writes :—

The principal grievance is the monopoly enjoyed by Liverpool steamship owners in dealing with unloading ships and delivering cargo from quay. The rates they are permitted to charge are one of the inflictions on the commerce of the port. The system is well known to be a source of considerable profit to shipowners. The work, when not performed by them, is undertaken by one of the extensive master-lumpers, by arrangement beneficial to all parties. The calling is a very lucrative one, making fortunes for the private individuals engaged in it.

Though the old price paid well, the stevedores were afterwards powerful enough to raise their price for the handling of cotton 10 per cent.

2. *Appropriated Berths.*—Great difficulty was experienced in getting quick discharge for ships at the docks, many of them having to wait days for their turn. At the same time, out of 22 miles of quay space, 6 miles were converted into appropriated berths for regular lines of steamers, and of the latter the greater part was let to past or present members of the Dock Board themselves. The Liverpool Chamber of Commerce in 1879 resolved—

That the space allotted to appropriated berths is out of proportion to the quay space available for the general trade of the port, and that much inconvenience is caused thereby.

Between 1879 and 1882 the appropriated space had increased 84 per cent., whilst the rental had only increased 35 per cent. Colonel Paris, a Liverpool shipowner, said in his pamphlet :—

Unless a steamer belongs to the family party, or has its influence, and so has its appropriated berth, she suffers more detention in loading and discharging in Liverpool than in any other port in the kingdom, and her loading and discharging are more costly. It may fairly be taken that any of our large steamers is worth in demurrage £70 to £100 per day. Such an one going to London can be discharged, loaded and despatched in four days or less. Coming to Liverpool she will be lucky to perform the same operations in fourteen days, in spite of all that her owners can do unless she has an appropriated berth.

The writer went on to say, when there is a berth to be appropriated instead of the Dock Board allotting,

Why not invite tenders and let to the highest bidder? I venture to affirm that they could thus obtain £50,000 instead of the £14,189 received. Further, I am disposed to view the Dock Board in the aggregate as a gigantic trade combination, more imperious, absolute and injurious to fair and open competition than any trade union could possibly be; more restrictive and depressing to the continual sturdy growth of the commerce of the port. That this large trade combination has, in effect, appropriated to itself, for its own purposes, the cream of the Liverpool Dock accommodation, leaving to the public that which is insufficient to meet its requirements, and depriving the public of even that when their own interests demand it. Thus might we not say that Liverpool is not the port of Liverpool, but the port of the Mersey Docks and Harbour Board Trades Union.

In August, 1883, a writer to the *Liverpool Courier* complained bitterly of insufficient cranes and accommodation at the docks harassing trade, and ended with :—

Mr. Hugh Mason has been eulogising the Board, asserting " that it is not possible for any railway company to be managed with greater zeal". We in Liverpool know better. It is freely asserted that till some of the present Sachems are shunted, and the present large staff controlled by a paid chairman of the type of Richard Moon, Sir Edward Watkin or Arthur Forwood, the Board will go on blundering and burdening the estate, becoming a bigger byword in the future than they even are at present.

The same correspondent, after pointing out several other extravagant expenditures, especially one connected with pumping apparatus, which he says cost £26,000 too much, ended with, "Yet Liverpool wonders why Manchester wants a ship canal"

Another correspondent in the *Liverpool Courier* wrote :—

The manner in which the old docks are sacrificed to the new works at the north end is little short of a scandal. The docks in the heart of Liverpool are left almost unused, their quays are grass-grown, and this simply because the Dock Board neglect to adapt them to the altered conditions of commerce. They are too shallow to receive large ships, their entrances are too narrow, and their quays are not equipped with the mechanical appliances indispensable for present necessities.

E. *The Liverpool Bar.*—There can be little doubt that the Dock Board received dock and town dues which ought to have been appropriated to the removal of the bar and other improvements on the river, but were actually spent recklessly, and all the time the bar was allowed to remain a peril to life and property. Yet, during the Ship Canal inquiry, some members of the Dock Board pleaded that if the bar was touched the entrance to the Mersey might become worse, and then ruin to Liverpool would follow.

Mr. John Laird, M.P. for Birkenhead, always advocated its removal, and strongly objected to dock dredgings being tipped to obstruct the mouth of the Mersey.

He did not blame Admiral Evans, the Acting Conservator, because that gentleman had also strongly objected to dock dredgings (amounting in one year to 213,000 tons) being emptied into the middle of the river, and had urged improvements on the bar. The Dock Board had full power to raise funds, and, along with the Conservancy Board, to deal with the question. Every manufacturing interest in the kingdom was deeply concerned and would applaud the spending of all necessary money. He moved for a report on the depositing of silt and the deepening of the bar.

Yet during the Ship Canal struggle, the advocates of Liverpool and the Dock Board frequently urged that all had been done that could have been done, and that the traders of Manchester had no cause for grievance. The main object of this chapter has been to show the error of this assertion, and that the Liverpool Chamber of Commerce and the merchants of Liverpool have been in the past constantly in antagonism with the Dock Board. No more bitter opponents of the Board's policy could have been found than Messrs. Patterson and Coke and Colonel Paris; yet these gentlemen were induced to go into the box and oppose Manchester freeing herself from the Liverpool toll bar.

The *Times* in an article on 18th October, 1882, graphically described the incubus on the trade of Lancashire, in the shape of Liverpool charges which the Ship Canal sought to remove :—

Five millions and a half of people are at the mercy of a combination holding a pass between them and the rest of the human race, and making use of their coign of vantage as the medieval barons did in the embattled toll gates thrown across the world's highways. City, port, dock and railway vie in extortion, and levy duties to the extent of human forbearance. Many millions of material and manufactures pass annually to and fro between the port and the industrious region at the back of it, and on every ton Liverpool has its profit. It cannot be expected that a large population, placed at the mercy of a single port, should sit quietly under it.   The more material conditions of the question are much in favour of a ship canal by one route or other, and the very idea of seeing a hundred acres of ocean-going steamers from one's own windows is so charming, that one cannot be surprised at Manchester being possessed with it.

The magnificent local patriotism and self-sacrifice exhibited by Manchester when she shook off the Liverpool toll bar has no parallel except perhaps in the Corn Law agitation, and it has been necessary at the risk of being wearisome to prove the grievances under which Lancashire laboured in order to show that it was misgovernment by the Dock Board, and self-preservation, not hostility or a desire to damage Liverpool, that prompted the Manchester Ship Canal.

CHAPTER V.

EARLY ATTEMPTS TO MAKE A SHIP CANAL TO MANCHESTER.

I consider the time has now arrived for Parliament to say once for all to the railway companies "hands off the canal". . . . England cannot in the face of increasing foreign competition afford to see her cheapest means of internal transit year after year closed against traffic.—PETER SPENCE, J.P.

UP to the year 1721 the growing trades of Lancashire had mainly depended for their supplies on the barge navigation of the Mersey. As far back as 1712 a scheme was elaborated by Mr. Thomas Steers to make a canal for ships to Bank "Key," Warrington, and so give Manchester direct water communication with Liverpool, taking a line very similar to the present Ship Canal. A map was published headed "A Map of the Rivers Mersey and Irwell, from Bank Key to Manchester, with an account of the rising of the water and how many locks it will require to make it navigable.[1] Surveyed by order of the gentlemen at Manchester by Thos. Steers, 1712." The plan shows the three locks then existing were to be replaced by nine. The object of the improved waterway is explained at the foot of the plan reproduced herewith.

Some ten years later the Mersey and Irwell undertakers obtained an Act to improve the navigation by locks, cuts and weirs. Carriage was cheapened, and the district benefited. After the Napoleonic Wars efforts were made to further improve the water communication with Manchester. Enterprising citizens in the Press, at public meetings, and in verse, advocated a ship canal to Liverpool; they realised the importance of cheap carriage both for raw material and manufactured goods.

The following lines were unearthed by Mr. James Crossley in reference to a ship canal movement. They appeared in a song entitled "Inland Navigation" in vol. xxxvi. of the *Gentleman's Magazine* for March, 1766. Speaking of the versifier, Crossley once facetiously remarked "he, apparently, didn't like Scotch".

[1] See Plan No. 1.

NO. 1.

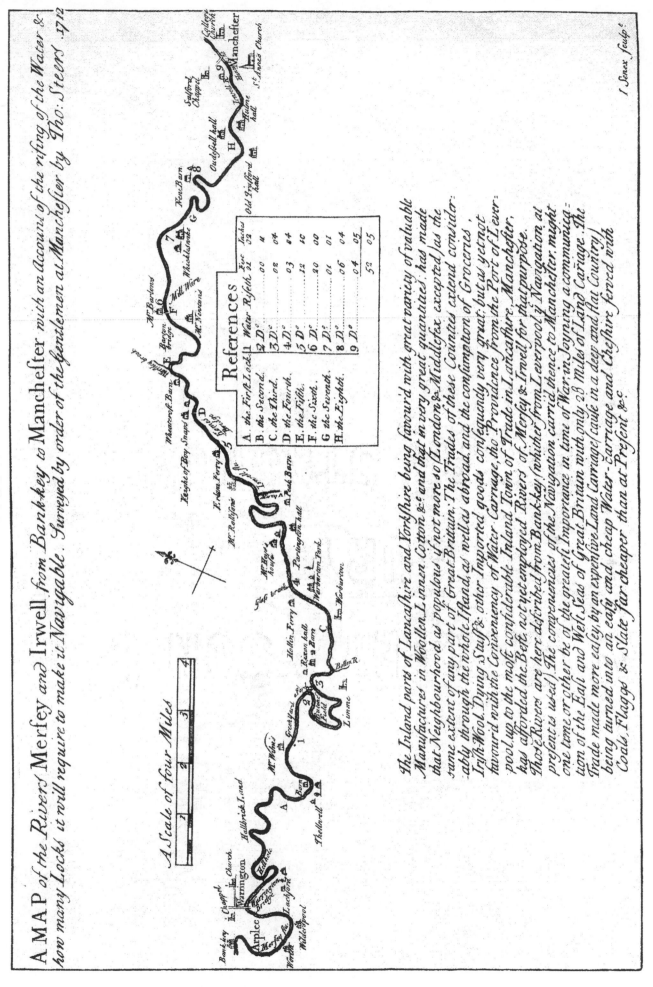

INLAND NAVIGATION.

In Lancashire's view, what a laudable plan,
    And brought into fine execution
By Bridgewater's Duke; let us copy the man,
    And stand to a good resolution!

If the waters of Trent with the Mersey have vent,
    What mortal can have an objection—
So they do not proceed to cut into the Tweed,
    With the Scots to have greater connection?

A free intercourse with our principal ports
    For trade must be certainly better;
When traffic's extended, and goods easy vended,
    In consequence things will be cheaper.

Our Commerce must thrive, and the Arts will revive,
    Which are now in a sad situation,
If we follow this notion, from ocean to ocean
    To have a compleat (*sic*) navigation.

'Tis this will enable our merchants abroad
    To vie with each neighbouring nation;
Who now, as they tell us, in fact undersell us
    For want of this free navigation!

In 1824 Mr. Matthew Hedley, a grocer of Manchester, came to the front as an advocate of a ship canal, and in 1825 a company was actually organised.

The proposed canal began below Padgate on the Dee, with docks at Dawpool; and passing along the Cheshire side of the Mersey, crossed the Wirral Canal, the Weaver above Frodsham, and the Grand Canal near Preston Brook, thence through Lymm and Altrincham to Didsbury, where it crossed the Mersey, and so on to the south side of Manchester. The plans were published by William Wales & Co. in 1824, the surveyor being Mr. Wm. Chapman.[1] The capital was to be £1,000,000 in 100,000 shares of £10 each. At the instance of a few supporters of the movement, a meeting was held on 3rd February, 1825, at the office of Mr. Wm. Norris, solicitor, in Old Exchange, Major Watkins in the chair. At that meeting it was resolved unanimously "that a navigable ship canal, capable of bearing vessels of 400 tons burthen and upwards, and to communicate with the Irish Sea direct from

[1] See Plan No. 2.

Manchester, would be of great public utility and advantage," etc.   Mr. Wm. Norris was appointed solicitor to the company, and Mr. Wm. W. Tait, of King Street, secretary.   At a subsequent meeting at the Albion Hotel (14th February), Mr. Matthew Hedley in the chair, the following gentlemen were appointed a Committee of Management: Major Watkins; Mr. John Marshall, of Ardwick; Mr. Francis Dukinfield Astley; Mr. Thomas Appleby, Bridgewater Yard; Mr. James Bremner, of Newmarket Lane; Mr. Richard Matley, Calico Printer; Mr. Charles Clegg, Timber Merchant; Mr. M. Hedley, of New Cannon Street; Mr. George Jones, of the Crescent, Salford; Mr. Edmund Wright, Calico Printer; and Mr. Leonard Cooper. It was further stated that from Dawpool at the mouth of the Dee (where Nature has bestowed what art could never have procured—a safe and commodious harbour) to Manchester was 45 miles, and the intended course of the waterway would run through a country suitable for the formation of a canal.

This scheme was brought before Parliament in 1825, but thrown out because plans and levels had not been prepared and proper notice had not been given. Curiously enough the 1883 Bill was subsequently rejected practically for the same reason.   The following lines, published in the *Liverpool Kaleidoscope* of 19th April, 1825, show that the cynical and sceptical attitude of the great port of the Mersey was the same then as now.

MANCHESTER GRAND SHIP CANAL.

One day, as old Neptune, delighting to rove,
In the blue tinted empire assigned him by Jove,
Resolv'd on inspecting his ample domain,
He summoned the Naiads and Nymphs of his train;
And, mounting his car, o'er the wide ocean skimm'd,
Bespangled with many a vessel well trimm'd,
And many a barque that was coasting it too,
'Till the Land's End of England appear'd in full view:
Here the monarch abandoned his car and his state,
And ordered the nymphs his return to await,
Whilst he plunged in the sea that to Liverpool led,
And at one stroke he made the far-famed Holyhead;
And here, as he cautiously rounded the coast,
Fearing lest in the quicksands he might have been lost,
He was often obliged on the bottom to creep,
To avoid the effects of the steam-paddles' sweep;
Nay, 'tis said, but I know not how true it may be
That his back got a scratch in his efforts to flee.

NO. 2.

PLAN OF THE DOCKS &c. AT DAWPOOL, FOR THE INTENDED MANCHESTER & DEE

SHIP CANAL.

Scale of Chains.
0 ... 5 ... 10 ... 15

Engraved by order of the Committee of the Manchester & Dee Ship Canal.

Feb.y 1825

INTENDED SHIP CANAL

SEA PIER

BEACH

FULL TIDE DOCK

□ □ Transport Buoys

OUTER PIER

LOW WATER at SPRING TIDES

Capstan
○ Capstan

HALF TIDE DOCK

Gully hole

○ Capstan   Gully hole

BASIN

Transport □ Buoy

○ Capstan

N.W. PIER

BEACH

Muscle Beds

DAWPOOL.

Transport Buoy ○

The course through Hoylaugh Swash

Transport Buoy

R   I   V   E   R   D   E   E

The material originally positioned here is too large for reproduction in this reissue. A PDF can be downloaded from the web address given on page iv of this book, by clicking on 'Resources Available'.

Be that as it may, a taut ship he espied,
And close in her wake he went on with the tide,
Till, finding his way rather cramp'd and confin'd,
He essayed to turn and see what was behind;
But vain were his efforts to set himself free,
For he only kept bruising his arms and his knee.
That the Monarch should thus be confin'd to a space,
Was an insult, he said, that no time could efface;
Compell'd thus to move in a different sphere,
The day seem'd to him like the length of a year.
At length he approach'd what the end seem'd to be,
For at least he observed that his arms were set free,
Then raising his head above water, he found,
That he'd got into what appear'd just like a pond.
Surpris'd and astonish'd at what now appear'd,
He mus'd with himself, as he strok'd down his beard:
"Can this be proud Liverpool, famed for her ships,
Her wealth and her docks, and her piers and her slips?
What mean these long chimneys—these smells that confound?"
Cried the Monarch as sternly he looked him around—
"Can these be her riggers, her pilots, her tars?
Can these be the men who, retir'd from the wars,
Employ'd now as porters, or boatmen in wherries,
To carry the passengers over the ferries?
Surely these squalid fellows could never have been
O'er the seas which fill up the vast spaces between
Both the Indies and England, and Ireland, and Spain,
Surely these never crossed o'er the wide spreading main?"

He paused, and survey'd the quick gathering crowd,
As they welcomed the ship with cheers long and loud;
But not knowing what sort of a fish had come in,
They survey'd the proud Monarch with wondering din;
Some thought him a porpoise, some thought him a whale,
While others observ'd he wanted a tail;
At length, one more knowing than others drew near,
And having survey'd both his beard and his spear,
Set him down as a customer, come from abroad,
And kindly he offer'd to show him the road,
Observing that, "if you want aught in my line,
I have calicos, muslins, and bobbin, and twine"

> "Avaunt!" cried the Monarch, and brandish'd his spear,
> "I pray you inform me, all you who stand near,
> Where am I? Who are you? that thus dare confine
> Old Neptune as if with a cord or a line."
> At the sound of his name down all fell on their knees,
> As the deity whom they most wish'd to appease,
> And implor'd his pow'rful protection and aid,
> In behalf of the Manchester Ship Canal trade;
> The Monarch, indignant at what he called treason,
> And contrary, too, to the dictates of reason,
> Advis'd them in future to stick to their Jennies,
> And in aping their betters not make themselves ninnies,
> "And as for your ditch there, why take it for granted,
> My protection in this case will never be wanted."
>
> Old Neptune on this disappeared from their view,
> Leaving doubts if this prophecy e'er should come true.
> How he got out to sea I have not heard it said,
> Or the way he crept back to his usual bed,
> But that he got home there cannot be a doubt,
> As the Manchester folk soon began to find out;
> For assembling his sea-gods, and Boreas to aid,
> With a good stiff north-wester he soon spoil'd their trade;
> By means of a land-bank he clos'd their canal,
> And to it and their projects he put a finale.

In the same session, strange to say, was brought up the Bill to make a railway from Liverpool to Manchester. It went before the Commons Committee on 21st March, 1825, and after a struggle of two months was passed by a majority of one. Difficulties arose about land clauses; it was defeated in the Lords and withdrawn.

In 1826 the preamble was proved. The cost of getting the Bill was £27,000. The railways urged they were combating the high charges of the waterways, just as the Ship Canal is now designed to be a check on extravagant railway freights. During the inquiry a learned counsel declared that a train would not be able to travel more than six miles an hour. A parallel statement was made before the Ship Canal Committee, *viz.*, that it would take at least two days for a ship to get up to

Manchester from Liverpool.   Herewith is a copy of the form of receipt that was given to the subscribers in 1825.

𝕸𝖆𝖓𝖈𝖍𝖊𝖘𝖙𝖊𝖗.   1825.

We acknowledge to have received from Mr.

a Subscriber for   *Shares in the*

intended "MANCHESTER SHIP CANAL,"

Pounds, being a Deposit of One Pound per Share, to be subject to the

Order of the Committee.

COPY OF DEPOSIT NOTE—MANCHESTER SHIP CANAL, 1825.

It was found in a drawer at one of the Manchester banks nearly sixty years afterwards, and is a curious document.   The owner, Sir Leader Williams, kindly lent me the original.   Popular enthusiasm was much roused at the time, and various poetical effusions were published.   I give one, with the music, which was sung by Mr. Hammond at the Manchester Theatre Royal in 1825-26.

> I sing a theme deserving praise, a theme of great renown, sir,
> The Ship Canal in Manchester, that rich and trading town, sir;
> I mean to say it *once* was rich, e'er these bad times came on, sir;
> But good times will come back, you know, when these bad times are gone, sir.
>
> In eighteen twenty-five, when we were speculating all, sir,
> We wise folks clubbed together, and we made this Ship Canal, sir;
> I should have said we *meant* to do, for we'd the scheme laid down, sir,
> That would have made this Manchester a first-rate seaport town, sir.
>
> Near Oxford Road the dry dock is, to caulk and to careen, sir;
> Our chief West India Dock is where the pond was at Ardwick Green, sir;
> That is to say they *might* have been there, had these plans been done, sir,
> And vessels might have anchored there of full five hundred tons, sir.

THE MANCHESTER SHIP CANAL.

As sung by Mr. Hammond at the Theatre Royal, Manchester.

Allegro.

I sing a theme deserving praise, a theme of great renown, sir, The Ship Canal in Manchester, that rich and trading town, sir; I

mean to say it *once* was rich, e'er these bad times came on, sir, But good times will come back you know—when these bad times are gone, sir.

CHORUS.

Tow row row.   Tol di rid - dy rol di rid - dy, Tow row row.

Instead of lazy Old Quay flats, that crawl three miles an hour, sir,
We'd fine three-masted steamships, some of ninety horses power, sir;
That is, had it been *made* we should; and Lord! how fine t'would be, sir,
When all beyond St. Peter's Church was open to the sea, sir.

At Stretford, Prestwich, Eccles too, no weaver could you see, sir,
His shuttle for a handspike changed, away to sea went he, sir;
I'm wrong, I mean he *would* have done so had it but been made, sir.
For who would starve at weaving who could find a better trade, sir?

Alas! then for poor Cannon Street, the hookers-in, poor odd fish!
Instead of catching customers, must take to catching cod fish;
That is, *supposing* it was made, may it ne'er be I wish, sir,
These cotton baits for customers, would never do for fish, sir.

Alas! too, for poor Liverpool, she'd surely go to pot, sir,
For want of trade her folks would starve, her custom-house would rot, sir;
I'm wrong, they'd not exactly starve or want, for it is true, sir,
They might come down to Manchester; we'd find them work to do, sir.

Success then unto Manchester, and joking all aside, sir,
Her trade will flourish as before, and be her country's pride, sir;
That is to say if *speculation* can be but kept down, sir,
And sure we've had enough of that, at least within this town, sir.

A kind friend has also supplied me with an extract from a broadsheet published on Tuesday, 27th December, 1825, and entitled "*The Manchester Times and Stretford Chronicle.* Price 6d." It is sarcastic and humorous, after the style of the *Free Lance*.

MANCHESTER AND DEE SHIP CANAL.

The Committee of the Manchester and Dee Ship Canal Company have great pleasure in announcing to Shareholders, that their hopes are not *de*-funct, nor their exertions *de*-creased; that everything continues to go on *swimmingly*, and that the difficulties, antici-pated, and propagated, by persons opposed to this *great undertaking*, are proved to be such as can easily be overcome. With regard to the procuring of a sufficient supply of Water, all doubts are removed, as the Committee are not only in treaty for, and confident of obtaining, the whole of the *Cornbrook Stream, in Hulme*, but have also the promise of the *waste Water* from no fewer than four PUMPS in the neighbourhood of Chorlton, all of which are situated within two miles of the intended line of Canal; the supply from these sources would doubtless be found amply sufficient, but in addition, the Committee intend to erect at *Daw*-pool, an Engine of *twenty ASS* power, which will of itself be calculated to raise water sufficient to create a second DELUGE. With respect to the passing of the Bill (alluding of course to the Act of Parliament, not the Attornies Bill), the Committee feel confident of success; in the lower House the two members for NEW-PORT are pledged to support it, and although in the House of Lords, LIVERPOOL will of course be hostile to it, the *Committee* have the assurance of MANCHESTER, PORT-LAND, and many other Peers, that their utmost and united exertions will be used in its favour.

As there are only a few Shares remaining unsold, those who wish to become SHARE-HOLDERS must apply early.

Ship-Tavern, Water-Street.

Some verses to be found in the *Liverpool Mercury* of 18th February, 1825, show the current feeling at the neighbouring port:—

HUMBLE PETITION OF THE LIVERPOOL CORPORATION TO THE MANCHESTER PROJECTORS OF THE GRAND SHIP CANAL.

Oh, ye Lords of the loom,
Pray avert our sad doom,
We humbly beseech on our knees;
We do not complain
That you drink your champagne,
But leave us our port, if you please.

Sweet squires of the shuttle,
As ye guzzle and guttle,
Have some bowels for poor Liverpool!
Your great Ship Canal
Will produce a cabal,
Then stick to the jenny and mule.

Your sea scheme abandon
For rail-roads the land on:
And to save us from utter perdition,
Cut your throats if you like
But don't cut the dyke,
And this is our humble petition.

The next attempt to make a ship canal originated with Warrington. Sir John Rennie, an engineer, was authorised, in 1838, to make a complete survey of the river Mersey, between Bank Quay and Runcorn, with a view of rendering it navigable for vessels drawing, at least, 5 feet of water at neap tides. He made his report on 21st April, 1838, concluding with—"Finally, I beg to repeat that, under all circumstances, the ship canal from Bank Quay and Runcorn Gap to Liverpool is decidedly preferable to improving the old river, and if my information as to levels be correct, a canal upon the same scale might be readily extended to Manchester".

He divided his report under the following heads, and as every subject is exhaustively dealt with, it is a valuable document for those who wish to study the early history and navigation of the Mersey, and of the Bridgewater Canal :—

1. History and course of the Mersey.
2. Rise of Liverpool.
3. Commencement of the canal.
4. Present state of the Mersey, winds, tides, areas, etc.
5. Conclusions from the state of the Mersey.
6. Mode of improving the river between Bank Quay and Runcorn.
7. Expense of making the river navigable for 5 feet, and 10 feet at neap tides.
8. Cost of canal between Bank Quay and Runcorn.
9. Canal preferable.
10. Improvement of river between Runcorn and Liverpool.
11. Plan of improving the river between Runcorn and Liverpool.
12. Cost of canal between Warrington and Liverpool.
13. Conclusion.

Bearing upon the position of Warrington, Sir John Rennie says :—

The largest vessels which now navigate the Mersey from Liverpool to Bank Quay, draw about 8 to 9 feet of water, and are from 80 to 100 tons burden ; but vessels of this class can only come up to Bank Quay at high water of spring tides ; and even then, unless assisted by a favourable wind, or by a steam tug, they cannot get up from Liverpool in one tide. Indeed, it not unfrequently happens that they run upon the banks, where they are obliged to remain until the next tide, or if the spring happens to be falling off they must remain until the next spring tide.

A comparison with the position of to-day will show how greatly Warrington has benefited by the Ship Canal. That the public and the Mersey and Irwell Navigation Company looked forward to having a ship canal to Manchester is evident from the fact that whenever railway companies sought to bridge over the waterway, clauses were inserted in the various Acts of Parliament that, if ever a ship canal were made, the fixed bridges should be converted into swing bridges to allow the passage of shipping. This precaution, in one instance, was omitted, but when the same company subsequently came to Parliament, they were compelled to repair the omission, and undertake to convert the existing structure into a swing bridge if called upon to do so. Without these precautionary clauses, the Ship Canal could never have been made. Parliament obliged the railways to comply with the spirit (if not the letter) of the law, and submit to bridges being raised. The London and North-Western Railway when constructing Runcorn bridge were also compelled to make it high enough for the passage of vessels.

The following *résumé* of the important swing bridge clauses may be interesting :—

The Grand Junction Railway Company (predecessors of the London and North-Western) was incorporated by an Act of 1834, under the authority of which Act that company carried their railway from Warrington to Birmingham across the river Mersey and across the Runcorn and Latchford Canal by means of a viaduct between Warrington and Lower Walton, known as the Walton Viaduct. The height of the rails from the surface of the river was only 25 feet. Nothing was said in this Act about requiring the railway company to assent to the conversion of any part of Walton Viaduct into a swivel or opening bridge for the purpose of allowing the passage of vessels through the railway in the event of the future improvement of the navigation of the Mersey towards Manchester.

In 1845 the Grand Junction Railway Company promoted a Bill to construct the "Huyton and Aston Branch". The plans for this branch included a bridge over the river Mersey at Runcorn.

The Conservators of the river Mersey required the insertion in this Bill of a protective clause in regard to Runcorn Bridge, and such a clause was submitted to them by the railway company and was approved. The letter of approval, however, which was sent by the Acting Conservator (Admiral Evans) contained an additional requirement of a remarkable and momentous character as will be best understood from the following full copy of the letter :—

No. 4.

GWYDYR HOUSE, WHITEHALL,
28th March, 1846.

SIR,

Having laid before my Lords Conservators of the river Mersey the clause you propose to insert in the Huyton and Aston Branch of the Grand Junction Railway Company, relative to a bridge over the river Mersey at Runcorn,

I am desired by my Lords to inform you that their Lordships approve of the clause ; but, in addition thereto, my Lords Conservators deem it to be necessary that you further insert a clause in your Bill to the following effect :—

"That in the event of any improvement hereafter taking place in the navigation of the river Mersey, eastward of the proposed bridge at Runcorn, for the purpose of establishing a communication for ships or vessels with Manchester, or any other place to the eastward of the present viaduct of the Grand Junction Railway over the river Mersey, near Warrington, the Grand Junction Company, or the then owners of that railway, shall consent to such arrangements for the passing through their present viaduct over the river Mersey near Warrington as my Lords Conservators shall deem it to be expedient to require."

I have, etc.,
GEO. EVANS,
Acting Conservator of the River Mersey.

To J. SWIFT, ESQ.,
    Solicitor for the Bill.

The Lords of the Admiralty instructed Mr. J. M. Rendel, a celebrated engineer of that period, to report to them on the project of a bridge over the river Mersey at Runcorn. Mr. Rendel reported on the 30th April, 1845, and recommended that if a fixed bridge were to be allowed it should be on condition that there should be a clear headway of 100 feet under the centre of each arch above the level of the flow of the highest of spring tides. The Lords of the Admiralty adopted this recommendation, and communicated its effect to the Secretary of the railway company by letter dated 14th May, 1845.

The Grand Junction Company's Bill was passed in 1846, and contained the two following provisions, inserted at the instance of Mr. Fereday Smith :—

(1) That if at any time thereafter any application should be made to Parliament for a Bill to render the river Mersey east of Warrington Bridge navigable for sea-going vessels, or to make a river or canal navigable for such vessels from any part of the Mersey in the said direction to Manchester, the line of which river or canal should cross the line of the Grand Junction Railway *at the Walton Viaduct or elsewhere*, the Grand Junction Railway Company should assent to a provision in such Bill for authorising a swivel bridge to allow of the passage of such vessels through that railway.

(2) That any bridge or viaduct for carrying the railway over the Mersey at Runcorn should be constructed according to plans and specifications approved by the Admiralty and the Conservators.

There was also a clause inserted in the Act as to damages to the river Mersey to be made good by the railway company, in which it was stated that it was the intention to carry the Huyton and Aston Branch over the Mersey at Runcorn by means of a bridge with a clear headway of 100 feet above the level of high water.

It will be seen that the swing bridge clause follows very closely the terms of the requirement in Admiral Evans' letter of the 28th of March, 1846. The "river or canal" mentioned in the clause is afterwards called (in the same clause) "the said Ship Canal or River".

The railway company did not construct the Huyton and Aston Branch authorised by the Act of 1846; and the construction of a bridge over the Mersey at Runcorn remained in abeyance until the North-Western Company, under the powers of their "Lines near Liverpool Act, 1861," constructed the high-level fixed bridge known as Runcorn Bridge, having a headway of 75 feet above high-water mark at ordinary spring tides. It is to be presumed that this reduced headway was sanctioned by the Admiralty and by the Conservators.

From 1825 the idea of a ship canal simmered in the public mind, and it was constantly referred to in the Press, occasionally in verse, as will be seen from the following extract.

The following lines, contributed by Wilmot H. Jones to the *Manchester Guardian* in October, 1840, were suggested by the arrival of the *Queen*, a coasting vessel, at Warrington :—

MANCHESTER AS IT MAY BE.

Ye Manchester merchants, let politics be ;
The *Queen* is at Warrington, up from the sea ;
And Forrest and Gill, like Columbus of old,
Have shown you the true way to gather the gold.
Your mayors and commissioner trade on our stocks,
And deepen your river, and dig out your docks,
Import your own cargoes close up to your doors,
And warehouse and bond upon Manchester floors ;
Your home and your foreign trade both shall increase,
And yours be the quarters of commerce and peace ;
To London stand next in the ledger of state,
And ships by the score shall scarce carry your freight.

The other song with a rattling chorus (as below) kept the enthusiasm alive, and was sung at many a convivial entertainment :—

THE SEAPORT TOWN OF MANCHESTER.

O dear, O dear, this a curious age is,
Alterations all the rage is,
Old and young in the steam are moving,
All in the cry—improving.

To Manchester there is news come down, sirs,
They're going to make it a seaport town, sirs,
Then 'sted of weavers, spinners, and tailors,
Nought you'll see but ships and sailors.
    Thus 'twill be, I'll bet you a crown, sirs,
    When Manchester is a seaport town, sirs.

When the first ship comes in sight,
The town will be all joy and delight,
Eating, drinking, dancing, singing,
And th' old church bells will crack with ringing.
They'll cover the bridge with tout and prigs, sirs,
Aldermen, too, in their gowns and wigs, sirs,
The heads of the town with all their forces,
The Manchester Mayor, too, drawn by horses.
                            Thus, etc.

They'll cover the river with boats and barges,
Men-of-war ships that never so large is,
Steamers back and forwards towing,
You may ride for nought and they'll pay you for going.
Sailors swearing, spars a batting,
Heve yo ho-ing, handspikes clattering,
Strange sails crowding every day, sirs,
Anchoring in Victoria Bay, sirs.
                            Thus, etc.

The Liverpool gents will be all undone,
Here there will be nought but fun done,
Pats, half wild, running their rigs, sirs,
Landing butter there, bullocks and pigs, sirs.
Then to make us jolly and friskey,
Meally potatoes and barrels of whiskey,
New laid eggs, a twelvemonth taken,
Then all will feed upon eggs and bacon.
                            Thus, etc.

Such lots of goods the boats will bring up,
Store rooms will, like mushrooms, spring up,
To hold the wares of every nation,
The town must have a transformation.
They'll make the Exchange into a storehouse,
Cotton and corn rooms out of the poor-house,
One for grocers to put their figs by,
And the Temperance Hall they'll make it a pig stye.
                            Thus, etc.

In time you will have trade enough, sirs,
Over the world you'll send your stuff, sirs,
Goods for every clime and nation,
Will all come here for embarkation.
Ringley coals, cabbages and carrots,
And in turn receive Poll Parrots,
Baboons, racoons and Spanish donkeys,
Jays, cockatoos, and ring-tailed monkeys

                                                Thus, etc.

In a few years—say perhaps twenty,
Man-o-war ships will arrive in plenty,
Then as the tide of time encroaches,
They'll run 'em about the streets like coaches.
Over the marshes, stones, or gorses,
Tars for jarvies, whales for horses.
But I'll be off—first make my bow, sirs,
For I really believe there's a ship coming now, sirs.

                                                Thus, etc.

In the Manchester Reference Library are a few numbers of *The Herald of Improvement or Manchester as it Ought To Be*.   Its object seems to have been the creation of bonding warehouses in Manchester.   This was bitterly opposed by Liverpool, and No. 2 article of the above periodical, dated May, 1841, showed that out of £4,500,000 of customs duty nominally paid by Liverpool, two-thirds were actually paid by Manchester and district.   It asserted that Liverpool produced nothing, but depended mainly for her existence on the milk of Manchester, from which she took off the cream.   In the same number we find—

THE WAY TO MEND TIMES IN MANCHESTER.

Tune : Rory o' More.

Behold! "Men" of Manchester, now is your time!
Though Liverpool gents cry out it's a crime!
Let's deepen old Irwell that vessels may glide
From Victoria Bridge o'er the Atlantic wide.

Chorus—
    Five feet let us dig—and the coast let us clear!
    One million, at least, it will bring us a year!
    Local strife let us drive to Old Nick in a flame!
    Ship-building's our study, navigation's our game.

The Liverpool gents cry out " Smugglers be wary !
Whate'er you may do, touch not th' estuary !
If you do this, my boys, by the big hill of Howth,
You'll self-murder commit, for you'll stop up the mouth ! "
                                            Chorus.

Five feet let us dig—make the crooked parts straight !
From New York let the steamers bring hither their freight !
Let Turks bring their coffee, dates, rhubarb, and figs ;
And Irishmen butter, eggs, pratees, and pigs !
                                            Chorus.

The Dons of Oporto will bring sparkling wine,
And herdsmen from Scotland their well fatten'd kine ;
The Lascars and Tartars will bring pure Howgua,
With Lapsang, and Souchong, Congou, and Twankey.
                                            Chorus.

Let the boats bring their codfish, fluke, haddock, and sole,
And even fat salmon with manorial toll !
To dock and to bond we must have working men,
And thousands employment are certain to gain !
                                            Chorus.

In 1840 Mr. H. R. Palmer, F.R.S., was instructed by the Old Quay Company to prepare a scheme for the improvement of the Irwell.[1]   About the same time Mr. John F. Bateman (afterwards water engineer to the Manchester Corporation) was also instructed to make an alternative report.   Mr. Palmer produced a most exhaustive scheme, and advocated a 400 to 600 ton ship canal, 12 feet deep, from Liverpool to Manchester, with six locks and draw or swing bridges where the railway crossed the waterway.   With the report was a plan showing how he proposed to straighten the navigation for sea-going ships.   Mr. Bateman was in favour of throwing an embankment across the Mersey at Runcorn Gap (where the river is 400 yards wide), with floodgates having a frontage of 100 yards to admit and discharge water.   By this means deep water would be secured up to Warrington. Both reports were published.

In the *Manchester Guardian* on 6th February, 1841, was an advertisement, stating that on 9th February, 1841, would be published No. 1, of the " *Manchester Gallery of Science and Art.*   Price 4d."   In its first number was an article " Facts Worth One Million Sterling to the Port of Manchester ".   This alluded to a conver-

[1] See Plan No. 3.

sazione at the Royal Victoria Gallery, when Mr. Thomas Ogden Lingard read a paper on "The Improvement of the Rivers Mersey and Irwell for Sea-going Vessels to Manchester". Mr. Lingard, who was agent for the Mersey and Irwell Navigation Company, read to the audience the principal portion of Mr. Palmer's report, and illustrated it by reference to a large plan of the river that was hung on the wall behind him. He advocated the amalgamation of the Bridgewater Trust and River Navigation, in order to give the greatest possible benefit to the trade of the district. The discussion on the reports lasted four days or sittings. Sir William Fairbairn, Sir John Hawkshaw and many eminent engineers took part in the discussions. Strange to say, three of the gentlemen present, Mr. Bateman, Mr. George White-head and Alderman Curtis, lived to take part in the Ship Canal movement forty-one years afterwards. The article in the *Manchester Gallery of Science and Art* of 1841, written on the subject, shows that Manchester at that date was groaning under trade disabilities :—

At a time when the artisans of Manchester are suffering great privations from the high price of provisions and the scarcity of labour, we find the Corporation of Liverpool applying last session for a loan of a million pounds of the public money to build new ware-houses to hold the property of the merchants of Manchester in bond. Where is the public spirit of this great manufacturing metropolis of the world? Does the whole district contain no local patriot to rid it from the wasteful outlay of time and capital in having the goods of its hourly consumption bonded in Liverpool? Why, the very fact of the Corporation of Liverpool wanting a million of money for the purpose of building additional warehouses ought to arouse the people *en masse*. The money that is worse than thrown away upon Liverpool by Manchester in one year would gladden the hearts of thousands of working families by an additional income of 10s. per week. Surely the merchants of Manchester cannot be aware that a vessel called the *Nemesis*, 650 tons burthen, 50 men, 168 feet in length, 29 feet beam, engine 120 horse-power, and drawing only 4½ feet water, has doubled Cape Horn and arrived safely at Ceylon. With 5 feet of water up to our very doors, have we amongst us neither science nor enterprise to imitate the noble example of the people of Glasgow, who by dredging have so improved the navigation of the Clyde, that where they had only 2 feet of water they have now 16½ feet, and vessels from China discharge their cargoes at the Glasgow quay? Are the people of Manchester so blinded to their own interests as quietly to look on whilst Liverpool increases her warehouses and accommodation for the reception of the bonded goods of Manchester merchants, so as to increase her asserted claim for vested interests? Would not the assessment alone of these warehouses (which bring no poor) materially reduce our poor rates.

Would not the wages of the bonded establishment, which would be many hundred pounds a week, be a great help in ready money to our shopkeepers, to say nothing of the deplorable fact that at the present time 25 per cent. of the assessment is rendered uncollect-

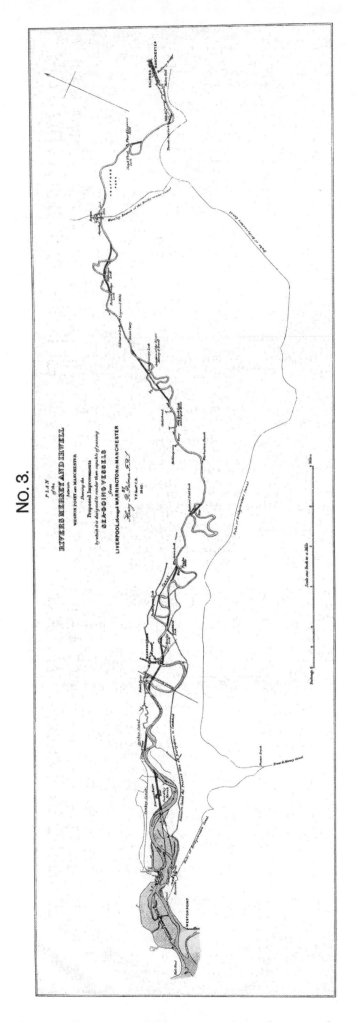

NO. 3.

The material originally positioned here is too large for reproduction in this reissue. A PDF can be downloaded from the web address given on page iv of this book, by clicking on 'Resources Available'.

able by empty houses, which would be filled in the event of our having bonded warehouses in Manchester? We would ask the spinners of Manchester, Bolton, Ashton, Hyde, Stockport, Oldham, etc., if, when a prospect appears of trade improving, they do not find themselves all at one time in Liverpool, and are they not satisfied of the fact that their presence has been to raise the market considerably, the brokers and speculators of Liverpool acting as one man, and taking advantage of the demand? We would ask them again, having a knowledge of this fact—*viz.*, that all the cotton arriving at Liverpool for the consumption of Lancashire must first come to Manchester—if it improves that cotton in quality, or increases its quantity by carting it first up the town of Liverpool and warehousing it, and when wanted for consumption re-carting it to the wharf to be sent up to Manchester, and further that large sums should be paid for brokerage to see this done? And further, would not all the trouble and expense of going to Liverpool to buy cotton and carting it up and down the streets of that town and the brokers' charges for the same be saved by the spinner of that cotton if it was taken at once from the import ships and sent direct to Manchester under bond, the duty to be charged there? This is performed every week on the Clyde, and not unfrequently on the Thames. Building sites are far cheaper in Manchester than in Liverpool, and employment would be given to our artificers, and when people had their property at their own doors they would look well after it, and thereby make an immense saving in leakage, carriers' samples and rats (*sic*). Manchester has now been asking Government for five years for what she ought long since to have demanded as a right—the privilege of having bonded warehouses. Liverpool has proclaimed the mighty advantage that privilege would be to Manchester by her energetic and determined efforts to prevent its being granted. The opposition of both Liverpool and London is not based upon sound policy, but upon selfish grounds; yet, so powerful will it be that Manchester will not obtain this important right without the most unflinching and determined perseverance. Since the advantages to be derived from the saving in dock and town dues only would, as shown by Mr. Lingard at the conversazione at the Gallery of Practical Science, afford an expenditure of at least two millions of pounds in improving the navigation, it is much to be hoped that our merchants and manufacturers will, with unity of hand, of heart, and of purse, follow the advice given in *Manchester as it May Be*.

Liverpool then, as afterwards, was death on the new scheme. A Liverpool paper declared that "meddling with the river would be an act of felony".

Singular to say, Dr. Clay, a well-known Manchester surgeon, took upon himself to combat Mr. Palmer's report, and in 1841 wrote *Mersey and Navigation Improvements Geologically Considered and New Plans Suggested.*

In the month of November, 1845, Mr. James Acland, who lived at a house (now in ruins) called "Acland's Retreat," in Cheetwood, issued a prospectus as sole promoter of what he called a "First section of the Mersey and Manchester Ship, Railway and Dock Company," capital £3,000,000 in 100,000 shares of £30 each, deposit £3 3s., three shillings per share being required for preliminary expenses, payable on allotment of shares. This gentleman was an Anti-Corn Law League lecturer, and was connected with the Stockport Press. He issued his address from his private house, and is believed to have had several public meetings to consider his proposals. The prospectus was full of facts and figures, and he got very wroth because a Liverpool editor "launched his puerile shafts of undisguised malice at the infantile port of Manchester". The scheme, however, soon died a natural death.

From 1845 onwards, the trade of the country improved, and things were prosperous till 1861, when the American War plunged Lancashire into dire distress. The same cause (scarcity of cotton) affected every other manufacturing country, so that when the war ceased there was a great chasm to fill, and spinners and weavers had a brisk time in supplying the wants of the world. This went on during the Prusso-Austrian and Prusso-French Wars; the producing power of the Continent was almost at a standstill, and the trade of England prospered accordingly for another ten years. Germany, then strengthened by the immense subsidy paid by France, became free to turn her attention to commerce, and soon became a competitor in the markets of the world; other nations followed suit, and people began to feel they must produce more cheaply if they were to hold their own. Traders sought economies, and found that the railways in England were charging about double as much as on the Continent; not only so, but the goods of foreigners were being carried into England by means of through rates cheaper than their home manufactures. Further, that Liverpool sat as a toll bar and levied heavy contributions on all imports and exports. Also that the cheaper carriage of goods by the internal waterways was being destroyed by railways getting hold of canals and navigations, and throttling them off by degrees. Notable instances occurred in Lancashire as before mentioned. The power thus acquired was exemplified on the Bolton Canal by the owners stopping it for eight months, because they would not repair sinkages from coal workings, whilst they repaired the adjacent railway and kept it in constant use, thus throwing the canal employees out of work, and making colliery proprietors and mill-owners on the canal bank carry by railway. These facts, and the decadence of Manchester in consequence of so many trades being driven away by disabilities, induced the writer to take up his pen and advocate in the Press (at first anonymously) the renewal of a ship canal scheme. He felt at least an effort ought to be made to stem the current that was reducing Manchester to the position of a second-rate city. If she had to trade with a weight on her

GEORGE HICKS, AUDITOR OF THE MANCHESTER SHIP CANAL, 1886
SEQ.

Franz Baum.

To face page 72.

shoulders, her manufactures must decay and her people be impoverished. It was absolutely necessary she must have as cheap carriage as her competitors. Not only were canals stifled but the Mersey and Irwell Navigation, once an important waterway, was now reduced to an almost unnavigable, huge sewer. Other people from time to time kept writing to the papers urging action and suggesting the possibility of a ship canal, which in years gone by had been favourably reported upon.

Correspondents pointed out that the cotton and other industries of Manchester were dwindling away. The sugar trade had departed; several machine shops were closed; some, like Sharp, Stewart's, were attracted to other localities, where they could carry on their business more advantageously. By this time, too, the Suez Canal was proving a success. Engineers and commercial men at first called it a mad undertaking, and shipowners said they would never use it. They were wrong. If a canal could be made through a desert 101 miles long, why could not one be made to the sea *via* Liverpool when traffic would be assured?

In the autumn of 1876 Mr. George Hicks, a Scotch gentleman of commercial training and progressive ideas, during a walk happened to notice the neglected river, with boats actually stuck fast in it, and reflected what a prostitution it was to see a river, which might be made so useful, utterly neglected. On his return home he wrote a letter (11th October, 1876), to the *Manchester Guardian*, expressing his regret that, whilst commerce was groaning under heavy freight charges, such a good and cheap avenue was lying idle and becoming a nuisance. He suggested it might be converted into a ship canal. This attracted the attention of a London engineer, Mr. Fulton. He wrote to Mr. Hicks to say he had just successfully restored the Nene navigation, and should be very glad to deal with the Mersey, expressing his willingness to come down, further examine the river, and then confer with any leading citizens Mr. Hicks could get to meet him. In December, 1876, Mr. Hicks had an interview with Mr. Fulton in London, and in February, 1877, he got up a petition to the Chamber of Commerce asking them to inquire into the possibility of making the river navigable for deep-sea steamers. In addition he got Mr. Fulton to make an engineering and Professor Boyd Dawkins a geological report, whilst he himself drew up a financial and commercial review of the aspects of the scheme. He then arranged with the Chairman of the Chamber of Commerce, Mr. Edmund Ashworth, that a petition and the reports should be brought before that body; and to prepare the public mind he wrote an article on "The Irwell Navigation Scheme," in which he dealt with the engineering difficulties that would have to be encountered.   On the 23rd April, 1877, a special meeting of the Chamber of Commerce was held to receive and consider Mr. Fulton's scheme.   A model of the proposed tidal navigation and docks was exhibited, and after a discussion that body passed a resolution that "it would be of the greatest service to the interests of Manchester and the trade of the district to have an improved waterway".   Subsequently Mr. Fulton's model was exhibited at the Royal Exchange, and at the request of Alderman Walmsley, Mayor of Salford, it was on view for some time in the Salford Town Hall.   Indeed that gentleman suggested that his son, Mr. Clement Walmsley, a young solicitor, should co-operate with the promoters and obtain signatures to a memorial to the Salford Corporation.   On 5th January, 1881, this was considered, and a resolution moved asking for a Government inquiry *re* the application of the Irwell for navigation purposes.   The motion was, however, negatived. In consequence of commercial depression and the Building Society crash all enterprise came to a standstill.   The idea of a ship canal to Manchester, though in general favourably entertained, remained in temporary abeyance   In October of the same year Sir William Harcourt, in an eloquent address, eulogised the energy of the Glasgow merchants, who, overcoming all difficulties and opposition by their splendid local patriotism, had brought the sea to their doors, and made Glasgow an important and busy port.   His text virtually was : "Heaven helps those who help themselves".   This speech caused quite a sensation, and brought a crop of letters in the papers; amongst others one on the 29th October, 1881, from "Mancuniensis" (Mr. J. W. Harvey), full of cogent reasoning and convincing argument. The idea "Go and do likewise" possessed the minds of many enthusiasts, and also of traders who were groaning under heavy and unjust charges for carriage.   They reasoned, "If Glasgow has achieved success, why should not Manchester follow suit and try and revive her industries?"   Amongst others who were stirred by the correspondence was Mr. Daniel Adamson, of Hyde.   From the time the question was brought before the Chamber of Commerce he had been deeply interested in the scheme for bringing a ship canal to Manchester.   The engineering difficulties even had a charm for him; he liked to try and master complex problems of the kind, and once he decided in his own mind that the canal could be made he threw himself into the work.   He had been approached early on by Mr. Hicks, who explained to him the commercial advantages and financial prospects, and he had consulted his old friend, Mr. Abernethy, the eminent engineer, upon it, who formed

WILLIAM J. SAXON, SOLICITOR TO THE MANCHESTER SHIP CANAL,
1883-92.

Higginson Bowdon.   *To face page* 74.

a favourable opinion.   His health, however, was bad; indeed it had been necessary that he should leave business and go abroad for some time.   Fortunately a change of air and rest did him great good, and his known indomitable energy pointed him out as just the leader who was wanted.   When it became necessary to draw up preliminaries the services of Mr. W. J. Saxon, of the firm of Grundy, Kershaw & Co., were enlisted, and there can be no doubt that to him is largely due the success of the subsequent Parliamentary proceedings.   An able and shrewd lawyer, resolute and determined in character, wise in counsel, of untiring energy, he threw himself heart and soul into the work, and a history of the canal would be incomplete if it did not bear witness to the Herculean task he achieved in fighting the Bill through, in face of the most powerful and persistent opposition that was ever offered to a Parliamentary Bill.   Poor fellow! there can be no doubt that his health suffered through his devotion to the work.   He died in harness; his name ought ever to be green in the memory of the people of Manchester.

Before Mr. Adamson consented to put himself at the head of the undertaking, several other gentlemen had been approached; among the rest, in 1878, was Mr., now Sir William Bailey.   Other gentlemen who met in 1879 at the office of Mr. Clement Walmsley were: Mr. O. O. Walker, M.P.; Mr. W. T. Charley, M.P.; Mr. Edward Walmsley, J.P., Stockport; Mr. J. H. Walmsley, J.P., Salford; Richard Haworth, Esq. and H. J. Leppoc, Esq.   They agreed: "That it would be of great advantage to Manchester, Salford and the surrounding towns to have a waterway to the sea for ocean steamers and other vessels; and that it was of the utmost importance that the proposed undertaking should be proceeded with".   It should also be mentioned that efforts were made to induce Lords Winmarleigh and Derby to head a movement for a ship canal; both were favourable, but declined to take an active part; however, they ever afterwards looked very kindly on the canal, and their Parliamentary services in vanquishing Lord Redesdale and his obstructive policy were invaluable.

CHAPTER VI.

1882.

DANIEL ADAMSON HEADS THE SHIP CANAL MOVEMENT— MEETING AT HIS HOUSE—PROVISIONAL COMMITTEE APPOINTED — RESOLUTION OF THE CITY COUNCIL— SCHEME ADOPTED—SUBSCRIPTIONS—OPINIONS OF THE PRESS—THOROUGH ORGANISATION.

You may have buildings here on the Pool Banks, worth more than £20,000, if God send peace and prosper trade.—Sir Edward Moore's advice to his son.

PRIOR to 1881, as has been recorded, efforts had been made to rouse public feeling in favour of a ship canal; odd letters and articles kept appearing in the local journals, mainly the result of Mr. Fulton's appeal to the Chamber of Commerce in 1877, which seemed to keep simmering in the minds of the commercial community. Trade was rapidly becoming worse, old manufacturing concerns were either giving up business or moving into districts where the rents and taxes were lower. The value of property had gone down and houses were emptying fast. Then there came more pronounced rumblings. People wanted to find a reason for the decay of Manchester. When things were prosperous and money was being made, little care was taken to go into details of expense, but when the reverse was the case, costs were carefully scanned. People's eyes had been opened by the differences between the Liverpool dock authorities and the railway companies. Liverpool said railway carriage to and from that port was the dearest in the country when, from the volume of trade, it ought to be the cheapest. The railways retorted that the docks in Liverpool were the worst equipped and the dearest of any in the country, except London. Manchester traders began to see they were working with a mill-stone round their necks, that they were handicapped in their business, and could not compete with other districts where goods and raw materials could be more cheaply imported or exported. The toll bar at Liverpool was strangling industries, and railways, by securing canals, were putting an end to cheap water carriage. "Argonaut" in December, 1880, wrote :—

Why cannot we pay the same attention to our waterways that America, France, Holland, Belgium and Hungary are paying to their waterways? Why should not the local project for the improvement of the Mersey-Irwell channel be taken up vigorously instead of being discussed in the half-hearted way it is from time to time? Steamers making ocean voyages would count the extra 30 miles to Manchester as a mere nothing, and would come up here for freight at the bidding of shippers. The freight which costs 10s. per ton would be virtually abolished, and the saving accomplished by the avoidance of the heavy Liverpool charges would more than pay the navigation tolls. This question of making Manchester a port is of very great importance. The freight involved in Manchester is far in excess of the amount computed by those who laid the project before the Chamber of Commerce a few years ago, etc.

"X." wrote to the *City News* :—

The immense value that a ship canal from Liverpool to Manchester would be to this district has been felt by me for thirty years. But I am far from being singular in my feelings. At times the idea has seemed to take possession of the mind of many people, but from the want of public spirit it has always been allowed to drop. I am afraid this will be so again. There is a painful want of unselfish public spirit amongst our leading wealthy men. So long as they are heaping up great fortunes, they do not care to put themselves to any great labour for the prosperity of the city. They come to town for business only, and leave it as soon as their pecuniary interests will allow. There is a terrible absence of sympathy with our common life, which shows itself in so few persons manifesting any desire to have their names associated with anything concerning the town. What will create public spirit for the promotion of a ship canal? An appeal to the selfishness or unselfishness of the public? I am afraid there will have to be an appeal to both.

The questionable transfer of the Bridgewater Canal into railway hands induced the author to urge in the papers the emancipation of waterways. In May, 1882, he wrote in the *City News* :—

Manchester does not keep pace with neighbouring towns as regards general prosperity. Walk through what used to be our busiest districts, such as Ancoats, and we find many shops closed and half of the workshops and mills empty. In the centre of the city the array of empty property is distressing, and the last census shows a decrease of population, and that 10 per cent. of the habitable houses are unoccupied. Large employers of labour have gone where taxes are light and land and labour cheap. In the future the prospect is that our city will cease to be a producing, and become entirely a warehousing and commercial centre. If this be true, a ship canal is an essential element of success, and I cast my lot most heartily with those gentlemen who desire it to be carried out with the view of reviving the trade of the city.

The above are extracts from three of the earliest letters that were written, and they were followed by many others generally showing the possibilities and advantages of a ship canal, and urging that the work done at Glasgow and other ports might equally well be done in Manchester if the city would rise to the occasion, determined to free itself from the monopolies exercised by the railways and at the port of Liverpool. It would be impossible even to give a *résumé* of the correspondence in the newspapers, but I purpose to attach an appendix with a few of the most important letters. Occasionally an opponent of the canal took up the cudgels on behalf of Liverpool or of the railway companies. There were also some able letters written by merchants of Manchester who did not see their way to support the canal, notably by Mr. James Angus, who wrote under the pseudonym of " Mercator ".

There can be no doubt that the success of the Ship Canal was largely due in its initial stage to the forcible and convincing pamphlet issued in May, 1882, by Mr. J. W. Harvey under the pseudonym of " Mancuniensis ". Mr. Harvey was a clerk with Mr. George Hicks, agent for the North China Marine Insurance Company. For years Mr. Hicks had been striving to bring before commercial circles in Manchester the necessity of cheapening carriage by means of a ship canal. His clerk threw himself heart and soul into the work, and being of a studious turn of mind, consulted all the books that had been written and the speeches that had been made on the subject, besides searching out a vast number of valuable statistics. These he gathered together under the heading of " Facts and Figures in Favour of a Tidal Navigation to Manchester, showing how to solve the cheap transport problem for the great import and export trade of Lancashire and the West Riding ". " Mancuniensis," by a comparative diagram, showed the immense population in the towns round Manchester. He maintained that their industries were crippled by dear carriage, and prevented from competing advantageously with foreigners, and even with cities, like Glasgow, which were situated near the sea. He demonstrated how inland towns were handicapped and their trades driven away by shipping dues and extravagant railway rates. Then he showed how rivers, canals and railways had in turn superseded one another, but through becoming monopolies had not advantaged the public. He pointed out how a toll bar at Liverpool sapped the vitality of Lancashire industries, and how the proposed navigation would result in immense saving, prevent goods being injured in transit, reinstate old industries and create new ones, develop important coal-fields and cheapen the food of the people. He traced

the various efforts that had been made in time past to bring a ship canal to Manchester, and how our ancestors had left the way clear by compelling the various
railways crossing the Mersey to give swing bridges when required.   That more
shipping accommodation was necessary was evident by the crowded state of the
Liverpool Docks, and the figures given made it clear that Manchester was even
better situated than Liverpool as a port of distribution.   The Clyde, the Tyne and
the Tees had all become prosperous by widening and deepening the rivers, and why
should not Manchester follow suit? especially when the Suez Canal had, in spite of
evil prognostications, turned out so well.   "Mancuniensis" then showed the rapid
strides made abroad both in improving the navigation of rivers and in making artificial
canals; the immense sums that were being spent upon them; and how goods were
being carried at ridiculously low rates as compared with what was being charged in
England.

This pamphlet had an immense sale, passing through many editions.   Originally sixpence, it afterwards was printed in a cheaper form and sold for one penny.
It was pregnant with facts verified by authenticated figures, and written in terse
and plain language that everybody could understand.   It had the effect of making
people think, and stimulating them to see that if Manchester and Lancashire meant
to hold their own, they must be up and doing, or other places would run away with
the kernel and leave only the shell.

The first paper to take up the Ship Canal cause was the *City News*.   In an
encouraging article of the 27th May, 1882, dealing with my letter, it pointed out
that cheap transit was as urgent a necessity as cheap production, and that in consequence of the existing railway and canal combinations, only a Ship Canal could
substantially reduce the cost of carriage from the manufacturing centres to the
distant markets of the world.   Further, that the excessive rates charged for
carriage between Liverpool and Manchester handicapped our industries.   Notwithstanding there were now five routes, the cost of carriage had been doubled,
and it was from 25 per cent. to 40 per cent. higher than the cost of carting by
road.   The editor asked its readers to study the valuable pamphlet recently issued
by "Mancuniensis," and endorsed his conclusions that a Ship Canal would cheapen
food, bring a hive of new industries into the district, give an opening for the
collieries of the neighbourhood to supply ships, create new markets and stimulate
old ones, raise the value of property, and in all probability effect a saving of at least
£1,000,000 a year to Manchester and district.   He prophesied fierce opposition

by vested interests in Liverpool and elsewhere, but counselled the promoters not to be daunted by early defeats, but to return to the charge till they won the day.

An extinct volcano, the *Lancashire Figaro*, in its issue of the 8th June, 1882, after some very complimentary remarks on several letters I had written to the *City News* said :—

> I trust Mr. Leech's pluck in advocating this scheme single-handed will meet its reward. The Corporation don't feel disposed to help the agitation. The railway magnates—monopolists, I ought to say—cannot be expected freely to give in an adhesion. Mr. Leech justly appeals to those capitalists who are immediately interested by their occupation as manufacturers, whose cost of production would be so materially affected. Mr. Leech ought to go further. He ought to appeal boldly to every citizen, for all—be they rich or poor—would be benefited by the carrying out of the proposal.

In the summer a few gentlemen, including Mr. Peacock (Beyer, Peacock & Co.), Mr. Lloyd (Hickson, Lloyd & King), Mr. Samson (Grundy, Kershaw & Co.), Messrs. Fulton, Leader Williams, Henry Whitworth, and others, took a trip down the river in a barge. They were met at Warrington by Alderman Davis, Dr. Mackie, Mr. Bennett (landowner), Mr. Darbyshire and Mr. W. H. Brooke (town clerk). After an inspection of the waterway all seemed impressed with its possibilities and that they had a practicable scheme before them.

In June, 1882, Mr. Hicks waited on me with a message from Mr. Adamson asking me to assist in the Ship Canal movement. He had seen my letters in the newspapers and knew that I was an enthusiast on the subject. I gladly consented.

The real start of the canal dates from a memorable meeting held at the Towers, Didsbury, on 27th June, 1882. It was summoned by Mr. Daniel Adamson, to whom the credit is due not only of giving this great enterprise an effective start but of courageously fighting its battles in face of tremendous odds till success was achieved and the Ship Canal Bill passed. Since the time when Mr. Fulton's scheme was before the Chamber of Commerce, Mr. Hicks had been vainly endeavouring to find a man of influence and determination to head the movement. At length, through the medium of a friend, he was introduced to Mr. Adamson, who, after going into the merits of the scheme with Mr. Fulton and Mr. Hicks, undertook to captain the enterprise. As a first step he invited the Mayors of Manchester and surrounding towns, the heads of the principal commercial houses in the city, the leaders of co-operative and labour movements, and several well-known

MANCHESTER TIDAL NAVIGATION

PLAN AND SECTIONS

ENGINEER HAMILTON H. FULTON M. INST CE 22 GREAT GEORGE STREET WESTMINSTER

No. 4.

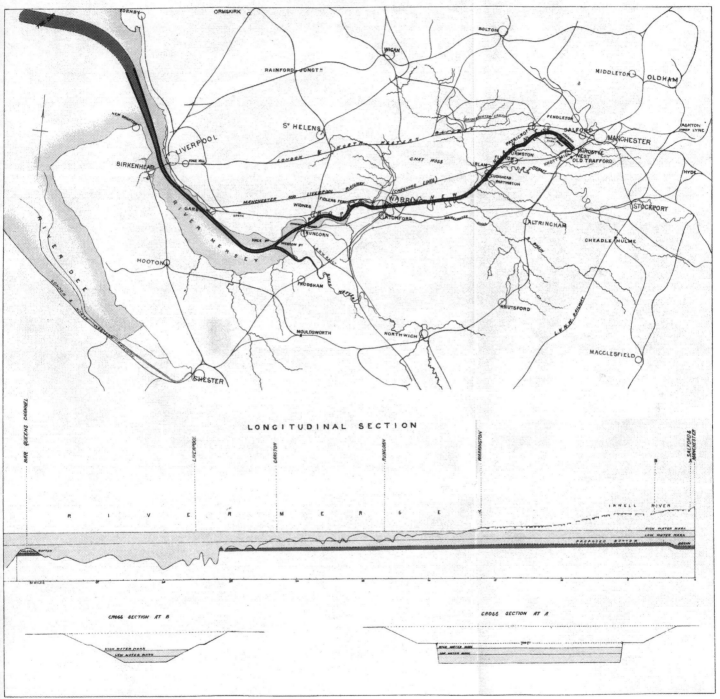

capitalists to his house, some to dinner and others to an evening meeting. There were present thirteen representatives of large Lancashire towns and fifty-five leading merchants and manufacturers. To meet them he asked Mr. Fulton, C.E., who for some years had been advocating a ship canal, and also Mr. Leader Williams, late engineer to the Bridgewater Navigation Company, who was known to have valuable local knowledge in respect to waterways. The historic evening meeting was attended by the Mayors of Salford, Ashton, Warrington, Macclesfield, Stockport, Rochdale and Stalybridge. Mr. John Rylands, Mr. Thos. Ashton, Mr. P. Spence, Mr. J. A. Beith, Mr. J. Thewlis Johnson, Mr. E. B. Dewhurst, Mr. Andrew Walker (Messrs. T. G. Hill & Co.), Mr. C. S. Agnew, Mr. Rd. Peacock, Mr. T. R. Wilkinson, Mr. W. W. Hulse, Mr. J. P. Higson, Mr. Henry Boddington, junior, Alderman Thompson, Mr. C. P. Henderson, Alderman Walmsley (Salford), Messrs. W. Richardson and S. Andrew (Oldham), Messrs. H. Bleckly, F. Monks and Alderman Davies (Warrington), Mr. A. C. Boyd (Dukinfield), Mr. J. H. Nodal, Mr. H. Dunckley, Mr. Mitchell (Co-operative Society), Mr. G. Hardman (Limited Liability Association, Oldham), Mr. Tomkins (Sharp, Stewart's), Mr. Hicks, Mr. Marshall Stevens, myself and others. The meeting was held in the large hall at the Towers.

Mr. Adamson was in good form, and with a strong Northumbrian burr reverted to the wonderful success attending the improvement of the Tyne, the Tees and the Clyde, and felt sure the Mersey was amenable to similar treatment with even better results. If the Suez Canal, situated in a barbarous country and where for 50 miles there was a solid cutting of the depth of 26 feet, could be carried out, there ought to be no engineering difficulties to stand in the way as far as the Mersey was concerned. His own impression was there ought not to be any interruption in the shape of locks, overhead bridges nor any lack of water for navigation. He advocated coming into Manchester at a low level and having an underground railway connecting with all parts of the city for the distribution of goods. He advised all present to read the pamphlet by "Mancuniensis," bristling with valuable facts and information, and he commended the scheme, believing it would be very advantageous to the constructors and a mighty blessing to Lancashire and Yorkshire.

Mr. Fulton then entered into a lengthy description of the engineering features of his scheme, explaining it would be tidal with 22 feet at low and 37 feet at high water spring tides, and that the basin at Manchester was to be of an area of $128\frac{1}{2}$

acres, 8,000 feet long and 700 feet wide.   The canal would be 228 feet wide at the top and 80 feet at the bottom, estimated cost £4,500,000.[1]

Mr. Hicks placed before the meeting the statistical aspects of the scheme. Four-fifths of the export trade of Liverpool passed through Manchester.   He believed Manchester would get 5,000,000 tons of traffic which at 3s. per ton, plus dues on ships and rents, and minus expenses, would leave a revenue of £750,000 per annum.

The Chairman then moved a resolution in favour of a tidal navigation, which was seconded in a cautious speech by Mr. Bleckly, of Warrington.

The Mayor of Salford (Alderman Husband) moved that a Committee be appointed to inquire into the best means of carrying out the project, to consist of Aldermen Walmsley and Davies, the Mayor of Stockport (Mr. James Leigh), and Messrs. Daniel Adamson, Henry Bleckly, Richard Peacock, John Rylands, W. Richardson and J. R. Pickmere, with power to add to their number.   The mover said he did not believe that there would be a grander sight under the canopy of Heaven than the docks at Manchester crowded with shipping from all parts of the world.   Alderman Walmsley moved "that the Provisional Committee be empowered to obtain a detailed survey by a competent engineer or engineers for the purpose of ascertaining approximately the cost of the construction of the proposed tidal navigation, and should the report of such engineer or engineers be satisfactory, that the Committee be empowered forthwith to form a company to be called the 'Manchester Tidal Navigation Company'".[2]

Another resolution was passed to establish a guarantee fund of £25,000 for preliminary expenses.   A few curious incidents connected with this remarkable meeting deserve recording.   One of the gentlemen asked to speak to a resolution was Mr. John Rylands, always an ardent supporter of the Ship Canal, but on this occasion he took the side of caution, and made a speech that threw a wet blanket over the meeting.   When he came to reflect on what he had said he was so vexed with himself that he came to town extra early next morning to see if he could do anything to qualify or nullify his speech of the preceding night, and expressed himself willing and anxious to give the Canal all the support in his power.

One of the most enterprising and courageous men of his day was the late Mr. Hilton Greaves, of Oldham.   He loved anything that had a dash of risk or adventure, and in business he recognised that difficulties were only raised to be

---

[1] See Mr. Fulton's Plan No. 4, in Pocket.   [2] See Prospectus, Appendix No. I.

conquered.   Therefore, when Mr. Adamson gave Mr. Hicks a *carte-blanche* Mr. Hilton Greaves was one of the guests invited by him.   Careless of appearances as he always was, Mr. Greaves, when the time of the appointment drew near, jumped on his horse and in mill costume rode from Oldham to Didsbury.   When he arrived at the Towers and saw a grand residence, with gorgeously liveried flunkeys in attendance, he was aghast, and, after a word with Mr. Hicks, he backed out and returned post-haste to Oldham.   But this little adventure did not daunt him.   At least once a week he used to invite me to his office, where, over a cup of tea we sat for an hour or two discussing Ship Canal affairs; and, when in the end he had his doubts removed, there was no more enthusiastic and liberal supporter of the Canal.   In its darkest days he cheered it on and he never lost heart.

Mr. Adamson's meeting was the means of introducing a gentleman who has played a most important part in Ship Canal history, and it was by the merest chance he got there.   Mr. Marshall Stevens (then in partnership with Mr. Nicholson at Garston) called on my agents, Messrs. Isaac Neild & Son, about renting some wharf land, with the intention of introducing water carriage, and thus cheapening freights from Liverpool to Manchester.   The Ship Canal project was casually mentioned, and my agent said his principal was much interested in it as a promoter, whereupon an interview was arranged.   I got an invitation for Mr. Stevens and took him with me to Mr. Adamson's meeting, after which he became a supporter and an enthusiastic worker.

The statement that the river would not allow the passage of large ships because of the bridges received a contradiction from a most unexpected source.   Captain Dutton, of the Allan Liner *Sardinian*, writing from Liverpool, and giving a sailor's opinion, said heavily-rigged ships were not a necessity, and that it was an easy matter to arrange for a 3,000 ton ship to pass under a bridge even with a height of 50 feet.

The first meeting of the Provisional Committee, to which my name had been added, was held on 7th July, 1882, at the rooms of the Frodingham Iron Company, in St. Ann's Square, and I well remember Mr. Adamson taking the chair.   In a spotless get-up and white waistcoat he quite beamed upon us.   In his opening address he said God had graciously restored him to health after a most serious illness, and that he was determined to show his thankfulness by helping to carry out the great Canal scheme, which he believed would benefit, nay even restore, prosperity to the county to which he owed so much, and in which he had made his money.

At the same meeting Mr. Henry Whitworth was appointed secretary, and Mr. Hamilton Fulton, C.E., of London, and Mr. E. Leader Williams, C.E., were instructed to make the necessary surveys, and to prepare a joint report and estimate, to be submitted to a future meeting of the Committee, and this, with the least possible delay.

At the next meeting of the Manchester Tidal Navigation Committee, held on 15th July, the name of Mr. Marshall Stevens and others were added to the Committee, and a hope was expressed that a guarantee fund of not less than £25,000 would quickly be raised to meet the preliminary expenditure.

The first Provisional Committee of the Tidal Navigation Company consisted of the nine gentlemen nominated at Mr. Adamson's house with the addition of:—

John A. Beith (Beith, Stevenson & Co.), Manchester.
A. C. Boyd, J.P., Dukinfield.
George Hicks, Manchester.
Bosdin T. Leech, Manchester.
C. P. Henderson, Jun., J.P. (G. & R. Dewhurst), Manchester.
Richard Husband, J.P., Mayor of Salford.
Henry Nevile (Agent, Clowe's Estate), Manchester.
Peter Spence, J.P., Manchester.
Frank Spence, Manchester.
Marshall Stevens, Garston.
Edward Walmsley, J.P., Stockport.

A list issued on 7th September, 1882, contains the following additional names:—

James Boyd, Manchester.
Colonel Bridgford, Manchester.
Edward Collinge, J.P., Oldham.
Wm. Henry Johnson (Johnson, Clapham & Morris), Manchester.
Gibbon Bayley Worthington, J.P., London.

When the Committee next met it was announced that further sums had been subscribed, and that Warrington had sent word they would support the scheme, believing it would be conducive to the interests of the town, and also prevent flooding by the river.

At each following assembly of the promoters adhesions of the leading men in Lancashire, accompanied by contributions to the guarantee fund, were announced, chiefly stimulated by the Manchester Tidal Committee's explanatory circular.

With the exception of the *City News*, the attitude of the Manchester Press was cautious.   They gave a full description of the scheme, but took good care to check

enthusiasm by pointing out the immense difficulties that would have to be sur-
mounted. The *Times* spoke favourably of Mr. Fulton's efforts for the last six
years to make a ship canal, but approved of more minds than one being engaged
on the job, and wondered at the prescience of our forefathers, who sixty years ago
secured a clause for swing bridges. Though in the past the whole scheme had
been considered visionary, they believed the time had now come when the people
had become clamorous for cheap communication by water, with a desire to copy the
example of Glasgow, Newcastle, and the Tees.

The *Morning Advertiser* of 24th July, 1882, did not wonder at Manchester
asking why Liverpool should levy a toll on everything that passed through. The
answer was easy: "Make a canal to Manchester broad enough and deep enough—
like that of Amsterdam and Bruges—so that American ships may sail up and discharge their cargoes without transhipping a single ounce". The editor doubted
the estimate of four and a half millions. "But even allowing the works cost
£15,000,000, which would probably be nearer the mark, a toll of 3s. per ton would
allow 5 per cent. if 5,000,000 tons passed through the waterway."

The *Liverpool Post* believed the scheme wholly chimerical; a promoter's
project; an engineer's dream, which, if even it were allowed to translate itself into
fact, would spread disappointment and ruin broadcast. Manchester was not as
prosperous as it used to be. Its recovery from the last commercial crisis had been
slow and doubtful. It did not figure as well in the census as had been expected.
Its political hegemony had been transferred to Birmingham; and so it had conceived the grandiose scheme of repairing all this by making a ship canal. "It will
not do: all nature and history are against it." It was a remission from half-barbarous
times when ships went inland to be out of the way of pirates, etc.

From many surrounding Corporations and from different trade societies the
Provisional Committee received the heartiest assurances of support.

The Salford Town Council, at its meeting in August, 1882, adopted a memorial
to the Board of Trade in favour of the tidal navigation scheme.

As early as July in the same year the writer had placed the following resolution
on the agenda of the Manchester Corporation, but, in consequence of the pressure of
business, it did not come on for consideration till the second Wednesday in September,
which was specially allotted for the purpose :—

That the question of widening and improving the rivers Irwell and Mersey, and so
making them navigable as a Ship Canal, be referred to the General Purposes Committee of
the Council.

The following is an extract from the Council's proceedings :—

Councillor Leech confessed to feeling some difficulty in bringing business before the Council which did not arise out of the proceedings of any of its Committees. But he saw that the Corporations of Oldham, Salford, Warrington, Stockport and other towns had already taken up the question of a ship canal, and he was afraid that as nothing had yet been done by Manchester with regard to it, an impression would be created that there was in the City Council considerable apathy on the subject, if not hostility towards it. This question of tidal navigation had been advocated for the last sixty years, during which various schemes had been spoken of for bringing sea-going vessels to Manchester, and at the present time a number of gentlemen, persons of influence in the city, feeling that Manchester was placed in a most unfavourable position in consequence of the heavy charges now imposed by the railway companies or levied at Liverpool, had joined themselves together to see if something could not be done to remove the load which weighed upon the commercial community and public of Manchester and neighbourhood. These gentlemen had called in the aid of eminent engineers, and it was hoped that the result of the surveys that were being made would convince the people of this district that the scheme contemplated was practicable. Either Manchester was losing thousands of pounds weekly, or the public mind was being disturbed without just reason, and he thought it was due to the public that the Council should now take cognisance of the movement. It seemed to him that the Irwell was at present practically wasted, and he wished members of the Council would go down and see for themselves how little was done with it, and how much better use might be made of the river. He had spoken to several engineers on the subject, and they all agreed that the scheme could be carried out and that it would be of vast importance to Manchester. Some members of the Council might think that this was a matter for private enterprise, and therefore did not concern the Corporation. That was not his view. He held that the Council existed for more than the discharge of routine duties. It was incumbent on them as far as possible to stimulate commerce, to protect the citizens from the exactions of other corporate bodies, and see to the removal of anything that militated against the welfare of the community. In support of this view he directed attention to the efforts that had been made by the Corporations of Liverpool, Glasgow, Newcastle-on-Tyne, and the benefits which were thereby secured to those towns. In the case of Liverpool the Corporation laboured hard to extend the docks until the business became so great that it was transferred to a special board; at Glasgow the Corporation, in conjunction with the local magistrates, laid the foundation of the prosperity of that city by the action they took to improve the navigation of the Clyde.

Again, at Newcastle, a great commercial interest had sprung up within the last twenty years as the result of improvements of the Tyne for navigation purposes, which though effected by the Conservancy Commissioners were strongly countenanced by the Corporation. The expenditure incurred by these Commissioners in improving and maintaining the river and in providing dock and other accommodation, amounted at the end of 1880 to £6,500,000 sterling, including interest. They were empowered to borrow £4,000,000, but had only

SIR BOSDIN LEECH, AUDITOR, MANCHESTER SHIP CANAL COMPANY,
1886-92; DIRECTOR AND CHAIRMAN OF LAND COMMITTEE,
1892 *SEQ.*

Wilkinson Bros.

To face page 86.

borrowed £3,500,000, and within the last twenty years they had paid £1,000,000 out of revenue. They had also at the end of 1880 a surplus profit of £56,800 after paying all charges. But this was not all, for as the result of this improved navigation, numerous works had sprung up on the banks of the river where, thirty years ago, there were none. After referring to what the Corporations of Bristol, Leeds, Hull and Southampton had done to remove certain burdens to which the people of those places had been subjected, he said he did think it was the duty of this Council to take up this question in earnest. One reason for doing so was the serious injury occasionally done by the flooding of the Irwell, the widening and deepening of which would to a great extent obviate the danger. The carrying out of such a scheme would also be some help to the solution of the difficulty felt by the local authorities in complying with the Rivers Pollution Act. It would also bring into use the river Irwell, which was at present practically useless as a navigable river. He submitted that the Irwell Navigation Company had signally failed to utilise the stream as it might and ought to be used. No works had been established along the banks of the river like those about Newcastle, and he contended that as a Corporation they were suffering a very serious loss from the want of proper facilities for the transmission of the refuse of their town yards to the farm lands on the banks of the stream. For nine months in the year the river was not used, and he submitted that it was high time that such a bountiful provision of nature was fully utilised. If the Bridgewater Navigation Company were not disposed to make use of the river, they were bound to give up their privilege, and let it pass into other hands that would employ it better.

Another reason why this matter should receive attention from the Council was, that it was now universally acknowledged that water carriage was the cheapest mode of transit, and he felt sure it was necessary to the commercial prosperity of Manchester that they should endeavour to secure this means of transit for goods to the fullest possible extent. He did not wish to say an unkind word about railways, but he believed as regarded Manchester they had made the best use of their opportunities ; they had formed what might be termed a "corner" and had brought their charges to a uniform sum, and the Bridgewater Navigation Company had raised their charges to the same amount. To show how the trade of Manchester was oppressed by the present charges, he would cite a few instances illustrating their effect on the cost of articles of daily use. A charge of 10s. per ton was levied to carry cheese and butter 1,000 miles in America, and 10s. per ton to bring the same articles from Liverpool to Manchester, a distance of 31½ miles, whilst it was brought from New York to Liverpool, 3,000 miles, for 15s. per ton. A charge of 4·48d. per ton per mile was levied on refined sugar to bring it from Liverpool to Manchester; whereas a town like Greenock, which had to carry it from London, only paid 1⅛d. per ton per mile. A fact like that showed the advantage of Manchester having access by means of a tidal waterway with the outside world. Again, 10s. a ton was charged to carry cotton goods from Manchester to Liverpool, and the merchant had then to pay 2s. 6d. per ton to get them on board ship, while the cost left for transit to Calcutta was only 6s. 9d. per ton ; 7s. 8d. per ton was exacted to carry corn from Liverpool to Manchester, the charge for bringing it from New York to Liverpool being only 6s. 3d. per ton. The effect of heavy charges like these was to drive away trade from Manchester, and merchants in different parts of the country were beginning to send their goods by way of Goole, Glasgow and other ports, in order to avoid the heavy railway charges and dock dues at Liverpool. Manchester, however, was not only burdened by heavy railway charges, but by being compelled to send her goods through Liverpool, which was one of the dearest ports in the world. It was frequently the case that the charges in Liverpool were double what they were in other ports. Raw sugar could be laid down at the refineries in Glasgow at 2s. 6d. per ton, whilst it would cost 12s. 6d. to bring it to Manchester. It was no wonder, therefore, that one large refinery in this city had had to close, and now stood empty. Shortly stated, Manchester stood at a disadvantage as compared with such places as Glasgow and Newcastle in respect to the charges for carriage to the extent of 4½ per cent. on foreign produce; from 3 to 4 per cent. as regards iron, and 2 to 3 per cent. on cotton goods.

We were at a disadvantage in many other ways in consequence of the great disparity between the charges which Manchester people had to pay and those which were levied in other towns. Many works were closed for these reasons. If there was any improvement in trade Manchester was the last to get the benefit of it. They got the overflow of other towns, and good trade left them the first. Mr. Fowler, the engineer who was rebuilding the Tay Bridge, stated the other day that iron could be imported into London cheaper from Belgium than from Staffordshire. This was owing to the energy of the Belgians, who, by means of waterways were enabled to get into the heart of Belgium and bring their iron and lay it down in London at a cheaper price than Staffordshire and North Country ironworkers could. Liverpool was not satisfied with the present state of affairs. She felt the difficulties which he had indicated, and application was this year being made to Parliament to go back to the oldest system of all—carriage by road. It was proposed to lay down a line of rails on the road between Manchester and Liverpool. But whilst this would help Manchester a little in the matter of carriage, it would not bring new trades into the city, nor give the impetus to existing trades that a navigable river or canal would, and it would not lessen the dues in Liverpool. Much attention had recently been directed to the desirableness of freeing the foods brought to the Manchester markets from unnecessary imposts, but the advantage to be derived in that respect would be as nothing compared with that which would arise if we brought ships with fish and fruit as was done in London. He could scarcely conceive the impetus that would be given if the scheme now under consideration was found practicable. Old trades would come back to us, new ones would be created, and we should be able to compete with foreigners. It used to be thought that we were ahead of Americans in matters of trade. Now it was a life struggle between them and us, and, if Manchester was to retain the proud position of being the second city in the empire, the obstacles which now interfered with her progress must be taken away. In further support of the scheme, he quoted the words of Sir William Fairbairn (see Chapter IX.), whose works now lay silent, he believed, to a great extent because of the difficulties to which reference had been made. He (Mr. Leech) hoped the City Council would take up this question with the vigour and earnestness which its importance demanded. Alderman Sir John Harwood seconded the resolution, and it was carried unanimously.

This meeting roused the indignation of the *Liverpool Courier* whose editor wrote :—

There are even yet Manchester people who believe in the ultimate realisation of the periodically revived scheme of a ship canal to relieve Cottonopolis from the extortions of the Mersey Docks and Harbour Board. Faith is a very valuable possession, and has accomplished much since the world began ; but even faith cannot accomplish things which are impossible, and the ship canal to Manchester is one of those things. From an engineering standpoint we have been taught that nothing is impossible except causing water to stream uphill, but there are other practical considerations besides the theory of engineers. The Manchester City Council includes some of the infatuated people who look confidently to the construction of the tidal canal in the near future. Could not the Manchester Council appoint a Committee to investigate the subject, and ascertain whether the cotton metropolis is not being victimised by a huge conspiracy of shipowners, cotton brokers and dock bond holders.

The Guarantee Fund on the 7th of September, 1882, amounted to £11,450. It was subscribed to on the understanding that the money was to be used for the payment of all preliminary expenses. If a company were subsequently formed subscribers would then be relieved of all obligations, and what they had subscribed would be repaid to them out of the funds of the proposed company.

The following were guarantors of £100 and upwards :—

Daniel Adamson, Dukinfield	. . £200	W. C. Crum, Manchester	. . .	£100
S. W. Clowes, Ashbourn	. . . „	E. B. Rumney, Manchester	. .	„
W. Rumney, Ramsbottom	. . . „	Michaelis, James & Co., Manchester	.	„
Wm. Robinson, Warrington	. . „	Hickson, Lloyd & King, Manchester	.	„
Rd. Peacock, Gorton	. . . . „	S. & J. Watts & Co., Manchester	.	„
G. & R. Dewhurst, Manchester	. . „	Horrocks, Miller & Co., Manchester	.	„
Geo. Benton, Stretford	. . . „	George Robinson & Co., Manchester	.	„
Grundy, Kershaw & Co., Manchester	„	S. L. Behrens & Co., Manchester	.	„
Lord Winmarleigh, Warrington	. £100	James Jardine, Manchester	. .	„
Colonel Blackburne, M.P., Warrington	„	George Fraser, Son & Co., Manchester	„ ,	
W. Cunliffe Brooks, Manchester	. „	James Boyd, Manchester	. . .	„
Sir Joseph Whitworth, Manchester	. „	Beith, Stevenson & Co., Manchester	.	„
The Mayor of Salford (Rd. Husband)	„	The Strines Printing Co., Manchester		„
Alderman Walmsley, Salford	. . „	De Jersey & Co., Manchester	. .	„
John Lowcock, Salford	. . . „	S. Schwabe & Co., Manchester	. .	„
P. R. Jackson & Co., Salford	. . „	J. R. Bridgford & Sons, Manchester	.	„
J. & N. Philips & Co., Manchester	. „	Oliver Heywood, Manchester	. .	„
John Rylands, Manchester	. . „	Peter Spence, Manchester	. . .	„
John Munn, Manchester	. . . „	John H. Garside, Manchester	. .	„
Edmund C. Potter, Manchester	. „	Bryce Smith, Manchester	. . .	„

Deane Stanley, Manchester . . £100	Councillor Leech, Manchester . . £100
Chas. E. Schwann, Manchester . . „	„ Howarth, Manchester . . „
Frank Spence, Manchester . . „	J. T. Emmerson, Manchester . . „
Jacob Behrens, Manchester . . „	William Butcher, Manchester . . „
James Holmes, Manchester . . „	S. Kershaw & Sons, Manchester . „
Chas. Moseley, Manchester . . „	Rd. Johnson & Nephew, Manchester . „
Ellis Lever, Manchester . . . „	The Broughton Copper Co., Manchester „
Thomas Bradford, Manchester . . „	Joseph Leigh, Stockport . . . „
Henry Samson, Manchester . . „	George Woodhouse, Bolton . . „
Wm. H. Johnson, Manchester . . „	R. S. Collinge, Oldham . . . „
T. R. Wilkinson, Manchester . . „	Sam. Platt, Oldham . . . . „
Lieut.-Colonel Sowler, Manchester . „	Wm. Richardson, Oldham . . . „
C. C. Dunkerley, Manchester . . „	Ed. Collinge, Oldham . . . „
Henry Marriott & Co., Manchester . „	A. C. Boyd, Dukinfield . . . „
Isaac Neild & Son, Manchester . „	Trustees of the late F. D. Astley,
M. Kaufman, Manchester . . . „	Dukinfield . . . . . „
Walmsley & Samuels, Manchester . „	Alderman Davies, Warrington . . „
Brockbank, Wilson & Mulliner,	Peter Whitley, Warrington . . „
Manchester . . . . . „	Edward Mucklow, Bury . . . „
Alderman Sharp, Manchester . . „	Thos. G. Stark, Ramsbottom . . „
Councillor Boddington, Manchester . „	Marshall Stevens, Liverpool . . „
„ Goldsworthy, Manchester . „	

In addition there were forty-three subscriptions of £50 each and £200 subscribed in smaller amounts.

The Provisional Committee were busily employed in strengthening the guarantee fund and considering the reports of the engineers, Messrs. Fulton and Leader Williams, who, finding they differed in opinion, agreed to send in separate reports. These the Committee submitted to Mr. Jas. Abernethy, C.E., an eminent London engineer. The Chairman, Mr. Adamson, was strongly in favour of a tidal canal, but after mature consideration the Committee felt they could not take the responsibility of rejecting the advice of such sound engineers as Messrs. Abernethy and Leader Williams, backed up by Mr. J. F. Bateman, and they passed a resolution adopting the lock instead of the tidal scheme.

In the autumn of 1882 Mr. Adamson went abroad to recruit. During his absence Mr. Peacock, as Deputy Chairman, presided at the meetings of the Provisional Committee held in the B. room at the Old Town Hall. Things were going flat, and Mr. Peacock at one of the weekly meetings said it was absolutely necessary to get some one experienced in organising and conducting agitations and

Parliamentary fights, and it was decided to try and find a suitable man. At the next meeting some names were mentioned by the Deputy Chairman, in particular that of Mr. Joseph Lawrence, who said Mr. Lawrence had been engaged in just the same kind of work at Hull. It was decided to make inquiries about the applicants, and I was one of the Committee deputed to do so. Subsequently when the merits came to be discussed, to the astonishment of the Committee the very mild Deputy Chairman said we must not be too particular; it would be hard to find the right man. We wanted a clever organiser, and it was the Committee's duty to look after him and keep him in check. The choice fell on Mr. Joseph Lawrence, who no doubt had been largely instrumental in successfully carrying out the agitation at Hull. The Committee found him to have an aptitude for organising which was perfectly amazing. He was not to be daunted, and he could write a book or arrange a large staff with equal facility: he seemed to spoil for a Parliamentary fight, and at buttonholing members he was perfection. With the Irish party he had special influence, and at a division he seemed to do just what he liked with them.

This brings us to another era in the Ship Canal history.

On 26th September, 1882, a meeting of the subscribers to the guarantee fund was held at the Old Town Hall, King Street, to receive the report of the Provisional Committee. In the absence of Mr. Adamson, the Deputy Chairman (Mr. Rd. Peacock) presided. The Chairman said Mr. Fulton's scheme would not have allowed time for a ship to come up on the tide; it would have placed the ship in a cutting of 70 feet, and have been very costly from the great amount of excavating to be done. On account of these objections, and because Mr. Leader Williams would give a wider canal where two ships could pass, the Committee had decided to adopt Mr. Williams' canal with locks. The object of the guarantee fund of £25,000 for plans had been achieved, and now it would be necessary to have a further fund of £100,000 for Parliamentary expenses, deposit, etc. He would like the Corporations to take the matter up, and make the canal a public trust. To substitute a ship canal for an idle and sewage-laden river was a subject of national importance; nothing more so had come before the public since the making of the Liverpool and Manchester Railway. As a Manchester trader he found it cheaper and better to send his goods to Glasgow for foreign shipment, though he paid £30 an engine more to get it by railway to Glasgow than to Liverpool. This showed the necessity of better and cheaper communication abroad. We ought to bring the sea to our city so that we could load and unload ships there, and be in touch with the world.   He moved: "That steps be at once taken by the Provisional Committee to carry out the scheme as recommended in the report," also "that steps be forthwith taken by the Provisional Committee to raise a fund of £100,000, and to apply for Parliamentary powers to carry out the scheme as recommended by the report".

These resolutions were duly seconded and unanimously carried.

The Chairman said a meeting was going to be held in Barnsley, presided over by the Earl of Wharncliffe, to consider the scheme, and he announced several subscriptions of £1,000 each to the preliminary fund.   He also said that the Manchester Corporation Committee were going to inspect the river.

These important resolutions brought out a host of communications in the London, Liverpool and Manchester Press.   Advocates of Mr. Fulton's tidal scheme expressed their disappointment that real salt-water could not come to Manchester, and that the canal would be impeded and disfigured by locks.

The *Liverpool Courier* wrote :—

The Lancashire cotton spinners have revived an old project to "corner" Liverpool. They are apparently afraid that the "corner-men" on the Liverpool 'Change are more than a match for the "corner-men" in the factories.   As they cannot defeat the combination of brokers they mean to attempt the complete annihilation of Liverpool.   Several projects had been mooted during the past century for making a canal, and securing to Manchester a share of the mercantile marine business of Lancashire, but they had soon been consigned to the limbo of popular illusions, and they anticipated a similar fate.   They had better spend their money in helping to straighten the river and get rid of the Pluckington Bank.

The London *Times* said :—

No doubt it would be a great gain to manufacturers, and in the long run it might prove a paying and a profitable work.   But we must confess to a misgiving that the first shareholders might find themselves paid rather by the glory and grandeur of their work than in solid coin.   The projectors say "Even if the outlay were fourteen millions—better say at once twenty—the saving to the manufacturing districts would be such as to secure a good return to the shareholders".   Have they considered the costs and difficulties of training walls, the competition of railways and the dangers of accidents?   Also as to capital that "millionaires are generally shy people".

The article ended by recommending a reconsideration of a canal to the Dee rather than to try and improve the Mersey.

The *Manchester Guardian* thought the reports were not documents likely to increase the popularity of the scheme. There was sufficient to show the tidal scheme must be put aside, but there was some question if Mr. Williams' scheme would stand the test of close examination. Was it clear we should not have a re-petition of the Amsterdam Canals with their smells? Then Mr. Fulton had altered his proposition of coming up to Manchester, and was content to bring his salt-water basin to Barton. Had this alteration been duly considered? The cautious article ended with, "Whether the plan adopted by the Provisional Committee will in the main do all this, or not, we must leave experts and the future to decide".

The *Pall Mall Gazette*, whilst thinking Mr. Williams' scheme was the best, deprecated the attempt to make a canal, and suggested that the people of Man-chester learned a lesson in modesty and took to making the best of their existing waterway.

Of course there were many criticisms, and these were dealt with in a subse-quent pamphlet by "Cottonopolis," entitled, "The Manchester Ship Canal: Why it is wanted! and why it will pay!" A second edition had a chapter on "Can the Canal be Made for the Money?" These pamphlets were admirably written by Mr., now Sir Joseph Lawrence, and were meant to prove the Canal case out of the mouths of opponents who had given evidence on previous commercial inquiries, to the effect that Liverpool was a dear port, and that sadly too heavy rates were being charged for the conveyance of imports and exports. Amongst others quoted to prove the Manchester case were Messrs. W. B. Forwood, Mr. Guion, Mr. Alfred Holt, and Mr. John Williamson, of Liverpool, and Mr. Findlay, of the London and North-Western Railway Company. "Cottonopolis" also gave quotations from leading periodicals to prove that it was water competition England required if cheaper carriage was to be secured.

But Manchester having put its hand to the plough did not mean to turn back, even though the canal did not meet with all the encouragement that could be desired from the local Press and from capitalists who were likely to give a preference to their own vested interests.

The Provisional Committee, strengthened by many valuable and voluntary recruits, at once set to work with vigour.. Their first object was to educate the people, and let the merits of the scheme be thoroughly known. They believed in the power and intelligence of the people, and wished to have their support as well as that of the middle class tradesman and capitalist. They arranged for Committee meetings in all the wards of the city, and also in the neighbouring towns. A certain number of the Committee dedicated themselves to this work, and engaged to go whenever and wherever they were wanted to address meetings or help Ward Committees.

Oxford Ward opened the ball on the 4th of October, Councillor Roberts in the chair, and it fell to the lot of Mr. Leader Williams and myself to address the meeting. With the aid of plans the main features of the scheme were explained by the former, and I showed how the creation of new industries and the cheapening of food would benefit all classes of the community, venturing to predict a saving of £1,000,000 a year to the district. This was followed by ward meetings all over Manchester and Salford, at which resolutions in favour of the scheme were passed and a canvass for funds instituted.

Meetings also were held in Oldham, Ashton, Stockport, Warrington and other surrounding towns, and the project was favourably received everywhere, except in Bolton. Here a resolution of support was opposed in the Council on the ground that, though the canal might advantage Manchester, it would be of little utility to Bolton, and further (as an Alderman said), because in the end the canal would be gobbled up by the railway interests, as the Bridgewater Canal had been.

At the October meeting of the General Purposes Committee of the City Council to whom the Ship Canal resolution had been referred, Alderman Harwood moved : " That the question of widening and improving the rivers Irwell and Mersey and so making them navigable as a ship canal, be referred to the Parliamentary Sub-Committee, and that they be instructed to consider whether the administration thereof should be vested in a public trust instead of private individuals, and further that they be authorised and instructed to obtain such plans and opinions on the subject and to incur such expenditure as in their judgment may be desirable, with the view to presenting a complete report, and that Messrs. Leech, Walton-Smith, Goldschmidt and Howarth, be added to the Parliamentary Sub-Committee".   He wished to strengthen the Committee, and asked that the Council should give their moral support to gentlemen who were spending their time and money to improve the river and give commercial advantages to the city.

Alderman Thompson, in seconding, threw out a suggestion that in view of a tidal navigation there should be one great and extended Manchester that would embrace Salford and the surrounding districts. If Manchester was to maintain its present position, they must not throw away any opportunities for cheapening the carriage of merchandise whether it were by water or by land.

In my contribution I said the promoters had no private interests to serve, their object being to benefit the commerce of the district.   They would gladly work hand in hand with the Corporation, and I hoped the new Committee would start their work with energy.   No time was to be lost, as rival schemes by the Bridgewater Canal and by a Plateway Company would also be before Parliament.

The resolution was passed unanimously.

At a subsequent Council meeting it was determined to oppose Liverpool laying the Vyrnwy water pipes on the bed of the Mersey, and to insist they should be placed at such a depth and in such a way "as will not cause obstruction to any proposed improved navigation".   In consequence there was an inquiry at Runcorn, and the Acting Conservator ordered the pipes to be carried at a considerable depth as suggested by the Ship Canal Committee.   The new Committee lost no time, and arranged to inspect the whole length of the river, dividing it into two parts.   On 5th October, 1882, accompanied by Mr. Leader Williams, they left Manchester in a barge drawn by horses and went as far as Warrington, and a fortnight later did the other half from Warrington to Runcorn in the same way. The engineer explained his proposition of carrying the Bridgewater Canal in a swinging trough over the Ship Canal, and other interesting details.   On the second occasion Mr. Abernethy, C.E., was of the party.   Apart from the information obtained these trips were very enjoyable.   The weather was fine, the company good, and there was something novel in being drawn by horses along an old-fashioned waterway and through a pleasant country, occasionally dropping a few feet down by an old-fashioned lock.   When we had left behind us the unfragrant part of the river an excellent lunch was served on board, and all were very well prepared to enjoy it.

These visits were the subjects of some innocent lampooning.   The comic paper of the day (*City Lantern*) called Mode Wheel "Mudwheel," and of the Irwell said :—

> Mud to right and mud to left,
> Mud oozed out of every cleft,
> Mud surged up and mud dropped down,
> Mud and slush caused many a frown.

It also issued a cartoon of the City Fathers at lunch, and many of the faces are quite distinguishable.

Ben Brierley too wrote *Ab-o'th'-Yate and the Ship Canal*, in which he gave "A Dream of 1892".   Herein he pictured what Manchester would be ten years afterwards :—

When "Owd Ab" tells his wife Sarah "They're gooin to bring th' sae here. What dost think abeaut that?" she replies, "Ab, hast' bin atin poork, as thy yead's gone a-woolgetherin? I know theau'rt subject to dreeamin when theau's o'er-weighted thy inside; but neaw theau'rt off at a corner, I think. Are thy brains gettin flee-blown, or summat? Theau couldno' talk crazier nonsense if theau're stark mad."

Then Ab goes to bed "but before mornin coome I'd lived ten year. Afore th' fust clod wur delved, ther a good deeal o' pooin deawn, an' shiftin away to be done. Th' Manchester Corporation did a wise thing for once, ut even th' ratepayers gan 'em credit for. They sowd th' Knot Mill Market to th' Cannell Company. Gooin lower deawn, rents had been doubled i' Lower Broughton; an' everybody livin theere wur gettin new furniture, after weshin th' slutch eaut o'th' owd, an' sellin it. That owd wyndymill facin Peel Park wur to be made int' a leetheause; an' owd Oliver Crummell wur to have a creawn put on his yead; but wur to be a creawn o' fire, for one o' Edison's lamps wur to be put on his yead, for t' guide ships to their harbour, an' sinners to th' Owd Church.

"Th' Cannell wur finished at last; an' neaw it coome to lettin th' wayter into it. This wurno' to be a straightforrad job. Liverpool 'gentlemen' said th' sae didno' belong to us, an' we must ha' no wayter eaut on't unless we paid for it. They couldno' prove their title to it, becose owd Noah laft no will. Th' biggest undertakin' o'th' nineteenth century wur creawned wi' victory an th' Teawn Hall bells—thoose ut wurno' cracked—rung a merry peeal."

A few days later some of the principal subscribers finding the Corporation barge *Eleanor* had not been dismantled, chartered her for a cruise down the river, and their trip is humorously described in the Press by one of the party on the lines of Artemus Ward when he gave the history of his famous cruise on the Wabash Canal. They had many adventures by land and water, among the rest one of the horses in crossing the river on a floating bridge tumbled in. They were much astonished by the primitiveness of the country and the people, and still more so that a fine waterway should have been allowed to go into disuse.

Salford was also enthusiastic about the idea of a Ship Canal. The Council on the 29th September, 1882, passed a resolution that their borough engineer, Mr. Jacob, should inquire into the proposed scheme and report. This he did most exhaustively. In closing this report he summarises thus:—

First, that the scheme which has been adopted by the Provisional Committee possesses, from an engineering point of view, all the elements likely to render the enterprise a successful one; secondly, that the canal will materially contribute to the solution of the floods question as regards Salford; and thirdly, that in my judgment the completion of the undertaking must be followed by the most beneficial results as regards the trade and prosperity, not only of Manchester and Salford, but of every manufacturing centre in their neighbourhood.

Alarmed by the enthusiasm with which the project was received, Liverpool devised a counterblast in the shape of a Lancashire Plateway, which it was said would effect the same end at a much less cost.   Practically this scheme was to place smooth parallel iron plates on the main roads, and draw trucks on them by means of steam traction engines.   Thus it was said goods could be cheaply taken from the ship's side into the user's yard without transhipment, and many of the railway and shipping charges would be avoided.   This idea emanated with Mr. Alfred Holt, himself a large shipowner and a practical engineer.   Liverpool merchants and shipowners saw the possibility of checkmating the new port and also of cheapening carriage, and they subscribed £75,000 to float the scheme.   It had, too, the powerful support of the London *Times*, in which several favourable articles were written.   A deputation of the promoters waited upon the Manchester Corporation, but received scant encouragement.   Eventually the Plateway scheme collapsed, in consequence of its being found much more costly than at first anticipated.   A preliminary prospectus was advertised with some of the best Liverpool names attached to it, but it was generally felt that every argument urged in Liverpool in favour of the Lancashire Plateway could just as profitably be used by the promoters of the Ship Canal.

About this time the *Manchester Guardian* published the following series of very instructive articles on "The Commercial Prospects of the Ship Canal":—

1. The Geographical Question.
2. The Coal Trade.
3. The Metallurgical Industries.
4. The Chemical, Glass and Pottery Industries.
5. The Cotton Industry.
6. The Cotton Industry.
7. The Cotton Market.
8.. The Woollen Industries.

They were admirably written by some one who certainly was not carried away by enthusiasm for the canal.   If he had had to rewrite them by the light of after events he would possibly have had to admit that his calculations in respect to cotton, etc., were far from being correct.

During October, 1882, the Provisional Committee was busily at work organising, and canvassing for subscriptions to the £100,000 fund.   They met weekly and their work was full of encouragement.

The following letter, enclosing a donation of £10, from a venerable old lady shows the remarkable enthusiasm felt for the Ship Canal :—

Mrs. —— is most deeply interested in the project of a Ship Canal for her dear native town of Manchester. She is now seventy-three years of age, and during the years 1823-24-25, being in the evenings the constant companion of her excellent father (a good, highly respected man), she heard a very great deal from him about the project, which was then hoped might soon be accomplished, but it fell through, greatly to the regret of Mrs. ——'s father, who over and over again said to her, "Manchester and Manchester trade will never really flourish until a Ship Canal be made". As she at that time read every document that came to her father on the subject, she longed for the Ship Canal to be made as much as he did, and under these circumstances it is a great delight to her to send the enclosed small sum of £10 towards the large fund required for the revived project. Besides being so old, Mrs. —— is a very great invalid, and is quite aware she may any day die suddenly, yet whenever she reads anything in the *Manchester Guardian* about the Ship Canal, she cannot help wishing our Heavenly Father may permit her to live to hear it is made and proving to her native town as great a benefit as expected.

SOUTHPORT,
18th October, 1882.

Surrounding towns sent resolutions in favour of the scheme, and in Manchester and Salford candidates for the Municipal Council, one and all, supported the canal. Building Societies promised their help. The Grocers and other trading Guilds recognised the benefit to their trades, and promised their support. Various Local Boards tendered their services to promote the success of the undertaking. Week by week came willing helpers to the cause, and the subscription list rapidly increased. Trades Unions and working men voluntarily offered their contributions to a scheme they saw must make work more plentiful. The numerous letters and articles that appeared in the London and Provincial Press attracted the attention of the whole of England to the efforts Manchester was making. One of the most able contributions appeared in the *Saturday Review*, and in its opening paragraph put the case very clearly :—

Within an area of 40 miles round Manchester there is a population larger than that of all Ireland, a population perhaps the busiest and the most productive on the face of the earth, to the increase of whose prosperity there appears to be no limit but that imposed by the conditions under which it exchanges what it produces for what it requires; and it is now crying out that these conditions are becoming so hard that they threaten the industry with paralysis. . . . Everything which tends to the greater efficiency and economy of industry will in the long run augment both wealth and population, and with the growth of wealth and population the traffic of the railways must increase.

With the view of increasing the circulation of Ship Canal literature, it was decided to publish a special organ, *The Ship Canal Gazette*, which was ably edited by Mr. Jas. W. Harvey and Mr. Joshua Bury, of John Dalton Street.   The first number was issued on 8th November, 1882, and the last on 15th August, 1883 ; it had a large circulation, and by its means a fuller report of speeches at various meetings was given than could be expected from the local Press.

An amateur poet called attention to the ship which forms a prominent feature of the City Arms, thus :—

> Mancuniensis !
> Some wonder from whence is
> The arms of thy city obtained,
> A shield 'tis with three bars
> (Suggesting the three R's),
> Through which all its wealth has been gained.
>
> Now these cause delays,
> And our merchandise pays
> Heavy dues ere it gets out to sea ;
> Quite a bar to success
> In our trade you'll confess
> When it's understood what these three R's be.
>
> Rail, Road, and River !
> But why did they ever
> O'er this shield put a ship on the ocean ?
> T'was perhaps to convey
> The idea, then, that they
> Of the " Ship Canal " had a notion.
>
> 'Tis no longer a dream,
> So success to your scheme,
> And away with the obstacle-dreamers ;
> We'll soon see the docks,
> The canal with its locks,
> And welcome the ocean steamers !

A well-known timber merchant, Mr. John Kirkham, also published some stimulating verse which did good service :—

SHIP CANAL SONG.

Written by Mr. John Kirkham.

To bring big ships to Manchester
　　Is what we mean to do;
If you delight in smaller craft,
　　Then "paddle your own canoe".
From Throstle Nest or Barton Bridge,
　　Just as it may suit you,
You'll sail direct to any port,
　　On board an *Ocean Screw*.

Chorus—Then love your neighbour as yourself,
　　　　We'll sing whilst sailing through
　　　　The Ship Canal by Liverpool
　　　　On board our *Ocean Screw*.

Monopoly has vexed me long,
　　It's done the like by you;
Down with the monster in the mud
　　And give fair play its due.
Then cotton, timber, corn, and beef,
　　Will come to us right through;
And ham and eggs and well-filled kegs
　　With savoury things for you.

You'll see great ships from distant climes,
　　And men of every hue,
Lay earth's vast bounties at your feet
　　And take your goods in lieu;
Then trade will flourish all around,
　　And peace and plenty too,
And every thrifty man, you'll see,
　　Will "paddle his own canoe".

Ye working men of Manchester
　　It much depends on you;
Come, put your "shoulders to the wheel,"
　　And "paddle your own canoe".
Hurrah! then for the Ship Canal,
　　Three cheers, my boys, for you;
Look up your rusting implements,
　　There's lots of work in view.

Ye merchant men oppressed by dues
  And inland charges too,
You must support this noble scheme,
  And "paddle your *grand* canoe".
Then docks and wharves and work's you'll see,
  Like magic rise to view;
You'll pass them all along the line,
  When on your *Ocean Screw*.

Some men will "rest upon their oars,"
  Like a faint-hearted crew,
You firmly grasp and "ply" them well,
  And "paddle your own canoe".
And now, my friends, my song is done,
  One parting word to you—
Take up its strains, and sing them out.
  "We'll paddle our own canoe".

Chorus—Then love your neighbour, etc.

On the 3rd of November a conference of the various Ward Committees was held to arrange for an extended canvass for funds. The Mechanics' Institute, Princess Street, was crowded by an enthusiastic audience which was addressed by Messrs. Adamson, Peacock, Peter Spence, Reuben Spencer and others. The Chairman, Mr. Adamson, appealed to the patriotism of the district in aid of the canal, and said if that was not forthcoming they might have to get M. de Lesseps to come over and do the work. Their object was first to make the canal, and then to make it pay. He hoped to see one large dock from Throstle Nest to Mode Wheel. Manchester must and would have the Bill, and the canal would be made in spite of all opposing difficulties. Mr. Peter Spence pointed out that Liverpool, with a population of near 500,000, had a smaller proportion of working men than any other city in the country. Of its people it might be said, "they toiled not, neither did they spin". Upon whom then did they live? Upon the Manchester people and others, by doing work which had been foolishly left in their hands.

Mr. Henry Sales said freightage represented one-third of the value of timber, and that if the canal were made, Manchester would be the depot for the North of England.

The meeting was adjourned, and the business completed at a subsequent meeting held in the Athenæum.

On the 13th November a mass meeting of working men was held in the Free Trade Hall, presided over by Mr. Henry Slatter, who was supported by Professor Boyd-Dawkins, Mr. Robert Austin, Mr. G. D. Kelley, Mr. James Mawdsley, and the leading Trade Unionists of the district.   The meeting was crowded, and resolutions were unanimously and enthusiastically passed in favour of the canal, one being to ask the adjoining Corporations to support the movement which, as a speaker said, "would lower the cost of production, would tend to cheapen the price of clothes and food, would exercise an enormous influence on trade, and increase the prosperity of Lancashire".   Another resolution called upon the members of Parliament in the district to give the Ship Canal Bill their strenuous support till it became law.

The Chamber of Commerce on the same day, at the instance of Mr. Peter Spence and Mr. R. C. Richards, agreed to support the Bill.   The latter said henceforth they must give up the idea of a tidal navigation and go in for a Ship Canal with locks, and stated on the authority of the Oldham Cotton Spinners' Association that a total saving of 5 per cent. would be effected in the import and export of cotton and cotton goods.

At the Town Meeting on the succeeding evening the Mayor said the requisition for the meeting was signed by 5,000 mercantile firms in and around Manchester —an unprecedented requisition.   The meeting was addressed by Messrs. John Slagg, M.P., Jacob Bright, M.P., Messrs. J. C. Fielden, Daniel Adamson, Richard Peacock and others.   The large room of the Free Trade Hall was densely crowded, as was also an overflow meeting in the Assembly Rooms, and the enthusiasm throughout the proceedings was intense.   Some incidents at this meeting are worth recording.   The late Mr. Tom Nash, the barrister, was fully aware that at least one M.P. was not as sound as he ought to be, but had at heart a wholesome fear of his constituents.   He went on to say, "He was glad to see that the great meeting of the Trades Unionists held on the previous evening had had a most remarkable effect in converting unbelievers to the scheme.   He was told that the heathen were coming in by shoals, and that there was a regular scramble for conversion.   There must, however, be no half-hearted support of the scheme, and no waiting to see how the cat jumped."   Every one knew for whom this was intended, and it caused some amusement.   Mr. Peacock, one of the most mild and gentlemanly of men, brought the house down by a story he told, perhaps a little coarse, but very much to the point.   Alluding to the taunt that the canal would not pay because it would cost too much, he told the tale of an Oldham spinner and a dog fancier.   Coming from the Exchange, the former saw the latter with a dog that pleased him, so he asked the price—"Five pounds," said the dealer.   Amazed the other said, "That's a hell of a price," to which the latter at once rejoined, "But it's a hell of a dog".   "So," said Mr. Peacock, "if the canal does cost an immense sum the advantages will outweigh the price."

At the same meeting Mr. Jacob Bright, M.P., made a most enthusiastic and encouraging speech :—

He believed it was as good a scheme as could be produced, and that it was in all respects practical.   It was quite true that neighbouring towns had not the same interest in this matter as Manchester and Salford, but at the same time they had a powerful interest in it.   Everything which they produced came through Manchester, and would go by the canal to Liverpool or the ocean.   Then the vast mass of the food and raw materials which they required would come by the canal to Manchester, and in many cases would be within carrying distance, so that there would be much more independence of railways than there was at present.   And if it be true—and nobody could deny it—that there might be industries in the neighbourhood of Manchester which could not possibly exist without the canal, the whole of Lancashire and the whole of England were at liberty to partake in those industries, and to profit by them, and they would find it much better than emigrating to the antipodes. The notion of a ship canal had been in the minds of men in this district for many years past. Only yesterday a friend of his told him that sixty-two years ago his father pointed out to him a man who was nicknamed "Ship Canal".   In those days that man was said to have ship canal on the brain.   He (Mr. Bright) supposed he had a brain somewhat better than his neighbours.   Therefore the canal which had been the dream of past generations in this city was, he sincerely believed, going to be realised in our time.   And when the day should come —and those amongst them who were no longer young might see the day—for the engineer told them that in four years the work might be accomplished, and when he knew that the best labour in the world was in our midst, he did not think that prophecy was a dangerous one.   When the day came when the smoke of big steamers should mingle with the smoke of their tall chimneys, he would not undertake to say that the external atmosphere should be purer on that account, but he would say that thousands of homes would again be prosperous which were now in a languishing condition, and this great community of Manchester might occupy a yet greater place in the position of the world.

Following on the two meetings came leading articles in all the papers.   The *City News* commenced with :—

The movement in favour of the Ship Canal progresses with astonishing rapidity.   Never before in South Lancashire has any great scheme been so quickly and generally approved as this.   The earnestness and warmth of public feeling, combined with the transcendant importance of the subject, has silenced all opposition.

It went on to say that the rich men of the district, with a few exceptions, had not rivalled the enthusiasm of the general public. Though they had accumulated their wealth in the district, they were hanging back, possibly because they had vested interests, and feared their railway dividends would fall with the success of the canal.

It would be an unpleasant record if the great Manchester Ship Canal is carried out without the sympathy and help of many of the men who have grown rich by our local industries. It would never do to let the canal fall through. Better let the Corporations subscribe the capital *pro rata* and make it into a public trust.

On the 18th November the Parliamentary notices for the undertaking were issued, and the necessary plans and sections were deposited.

At a special meeting of the General Purposes Committee, Alderman Harwood spoke against a private Ship Canal being allowed to pass, and urged the Mayor should confer with the Mayors of the neighbouring towns with the view of forming a public trust, as he believed a private company could not succeed. He urged that the Parliamentary Committee should quicken their steps as to the report, and moved that the Mayor should ascertain the views of surrounding municipalities.

At the meeting of the General Purposes Committee on the last Thursday in November, the Mayor moved the following resolutions which had been passed by the Parliamentary Sub-Committee :—

That the hearty and strenuous support of this Corporation be given to the project of improving the rivers Mersey and Irwell to the extent necessary to enable ocean-going ships to have direct access to Manchester. That the undertaking and administration of the canal should be so constituted as to be a trust for the benefit of the public in general. That subject to satisfactory provisions being introduced into the Bill, it is, in the opinion of the Committee, desirable that Manchester and other municipal bodies and local authorities in the neighbourhood of the proposed Ship Canal, should be authorised to contribute to the cost of the undertaking, and to take part in supervising the execution of the works and also in the general management of the canal, to such extent and in such manner as may be sanctioned by Parliament.

To this resolution Alderman King objected as regarded the authority to contribute. With a debt of £7,000,000, and a prospective £800,000 for sewage purposes, he was opposed to pledging the ratepayers to a contribution of £1,000,000, and possibly of £5,000,000. He twitted the promoters with having raised less than half of the £100,000 asked for, and moved that the contribution clause be deleted from the resolution.[1] Alderman Grundy seconded the amendment. He was willing to be friendly towards the promoters, but not to help them financially. Only the mover and seconder and Alderman Heywood voted for the amendment; all the other members of the Council voted for the resolution.

Nothing daunted, and with his usual tenacity, Alderman King returned to the charge on the following Wednesday, when at the Council meeting the proceedings of the General Purposes Committee came up for confirmation. He moved their rejection, and was seconded by Alderman Grundy, who remarked that it had been said the resolution committed the Council to nothing. This seemed to him a cowardly remark. Were some members trying to bamboozle people, and to make them believe the Council were committing themselves to something when they were really doing nothing of the sort? Was it not cowardly, he asked, to boast of a back door being left open out of which the Council could sneak? He held the Council would be morally bound to contribute if they passed the resolution.

Mr. Goldschmidt opposed any grant because it meant an increase in the rates, and because, as he said, the docks being in Salford and Stretford, merchants would take their warehouses into those districts. On the previous day Mr. Chamberlain, M.P., had doubted the right of Corporations to provide money for electric lighting ; he was quite sure a contribution to the Ship Canal would never be permitted.

The writer attempted to show that the civic bodies in Liverpool, Glasgow and Newcastle had, with great advantage, used their funds to provide docks and to improve the access to their ports. Also that the Corporation already traded in gas and water outside its boundary. He pointed out the many advantages to the city of a Ship Canal, and urged that a communion of interests might present a golden opportunity for amalgamation with Salford.

Alderman Bennett said, of all people, he would have expected Alderman Heywood to defer to the popular voice, which no one could deny was decidedly in favour of the Ship Canal. Did not Mr. Goldschmidt know that the Liverpool Docks were chiefly in Bootle, and a further extension was contemplated outside the city of Liverpool? It was a serious loss, and a monstrous waste, that the river should be used simply as a sewer. Manchester and the authorities round ought to join hands in making the canal, and if so, the capital might be borrowed at $3\frac{1}{2}$ per cent. For the sake of Manchester, for the sake of Liverpool, for the sake of the whole community of Lancashire they ought not to oppose a scheme required

[1] See Contribution List, Appendix No. II.

by the people.   On a division, six voted for the amendment and forty-six against. The dissentients were Aldermen Heywood, Grundy, King, Curtis, George Booth and Councillor Goldschmidt.

December was a month of unremitting work for the Provisional Committee. Its members were very busy attending meetings and collecting subscriptions. Personally, what with preparing for meetings and canvassing, I had little time left for my own business.   Canvassing especially was ever full of surprises and disappointments.   One of the first men I called on for the guarantee fund was Councillor Goldsworthy; he heard my case, and at once gave me £100, and promised to come on the Committee and work.   Another day, without solicitation, and to my surprise and pleasure, I opened a letter and found £100 cheque from Sir John Harwood.   Some days after my speech introducing the Ship Canal I was in the Health Committee, opposite Councillor Chesters Thompson, when that gentleman asked for a sheet of blank paper, and, without more ado, cut a slip off and wrote a cheque for £100, and threw it across the table with "Owd lad, take that for your canal".   Of course I was much pleased.   On the other hand, one had some very hard nuts to crack, and none more so than the county and borough members.   On some of them I called time after time, plying them with literature and statistics, but with small effect.   It was only when popular feeling forced their hands that a declaration was made, and then in some cases it was a qualified one.

Mr. Jacob Bright was a remarkable exception.   No man ever went more thoroughly into a cause, or worked more heartily for it.   He was always pleasant to deal with, and to be depended upon.   I was early brought into contact with him. Coming down the Reform Club steps in the autumn of 1882 Mr. Bright stopped me, and said he would like to be informed about the Ship Canal, its objects, and the benefits likely to accrue to the city.   He said: "As M.P. I have promised to give my assistance, but it will be more effective if I thoroughly understand the merits of the scheme and all its bearings.   Will you give me a long interview and post me up?"   This I expressed my willingness to do, and after chatting some time it was arranged I should see him in London, where we were both going. Accordingly I met him at the House of Commons, and we spent some hours in the smoke-room there discussing every phase of the question.   Just then one of my colleagues had been saying some extreme things, which enabled opponents to poke ridicule at the cause.   When I spoke of this Mr. Bright, to my surprise, said in a kindly way: "The world cannot do without enthusiasts, and you must not be too

JACOB BRIGHT, M.P. FOR MANCHESTER.

To face page 106.

hard upon them". At the end of our interview Mr. Bright thanked me, and expressed himself as being both interested and satisfied. He promised that he would give all the assistance in his power, and this he faithfully did. Manchester owes a deep debt of gratitude to Mr. Jacob Bright for his exertions on behalf of the Ship Canal.

Not only did large contributions come pouring in to the Provisional Committee, but welcome aid came from various workshops. Mr. Adamson's men clubbed together and sent the handsome sum of £430. When speaking to his men, Mr. Adamson mentioned that a leading paper had said the canal would cost £15,000,000, and ridiculed the idea. Strange to say, that sum was the actual cost, but of course for a larger and very different canal. At the Eagle Foundry, Salford, Mr. William Fletcher, the head of the firm, and a most energetic supporter of the canal, established a system by which he found the money, and the men repaid him at the rate of one shilling per week. Thus the working classes grew to have an interest in the scheme. The Co-operative Wholesale Society, on showing their members that the canal would save them £5,000 to £6,000 per year, obtained power to contribute £1,000 to the fund, and this sum was afterwards materially increased. A resolution to this effect was seconded by Mr. Belisha, of the Lancashire Supply Association, who was an early friend of the cause.

So much interest was taken in the scheme that two literary societies devoted special evenings to consider the question. The Manchester Scientific Students had the benefit of an explanatory address by Alderman Bailey, and the Manchester Statistical Society listened to a paper read by Mr. F. R. Conder, C.E. On both occasions there was an animated discussion, and a very decided feeling evinced in favour of the canal.

At the monthly Council meeting in Salford, on 6th December, Alderman Walmsley moved :—

That, in accordance with the recommendation of the River Irwell Conservancy Committee, the Council express its concurrence in and approval of the projected Ship Canal, whereby the navigation of the rivers Mersey and Irwell will be so improved as to enable vessels of large tonnage to come direct to the borough.

In his speech he said :—

There was also a side issue looming in the distance. It was a question they could not afford to lose sight of. It was the question of the amalgamation of the city of Manchester with the borough of Salford. Their borough, besides having its own interests, was an integral part of the greatest manufacturing and commercial centre of England.

The resolution was not enthusiastic enough for many of the members. On the other hand, Alderman McKerrow was disposed to put on the brake. Eventually the resolution was carried unanimously.

A day or two afterwards a public meeting was called in Salford in compliance with a requisition to the Mayor. The senior member, Mr. Benjamin Armitage, moved the mildest possible resolution; he could not be persuaded to move a stronger one. It was the fear of offending his constituency and not his love of the cause that took him to the meeting. This he himself made evident from the moment he commenced to speak:—

I confess that I do not come forward prepared to advocate this cause with the degree of enthusiasm which is shown by many friends who are around me. Again, when large sums of money lie unemployed, the investors want convincing of the profitableness and safety of the investment. With regard to this hesitation and shyness, some uncharitable things have been said, even meeting me in the form of a threat this week of the consequences at another election if I did not go fairly into the promotion of this thing. I am no more offended by this broad hint than any of you here, and could afford to laugh at it, but it is a sort of boycotting which is not very agreeable, and after saying that, I will not make any more complaint about it.

He went on to remark he was not sure the canal would both reduce freights and pay the shareholders, but that he had come to say he would give his recommendation and some degree of material support to the undertaking.

This wet-blanket speech was followed by one from Mr. Arthur Arnold, M.P., who made a good case as regards the necessity of cheapening carriage by the means of the canal, but took care not to commit himself.

I have been somewhat slow in giving encouragement to this scheme; but there is no harm, Mr. Mayor, in being slow, provided we are sure, and I thought till I was prepared to give the scheme material support, as many of my friends have done, it would not look quite the thing for me to give recommendations in its favour which might be likely to influence the opinion and judgment of others.

Fortunately the speakers who followed, by their resolutions and speeches, raised some enthusiasm in the meeting, but I recollect coming away in anything but a happy frame of mind, and the M.P.'s themselves, when they went to give evidence on the Bill, found they had provided their opponents with ammunition which they were glad to make use of.

The Press was not slow in expressing its opinion on the action of the Corporations of Manchester and Salford.

MANCHESTER-SUR-MER. A SEA-DUCTIVE PROSPECT.

(By permission of the Proprietors of *Punch*.)

The *Manchester Guardian* which had been unfriendly to the canal for some time previously now swooped down and condemned in a wholesale way the proceedings of the Parliamentary and General Purposes Committee and of the City Council.   Of the former's report they said :—

The main recommendation appears to us in a high degree impolitic and dangerous. Of the suggested contribution, the Committee has advocated a course the gravity of which every member of the Council and every responsible citizen ought seriously to weigh well before according to it his support.

Again :—

The surprising thing is that they allowed themselves to advocate any course which involves the participation of the municipality as a partner in what is nothing less than a gigantic commercial adventure.   No one can tell what it will cost, or how much it will yield on the outlay.   A project which involves the possibility of unknown additions to it is one which ought to be seriously discouraged by every one who has an interest in the continued prosperity of this community.   It is not the amount of the contemplated investment (about £15,000,000 is the extreme estimate), but its risk which induces the advocates to desire the Corporation to undertake the work.   That proposals so vast in their reach and importing principles into the administration of municipal finance so novel and so questionable should have been adopted without serious challenge, would have been nothing less than a reflection on the good sense and business capacity of the Council. That the governing body of a great Corporation, containing an exceptional number of members of tried character and business capacity, should have lent its hand at this stage to a scheme of this description, can only be regarded as an amazing proof of the contagion of a popular emotion, and the skill of an organised army of promoters.

Equally severe were the *Guardian's* two other articles.   There were to its mind six wise and discreet men left in the Council, who dare express their opinions, and they compared the prudent caution of the Salford Council with the rashness of the Manchester Corporation.   The conclusion of the last article runs thus :—

We venture to say that a gigantic undertaking was never more heedlessly entered upon than when the Manchester City Council determined to give the financial support to a vast engineering work, the very plans for which were not before it.   Certainly there will be need for plenty of prudence in the future to make up for the plentiful lack of it in the past.

The *Examiner and Times* and the *Courier* took up a fairly friendly attitude, particularly the latter.   The former advocated that the enterprise ought not to be a trust, but that the money should be raised and the work carried out by private capitalists.

The Council may pass resolutions in favour of the Ship Canal. They may give it the benefit of their best wishes, and assure the promoters of ready co-operation in carrying out the enterprise. But they should go no further. It is not expedient that the ratepayers should be laid under pecuniary obligations for the construction of the canal.

They were much opposed to impassioned advocates recommending the canal as an investment for the savings of working men.

The *City News* and the *Salford Chronicle* advocated the Corporation contributing to an undertaking that would be of such signal benefit to the city by stopping its decay and bringing new life to its commerce. The *City News* said if the Tyne ports and Glasgow had been so materially benefited by municipal help, why should the Manchester Corporation withhold help from an effort to remove the monopoly of Liverpool and the railways? It had been said there was no difference between making a railway and improving a river. The former was a private adventure, the latter was a natural means of transit belonging to the public; it ought not to become the monopoly of a private company, and if it wanted improving, the community should do the work, and profit by the result.

A few persons who in the earlier days of the canal scheme faintly praised it, now declare it to be a gigantic speculation, in which the risk is enormous. But the Council has declared it shall not be a commercial adventure, but a public trust.

At a meeting of the Liverpool Dock Board, Mr. W. B. Forwood thought Liverpool had preserved a dignified attitude towards the canal, believing it would not seriously injure the trade of the port.

If the canal were judged solely by its merits, it would quickly be relegated to the museum of exploded bubbles which amused the British public.

It seemed, however, an attempt was being made to make it a municipal and even a political question.

On the 20th December a Conference of Municipal Corporations called by invitation of the Mayor of Manchester was held, at which the Mayors of most of the neighbouring towns attended. The engineer of the Ship Canal, Mr. Leader Williams, was present, and explained the plans and engineering details. Some interesting facts were given. It was estimated more timber was consumed in Lancashire and the West Riding than in any corresponding area in the world. Annually over a million tons were used, valued roughly at £3,000,000. Two-thirds of this came through the eastern ports of England, and paid 8·67d. per ton per mile freightage. Grain from Liverpool to Oldham (station to station rate), cost 2¾d. per ton per mile, whilst grain was carried on American railways, a similar distance, at one-eighth of that figure.

Before closing a history of the year's proceedings, I should like to bear testimony to the exertions put forth by Mr. Henry Boddington, Junior, in furtherance of the canal.   He not only promised to devote £10,000 if necessary to the purpose, but at his own cost he published most useful plans of the proposed canal with full explanations.   By his permission I reproduce one that serves to show the original scheme before Parliament in 1883.[1]   His subsequent ill-health and retirement were a matter of deep regret to his colleagues.

[1] See Plan No. 6.

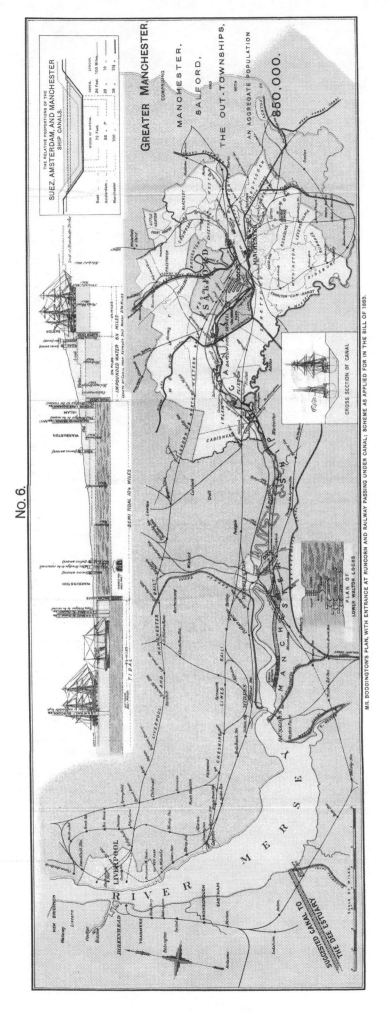

NO. 6.

THE RELATIVE PROPORTIONS OF THE
SUEZ, AMSTERDAM, AND MANCHESTER
SHIP CANALS.

GREATER MANCHESTER.

COMPRISING

MANCHESTER,
SALFORD,
AND
THE OUT-TOWNSHIPS,

WITH

AN AGGREGATE POPULATION OF

850,000.

CROSS SECTION OF CANAL

PLAN OF
LOWER WALTON LOCKS

MR. BODDINGTON'S PLAN, WITH ENTRANCE AT RUNCORN AND RAILWAY PASSING UNDER CANAL; SCHEME AS APPLIED FOR IN THE BILL OF 1883.

RIVER MERSEY

LIVERPOOL

SUGGESTED CANAL TO
THE DEE ESTUARY

CHAPTER VII.

1883.

CHRONICLE OF EVENTS—HOSTILE CRITICISM—MISHAP ON
STANDING ORDERS — LORD REDESDALE — SKETCH OF
THE LORDS COMMITTEE AT WORK—BILL THROWN OUT
—MANCHESTER IN EARNEST—LOCAL MEETINGS.

(To a Deputation of Liverpool Merchants and Shipowners.)

The deputation must not go away without his repeating that he considered Liverpool,
with the exception of London, to be the dearest port in England.—Sir Richard Moon.

TO all appearance the year 1883 opened auspiciously for the promoters of the
Ship Canal. Their organisation was working admirably. The canvass for
the £100,000 Fighting Fund had realised £63,000, and encouragement
and support were being afforded not only by surrounding municipalities and local
Boards, but by Chambers of Commerce, industrial institutions of all kinds, and by
the working classes through their Trades Unions. The Ship Canal Bill had been
deposited, also plans and sections as far as it was thought necessary, for Parliamentary counsel had advised that full powers already existed to make the length
from Runcorn to Liverpool, if only the plans met with acceptance by the Mersey
Commissioners, who, under the 1842 Act, were conservators of the river, and could
admit or refuse as they felt disposed. Further, it was known that the Lancashire
Plateway Company (opponents of the canal) who had also deposited a Bill, were
finding their scheme too costly, and were likely to withdraw it; this they did a few
days afterwards.

The horizon was not, however, cloudless. On the 21st December, 1882, Mr.
A. D. Provand, a Manchester and Glasgow shipper, published "A Criticism on the
Manchester Canal Scheme". It was an exceedingly well-written document, and
while it cast somewhat of gloom on the promoters, it was much appreciated by their

opponents. The Mersey Docks and Harbour Board bought thousands of copies and distributed them broadcast, one being sent to each subscriber to the £100,000 Guarantee Fund.

Mr. Provand objected strongly to the Manchester Corporation aiding the canal. He declared the statements made as to savings by the canal to be erroneous and deceptive; said a mild terrorism prevailed to compel unwilling people to subscribe; essayed to disprove the facts and figures given by "Mancuniensis," "Cottonopolis" and others; declared Mr. Adamson and other speakers were making extravagant and erroneous statements, and prognosticated the canal would be a failure. Further he maintained the cost of the canal and its maintenance would far exceed the estimates given; pointed out that a huge sum must be spent in equipment; pictured the united railways crushing the canal in the first year of its existence; doubted a sufficiency of water to work the canal, and prophesied it would neither cheapen traffic between Liverpool and Manchester nor earn a dividend. After deprecating the receipt of subscriptions from working men, Mr. Provand gave his idea of what ought to be done, and advocated a barge canal between Liverpool and Manchester.

This startling indictment fell like a thunderbolt on the promoters, just at the time they were doing their best to augment their subscription list, and it no doubt impeded progress. Mr. Lawrence, with wonderful celerity, issued an able pamphlet in reply, controverting Mr. Provand's statements, but there can be no doubt Mr. Provand's pamphlet supplied ammunition with which the canal was pelted for years afterwards.

He specially criticised the "four persons who are authorities respecting the canal". I happened to be one of them, and I freely admit that all the statements and predictions made in my speeches have not worked out exactly as I had expected. At the same time I firmly believe I was much nearer the mark than my critic, who must now see that many of his prophecies of twenty-one years ago have failed. Mr. Provand specially criticised my assertion that the canal would save 4s. per bale on cotton, would cheapen by 5 per cent. the food of the working man, and reduce the cost of the 4-lb. loaf ¼d. Whilst I embraced in my calculations not only cheapened carriage and port charges but also the effect of competition, contingent advantages, and a better supplied market, Mr. Provand would only admit direct savings (if any) on carriage. He wrote:—

If the whole cost of carriage were saved it would not amount to 5 per cent., but against this carriage there would be three or four shillings per ton for canal tolls and five or six shillings for steamer freight, pilotage and insurance coming up the canal. Any saving would be impossible, indeed, it is very unlikely that the canal would be able to do the work at the rates now charged.

In Chapter XXV. I shall show that the savings on the carriage of sugar are 70 per cent. and on wheat 64 per cent. and other food in proportion. In 1882 I obtained the cost of the 4-lb. loaf in Glasgow, Birmingham, Liverpool and London, and found the average was less than in Manchester, and at one of the meetings I held forth hopes that the effects of the canal would be to reduce considerably the cost of bread in Manchester. My justification may be found in a speech by Mr. Gerald Balfour, M.P., in the House of Commons (*City News*, 8th March, 1905), who, when replying to Sir Henry Vincent, said :—

There were some difficulties in making a statement of prices of the quartern loaf of the quality consumed by the working classes strictly comparable, but approximately the figures were as follows :—

	London.	Manchester.
March, 1902 . .	4½d.	4d.
March, 1903 . .	4½d. to 5d.	4d.
March, 1904 . .	5d. to 5½d.	4½d.
February, 1905 . .	5d.	4½d.

Time will not permit me to go fully into the question, therefore I will content myself with recording two other failures in Mr. Provand's predictions. Instead of the railways ruining the canal, and the use of the docks being limited to such goods as could be carted away, the railway companies are now its best friends. Mr. Provand said :—

It is certain the railways will not make connections. They will not spend millions to enable the canal to take away their own business.

On the contrary, the Lancashire and Yorkshire, the London and North-Western and other railway companies, have spent immense sums of money to connect with the canal, and they are reaping a good harvest thereby. Again, he was utterly wrong as to an insufficiency of water, and in his prophecy that in less than a month from the start the canal would have to submit to whatever terms the railways might impose upon it. "Mr. Provand's Epitaph," was thus written :—

> Who killed the canal ?
> I, said Provand,
> With my pen in my hand,
> I killed the canal.

An attempt was also made to ridicule the canal out of existence.[1]   Mr. McArdle, the Liverpool pantomime writer, said :—

> It looked well on paper
> But was likely to end in vapour.

He also made the following reference to it at the Rotunda Theatre, Liverpool :—

CINDERELLA.

Cinderella.   Talking of Manchester—what's this they're planning,
To make a seaport of it?
Prince.   The papers lately you've been scanning
Cinderella.   It's very rare indeed
I ever get a scrap of news to read.
Prince.   Well, you must know—the notion sets one laughing—
It sounds ridiculous, but I'm not chaffing;
They mean to break the great commercial rule
And make a ruin of old Liverpool,
Render her miles of docks no use—no less,
Let it become a howling wilderness!
Cinderella.   Ha, that can never be!   Her flag's unfurled
In every port and harbour in the world.
Prince.   In maritime importance she's supreme,
But certain folks in Manchester now dream
Of a vast Ship Canal, so cotton ships
Right up our Mersey can take regular trips,
And land their bales on quays 'longside the Irwell;
But if the waters of that ditch they stir well,
The awful scent, of which you have no notion,
Will taint the Mersey and stagnate the ocean.
Cinderella.   But they'll work out their scheme.
Prince.   It does sound prime,
But won't be worked out in *our* children's time.

Another correspondent, "Mancuniensis," puts the other side of the question in his ode to "Monopoly":—

MONOPOLY.

> Filled with traditions of the past,
> To the Dock Board there marchéd fast

[1] See Cartoons—"The Port of Manchester in 1950."   By permission of the Proprietor of *Tit-Bits.*   "Bock Again."   Re-action and how Manchester tried to become a Seaport.   By permission of W. T. Gray, Esq., Liverpool.

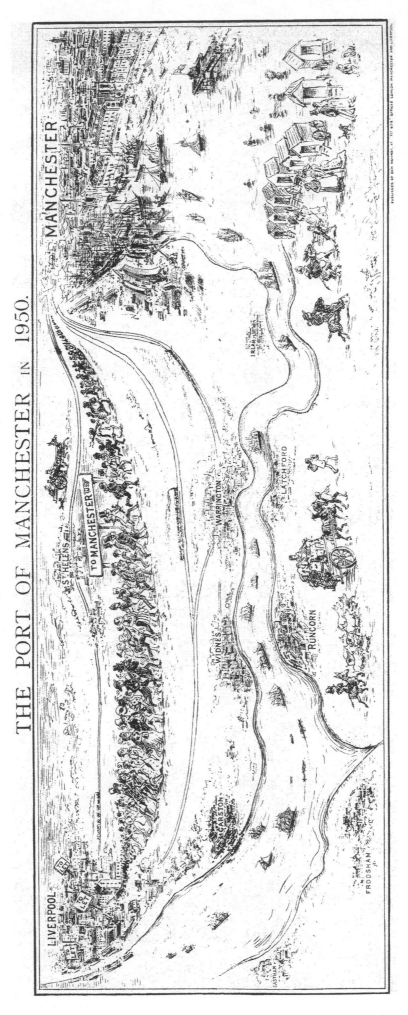

The material originally positioned here is too large for reproduction in this reissue. A PDF can be downloaded from the web address given on page iv of this book, by clicking on 'Resources Available'.

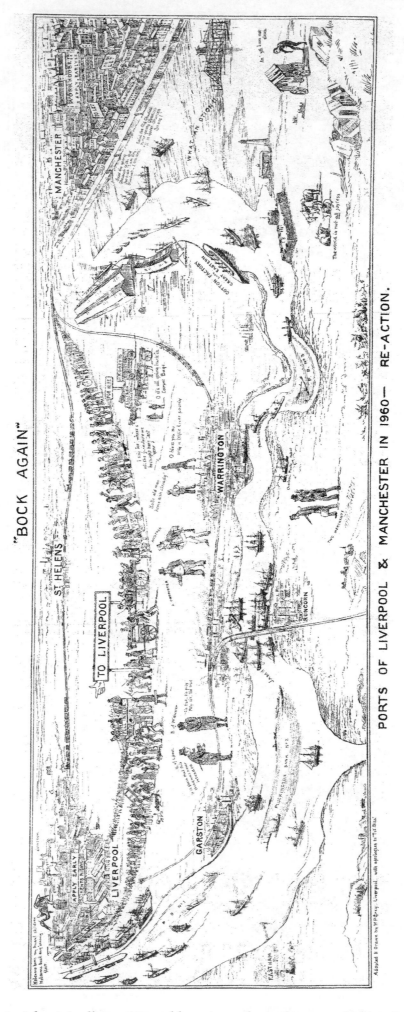

"BOCK AGAIN"

PORTS OF LIVERPOOL & MANCHESTER IN 1960— RE-ACTION.

The material originally positioned here is too large for reproduction in this reissue. A PDF can be downloaded from the web address given on page iv of this book, by clicking on 'Resources Available'.

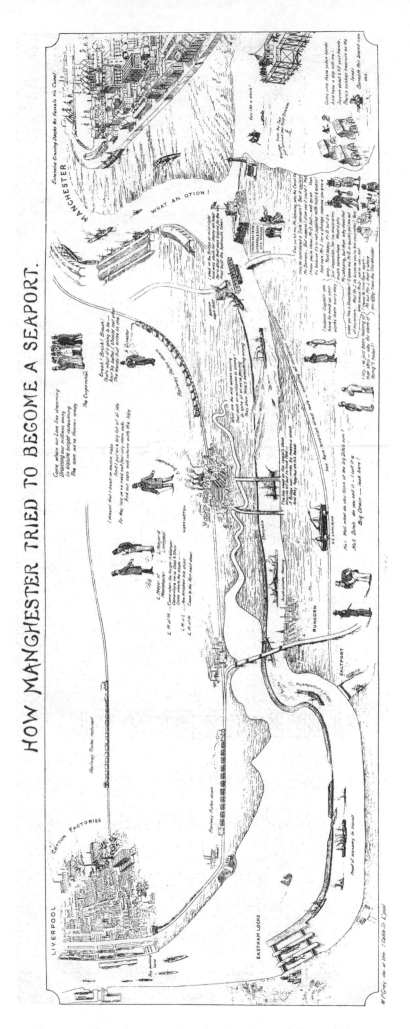

The material originally positioned here is too large for reproduction in this reissue. A PDF can be downloaded from the web address given on page iv of this book, by clicking on 'Resources Available'.

A Mersey Chief, who in a trice
A banner raised, with this device,
                    Monopoly!

His brow was sad, his eye beneath
Flashed like a falchion from its sheath,
And like a silver clarion rung,
In accents of his well-known tongue,
                    Monopoly!

"You know," quoth he, "our Mersey trade,
Is threatened by a strange crusade,
So let us with our friends combine,
And boldly fight e'er we resign,
                    Monopoly!"

Year after year, bold Harold [1] said,
"Do what is just to earn your bread,"
And still the friendly critic plied,
But loud that clarion voice replied,
                    Monopoly!

While commerce groaned a protest vain
(She could no longer hear the strain),
Still said the Chief, "'tis all my eye!"
And answered yet, without a sigh,
                    Monopoly!

"Yet pause awhile, lest sure revanche
Should crush you like an avalanche,"
This was the trader's last good-night,
Yet still replied the heedless wight,
                    Monopoly!

Poor trade, at last, crushed to the ground,
Oppressed by rates and dues was found.
The Chief yet held as in a vice,
That banner with the strange device,
                    Monopoly!

So they who buy and sell and make,
Themselves began the toll to take,
When, mindful of neglected prayer,
A voice wailed through the startled air,
                    Monopoly!

[1] Mr. Harold Littledale, a critic of Dock Board management.

> Peace! in Canning Place, cold and grey,
> Lifeless, yet fondly cherished lay
> A form, while from the sky afar,
> A voice fell like the morning star,
>   Monopoly!

At the commencement of the Ship Canal agitation Messrs. Walmsley and Samuels were the solicitors introduced by Alderman Walmsley, of Salford. With a stiff Parliamentary fight before them, the promoters deemed it necessary to join with them a firm that had had Parliamentary experience, and Messrs. Grundy, Kershaw, Saxon and Samson were selected. Henceforth, though both firms are named on the Bill, the latter firm conducted the whole of the Parliamentary business. Mr. Samson, assisted by Mr. M. J. Riley, drew up the brief, and I venture to say it was one of the most able and comprehensive documents ever placed in the hands of counsel. I begged permission to make a copy, which I prize highly. The *Book of Reference*, too, was a huge volume, and was one of the largest works of its kind ever produced. The first offices of the company were in Market Street, over the shops near the corner of Brown Street. For some months Mr. Henry Whitworth of King Street acted as secretary, but when the business grew, it was thought well to have some one who could devote the whole of his time to the work, and his son, Mr. A. H. Whitworth, was appointed. Messrs. Dyson & Co., of Westminster, were the Parliamentary agents, Mr. Coates of that firm having charge of the Canal Company's business, and most ably he did his work. The capital of the company was to be £6,000,000, in 600,000 shares of £10 each, shares not to be issued till one-tenth of the amount had been paid up. Qualification for a Director, 200 shares.

The first Provisional Directors nominated were :—

* Daniel Adamson.	* R. B. Goldsworthy.	* S. R. Platt.
* Richard Peacock.	C. P. Henderson, Junior.	* John Rylands.
Henry Boddington, Junior.	* Richard Husband.	* Peter Spence.
Joseph Davies.	* Joseph Leigh.	* Bosdin T. Leech.

The gentlemen who, on January 5th, 1883, became jointly and severally liable for a sum of £227,347, the necessary Parliamentary deposit, are named below.[1]

There was a provision in the Bill that if local authorities representing a

---

[1] The Provisional Directors marked with an asterisk, and in addition :—

William Richardson.	Hermann Hirsch.	John Taylor.
Alexander Butler Rowley.	Robert Bridgford.	Francis H. Walmsley.
Frank Spence.	William Henry Nevile.	Marshall Stevens.
George Hicks.	William Fletcher.	

ratable value of £4,000,000 applied to Parliament for the creation of a Trust, the promoters would not oppose it.

The first sign of opposition was the lodging of four petitions by the Mersey Docks Board, the London and North-Western Railway, and two others, alleging non-compliance with Standing Orders.

On 19th January, 1883, the Ship Canal Bill came before Mr. Frere, one of the Examiners on Private Bills. The opponents complained of non-compliance with Standing Orders, on the ground that the promoters had neglected to prepare and deposit plans and sections of the Mersey low-water channel, together with estimates; also that they had not made the Parliamentary deposit of 4 per cent. on the same.

Mr. Coates, for the promoters, brought evidence to prove that in consequence of the shifting character of the Mersey, it was impossible to deliver plans, and asked that the same roving powers should be granted as on the Tyne and Clyde. He claimed too that it was unnecessary to ask for new Parliamentary powers, inasmuch as the Mersey Conservancy Acts already gave the Conservators ample power to permit the promoters to dredge and also to erect training walls.

At the end of the second day's hearing the Examiner decided against the promoters, and threw the Bill out on Standing Orders. This necessitated an appeal to a Special Committee on the opening of Parliament, and the Provisional Committee at once determined to take this step. The omission of the plans was under the advice of their own Parliamentary agents. The opponents were overjoyed at this first check to the canal. The *Liverpool Mercury* wrote:—

This bungle at the outset is, we are afraid, characteristic of the whole jejune scheme. The promoters have got hold of "a big idea"; but if their capacity for dealing with it is to be measured by the case submitted in answer to the petitioners, we may pity the unlucky people who are induced to embark in the "magnificent undertaking".

Because the Chairman of the Ship Canal asserted the omission of plans was intentional, the same paper said:—

When promoters of enormous schemes adopt such crooked tactics, one would have thought their common-sense would have bid them not to blab. Yet here they are rather rejoicing in the trick, poor and unsuccessful as it was.

The *Manchester Guardian* commenced the year in an unfriendly way. In an article on 16th January it took as its text Mr. Provand's conclusion "that the project will benefit neither the producing nor mercantile classes, nor yet the shareholder". Speaking of the recent reverse it was not very cheering to say:—

The first practical movement that is made proves a fiasco. It will not be surprising if many observers regard the incident as ominous, and seriously doubt whether the promoters have the capacity to carry on successfully a work of such magnitude.

The *City News* considered the opponents had gained a triumph on a technical point, but that the lost ground would be regained, and believed the check might have a beneficial influence on the vigour and prudence of the promoters, and prepare them for the great struggles that awaited them :—

It is not bold and honest criticism that is objectionable, but bitter hostility under the guise of sapient friendliness. If fair argument is not strong enough to put down the movement, it ought not and will not be put down.

The Provisional Committee lost no time in attempting to repair their first disaster. They applied to all the Corporations, local bodies and Chambers of Commerce who had supported them, asking that they would petition in favour of the Standing Orders being dispensed with, and that the Ship Canal Bill should be dealt with by the Commons on its merits. In addition to this, a monster petition was signed in Manchester by nearly 200,000 persons. It required six porters to carry this to the Private Bill Office of the House of Commons, where it was presented by Mr. Jacob Bright.

Another mammoth petition from three hundred other Corporations, local boards and public bodies, merchants, bankers, manufacturers and ratepayers of Lancashire, Cheshire, and Yorkshire, was deposited by the same gentleman ; it was $2\frac{1}{2}$ miles long and weighed seven hundredweight. The Manchester Chamber of Commerce also sent a strong petition. There were twenty petitions against dispensing with the Standing Orders, including one from Liverpool, the Dock Board, various railway companies and others.

As a rule the boroughs asked their members to support the Ship Canal petition. Bolton and Warrington were exceptions. The former refused the invitation of the Mayor of Manchester to a conference, and when a deputation addressed a meeting there, they were told that the scheme had been got up by lawyers, engineers and interested parties, and that it would not benefit the borough. As the promoters were anxious to secure the goodwill of Bolton, it was arranged that Mr. Adamson and some of his co-Directors should meet the Corporation of Bolton and their local M.P.'s. I was one of the party who went. Unfortunately defeat and disaster seemed to dog our steps whenever we tried to please Bolton. On this occasion the deputation duly started by railway for the conference. On the way we got into earnest conversation, and at the first stop one of our party looked out of the clouded window and said, "This is Farnworth where they take tickets". Directly after leaving, to our annoyance we found we were passing through Bolton and speeding on to Darwen, the station beyond! We shouted our loudest to try and stop the train but without effect. We were obliged to disappoint the Bolton Corporation, the very people we hoped to propitiate. There was no help, so we had to make the best of it. We telegraphed the mistake from Darwen, and said we were returning by road (there was no train); but it took some time to get traps, and though we drove as fast as we could, we arrived just in time to see the last of the wearied Bolton Aldermen, and tell him our tale of woe. In turn he told us how much they were disappointed, as their Members of Parliament and Aldermen had waited for us till they were tired.

When Warrington was asked to petition, to the surprise of every one, a difficulty was raised by Mr. Bleckly (previously a warm supporter of the canal), and though the petition was adopted, two parties were formed in the Council; and this division has since militated against the existence of harmony between the Corporation and the Ship Canal. Even the right of members of the Town Council who were also Ship Canal subscribers to take part in Ship Canal debates, has been questioned many times.

When the petition for dispensing with Standing Orders was discussed in the Salford Council, Councillor Dickins opposed it.

To send such a petition as was proposed would make people outside the Council believe that Salford was taking great interest in the matter. He would like to ask how much of the £60,000 subscribed had come from Salford. Outside the £1,000 his Worship had given, he did not think £2,000 had been subscribed in the whole of Salford.

The Council, however, adopted the petition by a large majority.

The Manchester City Council raised no difficulty, and immediately adopted a petition for a suspension of Standing Orders.

At a meeting of the Chamber of Commerce on the 5th February, Mr. Peter Spence moved, and Councillor Leech seconded, a resolution that the Chamber should petition Parliament to dispense with the Standing Orders as regarded the Ship Canal Bill, and they were supported by Alderman Bennett. On this Mr. H. M. Steinthal moved, and Mr. W. Fogg seconded, an amendment to the effect that no petition be sent. The latter ridiculed the idea that dear carriage had anything to do with failing trade, said the promoters had even been to the lowest classes to get petitions signed, and asked whether the Town Hall was not a monument of corporate folly if they believed in a decaying Manchester. This roused Mr. Daniel Adamson, who made a sledge-hammer speech, and told Mr. Fogg he had got into a great fog himself, and that he had dragged them through the dirt among the dustmen in a way that was uncalled for.   Mr. Jacob Bright restored harmony by one of his happiest speeches, and among much excitement it was decided by 39 votes to 21 to petition Parliament.

The action of the various public bodies was severely criticised in the Press and elsewhere.   Mr. Moon, of the London and North-Western Railway, addressing his shareholders, said :—

He could not imagine that any such outrage on the Standing Orders would be allowed as was proposed by the promoters of the scheme.   The Examining Officer had already refused to sanction the Bill, and he did not believe it would go any further.

The *Examiner and Times* held that the Examiner was right in reporting against the promoters, because the most important part of the project had not been disclosed, and thought they would be unwise to base new hopes on an appeal to Parliament.

The *Guardian* attributed the decision of the Chamber of Commerce to the eloquence of Mr. Adamson and Mr. Jacob Bright, but said :—

The speech of the former was in the main merely the declamation of an advocate, and however much we may admire his zeal, we must point out that his remarks left the Examiner's decision practically unassailed.

Sir Edward Watkin, speaking to the Manchester, Sheffield and Lincolnshire shareholders, said :—

His company had not opposed the Bill on Standing Orders and they did not intend to do so.   If Manchester would find capital for a canal, it was not for them—if they approved it or not—to oppose the scheme on its merits, still less on Standing Orders.   He had no jealousy on the question of competition between water and railway carriage.   They would, however, bitterly oppose the Corporation finding capital for the scheme.   He felt very strongly that nothing had been done to remove the bar.   Was the great commerce of this district to be dependent on the bar at Liverpool?   They must have a harbour accessible at any time, and if the bar could not be removed, he recommended a new entrance by the Dee.

On the 2nd March the case of the Ship Canal came before the Standing Orders Committee, with Sir John Mowbray as Chairman.   After a lengthy deliberation in private, they agreed to dispense with the Standing Orders on condition that the portion of the Bill which related to the creation of a navigable channel between Eastham and Runcorn should be struck out, and the promoters left, as they desired, to arrange with the Mersey Conservancy in regard to this vital portion of the scheme.

In addition to Sir John Mowbray, the Committee consisted of Sir E. Cole-brooke, Mr. Cubitt, Mr. Monk, Mr. Mulholland, Mr. Dennis O'Connor, Mr. Playfair, Lord Arthur Russell, Mr. Whitbread and Mr. Yorke. Intense interest was felt in the Lobby by those awaiting the result, broken only by the Committee sending to consult Sir Erskine May's book on procedure. In the evening, to commemorate their success, Mr. Adamson, Mr. Stevens and the officials dined with Mr. E. J. Reed, M.P. for Cardiff.

This move forward was gratifying to the promoters, but as usual was received with mixed feelings abroad. The Bill next had to go before a Committee in the usual course. If it passed the Committee it would have to go before the Standing Orders Examiner in the House of Lords.

The *Manchester Examiner and Times* thought the Committee would have been more than mortal if they had not been influenced by the unprecedented number of petitions for the Bill. It would now remain for the promoters to make terms with the Mersey Conservancy Board about the estuary, the members being the President of the Board of Trade, the First Lord of the Admiralty and the Chancellor of the Duchy of Lancaster.

The *Guardian* deemed the result an additional testimony to the proficiency of the art of promoting; petitions had lain at every street corner, and had been signed by persons quite incompetent to form a judgment on the case, but there was no doubt the petition of the Chamber of Commerce had had great weight. But without the omitted portions the Bill would be of no use at all, and Mr. Frere said to deal with them was outside the power of the Conservators. Even if the present scheme were doomed to failure, a close examination of its details in Parliament might throw light on other possibilities, and hasten the advent of a really useful, if less brilliant project. In another article, the Editor dealt with the personal asperities which had grown up between the promoters and opponents of the canal, and alluded to charges of inconsistency made by Mr. Adamson against the *Guardian*. He noted a tendency to irritation on the part of the advocates of the Ship Canal, which might be pardonable, but still was injudicious. He then proceeded to deal with Mr. Provand's pamphlet and the reply, criticising the latter very severely. He objected to attempts to carry the Bill by a series of surprises, and ended: "Such tactics may be suitable in cases of private rivalry, but they are uncalled for in connection with a scheme which is avowedly promoted in the public interest alone, and which, if sound, has everything to gain by publicity and searching scrutiny".

The Bill was read for the first time in the Commons on the 6th March and the second reading was fixed for the following week. In two or three days it had also to come before the Standing Committee of the Lords, consisting of forty members, and it was known the Chairman, Lord Redesdale, was dead set against dispensing with Standing Orders in case of informal Bills, and that the Lancashire peers who presumably would be in favour of the Bill could not act. Then, if the Lords Standing Committee did throw it out, all that had been done would be labour in vain, even if the Bill succeeded in running the gauntlet awaiting it in the Commons Committee. Further, all the costs would be lost, and these were very heavy.

On the 9th March the Manchester Parliamentary Sub-Committee advised that a friendly petition should be presented against the Bill, so as to give the Corporation an opportunity to be heard on clauses. This is a position often adopted but which cannot be too much reprobated. It means piling up legal costs unnecessarily. The Ship Canal Committee would gladly have consented to any reasonable clauses.

On the 16th March when the Bill came for second reading in the Commons, Mr. Raikes asked for an explanation of a bifurcated and incomplete Bill being allowed to go forward, and called attention to the fact that it sought powers for municipal bodies to subscribe capital towards a private adventure. Sir John Mowbray said the decision of the Standing Orders Committee was unanimous, and that Mr. Raikes was too late in objecting now. Other members, including Mr. Bright, spoke, and the Bill was read a second time amid cheers, and referred to a Select Committee.

When the Bill came before the Standing Orders Committee of the Lords, only six out of its forty members were present. Out of these it was known that Lords Devon and Penrhyn were usually followers of Lord Redesdale, so the promoters' chance of success depended on Lords Monson, Camperdown, and Cork balancing them, and going for the promoters. An unusually long inquiry took place. Mr. Coates manfully fought the case. He showed the enormous interest taken in the Bill, and the value attached to it by all classes in Lancashire; never before had so many and such influential petitions been presented in favour of any Bill, and he made it plain that the Mersey Conservators could prevent a stone being touched in the estuary unless they had first been perfectly satisfied. Mr.

Coates had to fight his case single-handed against the Parliamentary counsel of the various opponents, but so well did he manage that it is believed the Committee were much impressed and were evenly divided.   Lord Redesdale, the Chairman, said :—

> After due consideration the noble Lords assembled here on the Committee to-day are desirous that the matter should be heard again before a larger Committee, and considering the state of the session and of the attendance in town at the present moment, we desire that that shall not be till after Easter.

On the 26th March the *Guardian* wrote one of its most scathing articles on the canal.   Of the Bill it said :—

> It becomes increasingly desirable that close attention should be given to its details. This is the more so as certain prominent members of the House of Commons, whose voices are likely to be influential, and who might have been expected to maintain a judicial tone with respect to the Bill, have already—somewhat hastily, as it appears to us—committed themselves to advocacy of the scheme.   Those observers who have ventured to criticise the project, to expose some of the rash statements of the promoters, and to call those who were allowing themselves to be carried away by an organised agitation back to the exercise of sober judgment, have been stigmatised as persons opposed to the trading interests of the district, whom history will regard with contempt.

It went on to say that such opponents, while anxious to cheapen traffic, saw germs of future disappointment in the scheme.   The article stated that the promoters had promised to bring the cotton market, the grain market, and the timber, as well as other markets, to Manchester, besides cheapening the working men's food, removing pollution from the river, and earning dividends for the shareholders.   It then went on to adopt Mr. Provand's "Criticism," and to show why all this was impossible, chiefly because there would not be dockage sufficient to do the business, and because of the cost of the canal :—

> We might have trusted the watchfulness of such bodies as the Manchester City Council and the Manchester Chamber of Commerce, but unfortunately the former has already shown itself to be too susceptible to hasty enthusiasm, and the latter has actually petitioned Parliament to abandon the most efficient safeguard against impracticable measures.   There is, therefore, some force in the suggestion of a correspondent that a special Vigilance Committee should be formed.

There can be no doubt that in many speeches, and also in letters and articles contributed to the Press by advocates of the Ship Canal, statements were made that had better have been omitted.   They often emanated from parties not recognised by the Provisional Committee, who regretted such injudicious statements.   But when gentlemen were working earnestly and contributing both of their time and money in the belief they were serving—with a single eye—the best interests of their city, it was hard to bear not only the indifference of merchants with vested interests and bitter attacks from abroad, but stinging articles from the Press of one's own city. Mr. Adamson, smarting under the suggestion of a Vigilance Committee, replied in a letter dated the 27th March.   After appealing to the *Guardian's* sense of fairness, he said he believed the subscribers to the Parliamentary Fund were actuated by no other desire than to advance the general welfare and prosperity of the community.

The only justification for a Vigilance Committee would be to exercise a species of espionage on a body of gentlemen whose sincerity and singleness of purpose, I hold, are beyond the shadow of suspicion, and who are prompted solely by a desire to do public good, with the intention of offering the boon (if obtained) to a public trust.   I consider the approval extended to the very mention of such a suggestion is a poor reward for the time, money and labour the Ship Canal subscribers are devoting to the development of a movement for securing cheap transit to our already heavily handicapped industries.

The eve of a Parliamentary fight was no time to deal with the little holes that Mr. Provand tried to pick in the scheme.   But if a Vigilance Committee was wanted it was—

as much to protect the promoters of the canal as the public, since I fear the Ship Canal Committee could hardly fare worse in the criticism it receives in your columns under such a régime than it does at present, and which, I am going to say, is strangely inconsistent with the generous, progressive and enterprising tone you adopted a few months ago.

The *City News*, reviewing the position, wrote an encouraging article :—

It is unfair for local opponents to attempt to force the promoters at this critical stage of the measure to expose the details which are prepared for use in the approaching examination before the Select Committee. . . . The promoters of the canal are charged with all the extravagant predictions and unfounded statements which have been spoken or written upon the subject.   Praise and condemnation always exceed the bounds of discretion in popular movements.   There is much chaff carried along with the grain.   A distinction should in fairness be drawn between the utterances of private individuals and the statements of responsible promoters. . . . It is certainly a new phase in the courtesy of the city to talk about a Vigilance Committee to watch men spending their money, time and labour in what they consider to be for the public good.

In view of the coming Council meeting, Alderman King wrote a long, and from his point of view, an able letter, to protest against the Corporation giving any aid to the canal ; and at the meeting of the General Purposes Committee he repeated his objection to the ratepayers being taxed to support the canal. "If," said he, "the Council adopted the principle, they would be establishing a precedent for making railways or building Atlantic vessels out of the rates." He moved there should be a petition against the Bill, and especially against this clause. It was pointed out the clause was only an enabling clause, and the report of the Parliamentary Sub-Committee was passed.

On the 14th April the Bill again came before the Standing Committee of the House of Lords. It was known there had been a special whip, and fifteen out of the forty noble Lords turned up. After the Parliamentary counsel had made their speeches, the room was cleared, and on the re-admission of the public Lord Redesdale said, "We are of opinion the Standing Orders should be dispensed with".

It is understood Lords Carrington, Devon, Sydney, Monson, Cork, Morley, Sudeley, Lansdowne and the Duke of Somerset voted for the Bill, and Lords Redesdale, Hawarden, Longport, Lathom, Donoughmore and Henniker against. Lord Powys did not vote.

The suspension of Standing Orders was reported to the Lords, and the next move was the appointment of a Select Committee of the House of Commons. The report of their proceedings forms the subject of a separate chapter.

In Manchester the Committee's proceedings were followed with the greatest interest, and the action of the various local Members of Parliament was the subject of much criticism. Mr. Houldsworth, M.P., wrote to the *Examiner:* "I am not aware of saying 'I believe myself in the success of the scheme'." Strong letters appeared in the Press respecting the attitude of Messrs. Slagg, M.P., and Armitage, M.P. They were told "it was utterly inexcusable to plead ignorance of the details of an undertaking in which their constituents were so vitally interested," and it was hinted to Mr. Agnew that his cash would be more valued than his sympathy. This reminds me of a good story. At one of the many meetings, Mr. Agnew, M.P. (who was never an enthusiast, but now and then made a speech just to show he did not scout a popular movement) was making an airy speech calling on the people to support the canal, the while with one hand in his waistcoat pocket, when an old fellow in the far corner of the room called out that "Mester Agnew should take his hand out of his waistcoat and put it into his breeches' pocket, and bring out some brass hissel for the canal".

During the Parliamentary fight the promoters made the Westminster Palace

Hotel their headquarters.   Here all the officials were to be found in conference at the end of each day's proceedings.   The Committee relieved one another, but the Chairman never left the case.   It could not be expected that barristers of the first rank could always be in attendance, but at odd times they were all absent, and then Mr. Adamson became furious.   On one occasion he declared he would telegraph for a new lot of counsel from Manchester, and if Mr. Goldsworthy had not used strong means to stop him, I believe he would have done so, for he was a very determined man.

Some of us much feared a rupture, for we had got a leash of barristers we could not replace ; the fiery, eloquent Mr. Pember, the genial, scientific Mr. Michael, and the cross-examiner *par-excellence*, Mr. Balfour Browne, who was like a terrier when he once found a hole in the armour of a witness.   Then there was the delicate and juvenile-looking Mr. Cripps, and the solid lawyer Mr. Pembroke Stephens.   Arrayed against them were a host of counsel, and as they all helped one another, there were very few weak places in our evidence that were allowed to pass unexposed.

It was very monotonous to sit day after day in either a stuffy, or, if the windows were open, a draughty room.   Occasionally a little sparring between counsel, or a humorous witness caused a welcome break, as when Mr. Marshall Stevens confounded the whole of the opposing barristers and let them drop into a hole, or when the burly good-natured manager of the Co-operative Society, Mr. Mitchell, crossed swords with Mr. Pope, Q.C., but these proceedings will be recounted in another chapter.

It was no child's play to be for days under examination and then to be crossexamined by such adepts as Messrs. Pope, Bidder, Littler, etc.   Mr. Leader Williams and Mr. Marshall Stevens made admirable witnesses, sometimes even being able to nonplus the cross-examiner.

Naturally the passing of the Bill in the Commons was received with great jubilation.   Both sides admitted the great fairness and tact with which Sir Joseph Bailey had conducted the long and tedious inquiry.   Mr. Pember's summing up of the vast mass of evidence on both sides was a masterpiece of brilliant and acute reasoning, and the promoters undoubtedly owe much to him for the energy with which he worked up their case.   His concluding speech drew such a picture of Lancashire suffering grievously from foreign competition, and anxious for the canal as a remedy, as no doubt affected the minds of the Committee.   To give an idea of the costs of a Parliamentary fight, it is said the leading counsel for the Ship Canal had 500 guineas on his brief, besides a heavy refresher and consultation fee.

When on the 11th July the Bill came before the House for a third reading, Mr. Whitley, M.P. for Liverpool, got it deferred by asking for information as to the conditions under which Admiral Spratt was Conservator of the Mersey, and Sir John Coode had been called upon to examine the plans. This was taken to be a device to delay the Bill, and was an expedient which was not effectual.

After the following pronouncement, which Sir Joseph Bailey, as Chairman of the Committee, went out of his way to make, "That it appeared from the evidence that if the scheme can be carried out with due regard to existing interests, the canal will afford valuable facilities to the trade of Lancashire and ought to be sanctioned," it was expected by both sides that the fight in the Lords would be, to a great extent, a matter of form. There were a few, however, who distrusted Lord Redesdale; they knew he was unsympathetic, if not hostile, and they could not forget that at its second reading he had tried to block the Bill. When a Committee was asked for, he had declined to move it on the ground that it was too late to get one together, and the Earl of Cork had been compelled to take the unusual course of moving for a Committee. Of course the selection of it was, to a great extent, in Lord Redesdale's hands, and he knew the tone of mind of those whom he selected. Five Lords are necessary for a Committee, and only four Commons. Is this a measure of their capacity? Anyway in this case, unfortunately, the Duke of Bedford had to retire in the midst of the inquiry, and rumour says the Committee were evenly divided, and that as usual in such cases the *status quo* had to be preserved. The unusual action of Lord Redesdale was portrayed by *Punch* as follows :—

Punch's ESSENCE OF PARLIAMENT. 4TH AUGUST, 1883.

Extracted from the Diary of Toby, M.P.

Tuesday.—Little row in the House of Lords to-night. Manchester Ship Canal down for second reading. Lord Redesdale doesn't like ship canals.

"Never had them in my day!" he growls. Content then with ordinary and proper thing broad enough for canal boats. If this thing goes on, have England cut up into mince meat in a few years. Make a sort of Holland of the island. Never be able to drive half a mile without coming across ship in full sail. Have steamers pouring smoke into your front bedroom window, and get hit on the head with maintop mizzen boom when you look out to see where smoke coming from. Have enough of ship canals at Suez. Have no more of them here as long as I am "Chairman of Committee".

So puts his foot down on proposal. Warns House if they agree to second reading he won't undertake to find Committee. This would have been enough at one time; but House sadly changing. Growing quite radical. Dares dispute what Redesdale says. When he got up in defiance of Rules to make second speech, there were cries of "order"! The stout Earl aghast.

"I am," he gasped, "standing up for order." "You'd better sit down for it," Lord Granville smilingly said.

Redesdale mechanically felt in trousers' pocket for his ruler. Attempted to draw it out. But Lordships only smile, and with a scowl at unoffending clerk at table, he resumed his seat.

"Take me away, Toby," he said a little later, in plaintive tones, that brought tears to my eyes, and nearly made me howl. "Take me away, and if it can be conveniently done, bury me in Westminster Abbey. The constitution is in danger; the throne is toppling to a fall; the sunset of the empire is at hand, and the House of Lords has shouted me down."

With the verdict of the Commons before them, *viz.*, that the canal would be of great use commercially, the Lords Committee evidently entered on the inquiry with the idea of keeping the evidence within a narrow compass. The Earl of Camperdown, the Chairman, specially pressed the promoters to curtail both their evidence and the speeches of counsel, and made it plain that the Committee were disposed to let the Bill go through, provided they could be satisfied on one or two points. So convinced were both sides that the passing of the Bill was a foregone conclusion that they were negotiating about clauses. Great was the disappointment when, after only a quarter of an hour's discussion, the Bill was rejected. It was rumoured that Lords Camperdown and Aberdeen were in favour of the Bill, and Lords Devon and Caithness against it.

After hearing Mr. Pember's reply, which to suit the Committee he had cut very short, I went back to my hotel expecting to return in time to hear the result. But on my way back I met Mr. Marshall Stevens looking very gloomy, and in the words "they have slated our Bill," he let me know its fate. Disappointment, nay almost despair, sat upon our faces. Mr. Adamson was so depressed that he decided the best course was to return at once to Manchester and discuss what was to be done afterwards, and without delay we all returned home in a very subdued frame of mind.

The only crumb of comfort to be gathered from the decision was contained in the words "not expedient to proceed this session of Parliament," giving some reason to hope that the Bill was rejected because it was incomplete, and that a complete Bill next session would have a chance of succeeding. No doubt the

Committee delivered a staggering blow to Lancashire, and this induced people to ask what qualifications the noble Lords had to deal with the weal or woe of a great commercial community.   This was answered by "A Sketch of the Lords Committee at Work," written by a member of the City Council (Alderman J. W. Southern), not then in Ship Canal harness, but who has since taken a leading part in guiding its career.   The article is too long to give *in extenso*, so I must content myself with a few quotations :—

There must be many people in Manchester who never enjoyed the felicity of watching the operations of a Committee of live Lords.   I sat for the whole of a day watching the proceedings of the Lords Committee on the Ship Canal, and had I not been made aware that the five gentlemen at one end of the room were real Lords, I might not have discovered their real superior constitutional endowments.   The Lord nearest the door is a very young man, about twenty-four, and I should think very good natured and kindly disposed.   He has a thin face, rather expressive mouth, and jet black hair, carefully brushed.   He smiles a good deal, and seems quite interested in what is going on.   Evidently he has not yet learned the Olympian attitude of sublime indifference which becomes a Lord.   For this you must look to his next neighbour, who is a commonplace looking personage of about sixty.   If I had met him in a railway carriage, and been asked to guess his business, I should have said a cheesemonger, and the pursed mouth and elevated nose I should have referred to some recent experiences with a cheese somewhat too far gone.   He never speaks—I was going to say never opens his mouth : that would have been incorrect.   He often yawns, and when the attention of the Committee is requested to some plan or diagram, he ostentatiously turns his face to the window, yawns, and stretches his arms.   The next noble Lord is the Chairman.   He is a comparatively young man, say about forty.   It is proper he should be Chairman.   He seems to have common-sense enough for an ordinary man.   What an *enfant terrible* such a person must be running in and out among the beings of a superior order.   He is a trifle caustic, is not to be imposed upon by secondhand evidence, is rather tart with both witnesses and counsel, and seems particularly desirous not to have too much evidence brought before the Committee.

To the right of the Chairman is an elderly man, perhaps sixty-five or seventy, maybe even more.   He looks a very superior person.   He has just the look a man ought to have who had been allowed all his own way when he was a child, and had been in the habit of patronising common people, and receiving the reverential obeisance of rustics ever since. His eyelids are rather heavy for his eyes and this gives him a sleepy look.   But that you know he is endowed with hereditary wisdom, you would say his prevailing expression was a compound of apathy and vacuity.   The only remaining member of the Committee is a powerfully built and military looking man.   He seems taciturn and grave.   He never utters a word, nor takes a note, but attends with impassive face to everything that is said.   I should say he belongs rather to the fighting than to the law-making type of man.   There is a look of cool self-control that would be invaluable on a battlefield.   Pity for such men that times change and court rapiers take the place of stout claymores, or councils of war give place to Committees on Private Bills.

As I reflected afterwards on the experience of the day I could not help feeling that the whole business was a monstrous and most expensive farce.   Here were five men, none of them in any way distinguished for wisdom, engineering skill or intellectual power.   They could not pretend—no one could put forward the pretence on their behalf—that even had they sat like the Commons Committee thirty-nine days and heard all the evidence that Sir Joseph Bailey's Committee heard they were better endowed or more capable than that Committee of arriving at a wise conclusion.   But they sit a few days—they limit the evidence they will hear—they disregard the wide conditions imposed in the public interest by those who represented the public.   These men who represented nobody and nothing but an obsolete and antiquated feudalism are permitted to refuse the just desires of a large and industrious population fortified by the approval of many municipal and commercial bodies, to override the decision of a painstaking and representative Committee of the Commons, and to cast to the winds the efforts and the hard-earned means of the promoters.   Our eyes are blinded by custom or we should never submit to such an absurdity.

Of course there was great jubilation in the enemy's camp.   Sir Wm. Forwood said the expenses of the unsuccessful promoters were £57,000, whilst Liverpool had spent £5,000.   Mr. Moon, Chairman of the London and North-Western Railway Company, remarked that they had had to fight the Ship Canal, probably one of the wildest schemes that had ever entered into the mind of man to conceive.

At home too there were Job's comforters.   Our leading paper discovered—

That the plea for a canal was founded on plausible fallacies.   The decline of industries and in the value of property had little to do with Liverpool and railway exactions, or our insular position.   It was mainly due to property jobbers and building societies, who hoped to make fortunes rather by land speculations than by industry.   If the Ship Canal is simply to transfer money from the railway companies to the landowners and property jobbers, how will the once flourishing industries of Manchester be benefited?

Another alluring statement of the promoters was that manufactories would spring up along the banks of the Ship Canal.   If the argument is sound, Oldham will remove to the banks of the canal, and Manchester itself will be found "stepping westward".   Such an outburst of industry along the banks of the canal as that dreamt of by the promoters would entirely destroy the value of an enormous amount of residential and rich agricultural property on the Cheshire side of the city, and make the city itself almost uninhabitable by poisoning the only fresh breezes we at present get.

Comment on this is scarcely necessary.   Events have proved how erroneous were the statements used in order to throw cold water on the scheme and discourage its promoters.   Trafford Park has become a hive of industry, and neither has the value of property been destroyed nor the city poisoned.

The City Council on the 27th August, in passing a vote of sympathy with the promoters, "deplored the vast expenses which had been incurred without adequate result, and considered the present to be a striking instance of the hardship of the present mode of procedure upon Private Bills".

After a few days' breathing time the Provisional Committee were called together, and they decided to hold a conference on the 28th August, to which representatives from the various local bodies who had supported the scheme were to be invited to attend, the object being to decide as to the renewal of the application to Parliament, and to discuss the scheme with the view to arrive at such an agreement as would secure for it the approval and co-operation of all classes; also to have a report on the financial position.

On the day named a large and enthusiastic meeting was held in the Town Hall, Manchester. Mr. Adamson made a stirring speech and ended with :—

If they got a canal their children and their children's children would bless them for increased facilities for earning their bread. There must be a development of trade to provide for an increase in population. Unless they made reasonable provision for the future there would certainly be disaster. If they entered into this subject with heart and soul, they would overcome every obstacle, and the canal would be made.

He moved the financial statement and report, and stated that so far their assets were £67,000 and their expenditure about £57,000.

This Alderman Bailey seconded. After a resolution had been passed deploring the rejection of the Bill, Alderman Harwood moved :—

That having received from the Provisional Committee its opinion and recommendation that if sufficient funds are forthcoming an amended scheme should be submitted to Parliament, and such assurances of renewed support having been given to this meeting as to justify the belief that a second Parliamentary fund of sufficient amount to carry the measure through Parliament will be provided, this meeting resolves that the Provisional Committee be authorised and empowered to promote a Bill in the ensuing session of Parliament for the establishment of a Ship Canal to Manchester, in such manner as may be found most expedient or desirable.

He said that having set their hands to the plough it was not the habit of Lancashire men and Yorkshire men to go back.

Dr. Mackie, of Warrington, seconded the resolution, and the Chairman announced that about £5,000 had been already promised for a renewed struggle, towards which Mr. John Rylands had given £1,000, and Mr. Jacob Bright £500. At this meeting there is no doubt the Chairman, carried away by his strong feelings, made remarks that had better been left unsaid, and which left sores.   Speaking of Lord Camperdown, "he must say that he had rarely met a more fractious, disturbing, interrupting Chairman than Lord Camperdown"   During the inquiry Mr. Pope had in his cross-examination, and subsequent speech, roused Mr. Adamson's indignation, especially when he likened the canal to a bubble which would burst if pricked.   So the latter took the opportunity to retort in these words :—

The great, renowned, big, infallible Lancashire Pope might misstate the figures and misrepresent the facts, but he would never be able to alter the fact that the enterprise was one of the best paying that had ever been placed before the country, and when he did prick the great canal bubble, the contents would overwhelm him like an avalanche, and nothing would be visible of him save the dishonoured wig of a Queen's Counsel.

Fortunately Mr. Pope, knowing his man, did not treat the matter seriously, but he now and then humorously alluded to it in his speeches.

Having determined to make renewed application to Parliament in the ensuing year, the Provisional Committee set to work with courage and determination to raise fresh funds, to educate the people by circulating literature, to consider the various suggestions made to improve the engineering details, to get experience from the different towns that had improved their navigations, and to arouse the enthusiasm of Manchester and the surrounding boroughs.   They asked the latter to enlist their members in the service.

Whilst the Local Committees were collecting funds on the outskirts, a Central Committee divided the city into wards in the hope of converting the wavering merchants and of securing subscriptions.   Many were the rebuffs the canvassers received, and it is amusing now to think of some of the scenes that occurred.   Mr. Ben. Armitage, M.P., was on my list, and I well remember the many hours I spent in trying to get him out of his shell.   He wanted to please his constituents by supporting the canal, and yet, as his immediate coterie of friends evidently were not favourable, he could not summon courage to throw aside their leading strings. All through he was an example of the man sitting on the fence.   One of the calls made by a friend and myself was on an ex-M.P., a rich home trade merchant, whose firm had subscribed £100 to the preliminary fund.   It was a hot day and we found him in his shirt-sleeves.   We put our case before him with all the ability in our power, and he kept throwing in jocular remarks, but when we came to ask for money he shrugged his shoulders and said he was too poor to help—that he was largely interested in railways—and from this position we could not move him.   To compensate for our disappointments we often received the heartiest encouragement; comparatively poor men would subscribe £100 when we did not expect a fifth part of the amount.

Meetings were arranged in the various wards of Manchester and Salford and in the neighbouring towns, and a systematic canvass for funds instituted. It was my lot to visit outside towns, and I well remember one amusing incident. Arrangements had been made with the local secretary at Littleborough for a meeting to be held at 7.30 in a large hall. But in consequence of some mistake about dates, when with a colleague I arrived, no secretary met us and no room had been prepared for the meeting. On hunting up the official, we found he had arranged for a room and speakers at a later date. It was a complete fiasco. We did not like coming on a fool's errand, so, hearing there once had been a bellman in the place, we hunted him up, paid him well and sent him round the town to say some gentlemen from Manchester had come to explain all about the Ship Canal, giving time and place, admission free. After all we had a good meeting and an attentive audience, who heartily passed resolutions in favour of a renewed application. In fact the bungle was turned into a success.

When a cause is struggling against adversity people are to be found very willing to sneer at it. So it was with the Ship Canal. Many people were ready to say "I told you so," and wrote comical letters to the Press. It was felt necessary not only to combat every adverse statement, but to put forward convincing arguments as to the benefits of a canal. This work largely fell into my hands, and the midnight oil burned over it no one will ever know. One day I had an unexpected reward that greatly encouraged me. I knew Colonel Shaw, the American Consul, but he did not know me. However, as we were travelling together to Bowdon, he began talking to a gentleman about the canal, and said he could not see his way clearly. Much that was said about it was extravagant and foolish, but said he, "there is a writer whose letters I read with interest, because he seems to have studied his subject, his name is Leech, and I should like some day to meet with him, and have some further explanations." The gentleman he was speaking to at once said, "And he sits next to you, let me introduce him". The Colonel shook hands and said a kindly word of compliment. We had many a talk afterwards, and the Colonel eventually did the undertaking good service both in England and America.

The visit of the Iron and Steel Institute offered an opportunity for Mr.

Adamson and several engineers interested in the canal to examine the improvements on Tyne and Tees.   A large number of subscribers, at the invitation of the Tyne Commissioners, visited the Tyne.   The heartiest goodwill was shown to the Ship Canal, and encouragement and promises of assistance given.   They afterwards visited the Tees with the same result.   Mr. Pember accompanied the party, in order that he might make himself personally acquainted with the waterways and harbours about which he had to say so much.

At their last meeting in October, the Salford Town Council passed a resolution expressing their regret at the failure of the first Bill, and congratulating the promoters upon their decision to make a further application.

On 31st October an immense gathering of the friends of the canal took place in the Free Trade Hall to support the Provisional Committee in their second application to Parliament.   Long before the doors opened many hundreds were waiting for admission, and it is supposed that nearly 2,000 people were turned away. Messrs. Jacob Bright and John Slagg, M.P.'s for Manchester, were there, as were also a large number of leading citizens and representatives from Limited Liability Companies and Trades Union Societies.   Several speakers, fresh from their recent visit to the Tyne and the Tees, recounted the wonderful results of energy and perseverance in opening out what used to be impassable rivers, and in making them navigable for large ships.   Mr. Adamson made an appeal to Lancashire to come to the rescue, saying he believed that county possessed as much foresight, backbone, pluck and endurance as the men of the North, who had been successful in bringing a vast volume of trade into their district.   Mr. Jacob Bright urged his hearers to reject the idea of a barge canal with which Liverpool wanted Manchester to be satisfied.   He gave facts connected with the industries of Sheffield and Warrington to show how the trade of inland towns was crippled by heavy charges, which all acted as a premium to our German competitors, and he hoped a Ship Canal (the dream of our forefathers) would be secured.   Mr. Bright then alluded to the hostile attitude taken up by the Manchester Press.   " I don't believe that our great Manchester morning newspapers, which have not yet been very friendly on this subject (Hisses and Hear, hear), would deny that it has to be a waterway or nothing. Some of our friends are afraid of a canal that shall bring ocean-going ships; the proposition is too bold for them, and they ask us to consider something else."   He concluded a most telling speech, which had roused the audience to a high state of enthusiasm, with, " I believe in your success just as Mr. Adamson believes in it, because I believe in the enterprise of Manchester men. We cannot succeed without a great effort. Parliament must not have the notion that we are lukewarm. Let Lancashire men, and especially the people of Manchester, Salford, and the neighbouring towns, do their duty, and this great enterprise will be carried to a successful conclusion" (Loud cheers). Alderman Thompson, the Mayor of Salford, Dr. Mackie, of Warrington, Alderman Bailey and others spoke, and the meeting (a most enthusiastic one) broke up with the firm resolve to pass the Bill.

Mr. Houldsworth, M.P., wrote regretting that he could not be present at the meeting, but said that he should support a Bill if it was applied for. "Of course I reserve to myself the right to consider from time to time any facts which may be brought forward against it."

By the end of November the Ship Canal Company issued the statutory notice necessary for an application to Parliament, and the Bill was deposited in due course.

On 26th November a public meeting was held in the Town Hall, Salford. Resolutions in favour of the canal were carried unanimously Mr. Armitage, M.P., and Mr. Arnold, M.P., both spoke, and the former not only gave his adhesion to the cause, but promised a subscription towards the Parliamentary Fund.

Mr. Adamson had great faith in the judgment and ability of M. Lesseps, and more than once he had said in a jocular way that if Manchester would not make a canal he should have to fetch M. Lesseps. It so happened the latter was in England, so an invitation was sent to him to inspect the plans and works, and attend a meeting in the Free Trade Hall. This he did on 19th November. Mr. Adamson, the Chairman, reminded the meeting that fourteen years ago that very day the Suez Canal was opened, that it brought with it peace, happiness and prosperity, and that, though the cry was raised it would never pay, figures proved the contrary :—

						Ships.	Tonnage.	Revenue.
1870	.	.	.	.	.	486	435,911	£206,000
1882	.	.	.	.	.	3,198	7,122,125	2,421,824

He concluded by presenting a congratulatory address. Mr. Jacob Bright made an appropriate speech, and said that nine years after the canal was begun the great engineer, M. Lesseps, had the satisfaction of steaming through.

In responding, M. Lesseps said he had not, from lack of time, been able to study the Ship Canal, but he should do so on his return with the greatest interest. M. Chas. Lesseps accompanied his father.

About this time Mr. Harcourt Thompson, C.E., tried to propound a scheme by writing to the papers and getting the assistance of the *Manchester Weekly Post*. He proposed to make a canal 800 feet wide, in the present course of the river, place locks at Warrington and pump up salt water from thence to supply the higher portion.   At some of his meetings he said he had shown his plan to Mr. Lyster, engineer of the Dock Board, who approved of it.   This was not felt to be a sufficient recommendation, and the scheme eventually died a natural death.

Unfortunately, after the rejection of the Bill, difficulties arose amongst the Provisional Committee.   When the question of a renewed application to Parliament came before them, there were some of the older members who had a strong feeling that there ought to be a reorganisation of the staff, a reduction in expenditure, and a line of policy adopted that would make friends, and draw rather than repel many of the influential people of the district whose help would be essential when capital had to be raised.   They felt it was absolutely necessary to husband their resources, and to be very careful not to alienate by hard words those who could not see eye to eye at once with the promoters.

This change in policy did not commend itself to the Chairman or Organising Secretary, who preferred the bolder policy of carrying the war into the opponents' camp, and taking the consequences, trusting to the public for support.   My own case was a typical one.   In 1883 I was a Director named in the Bill, and signed, with others, a joint guarantee for £229,905 on its deposit, but when asked again to take the same position in 1884, I demurred unless I could be assured of a change in policy. This the Chairman resented, and he wrote me a characteristic letter, and afterwards met me on the Exchange, which we paced for nearly an hour talking matters over. He wanted me to resume my place on the Directorate unconditionally, which I declined to do.   I assured him of my unabated faith in the canal, and of my intention and willingness to work for it to the utmost of my power, but said that though ready to assist in pulling the coach out of a rut, or pushing it up a hill, I would not take a seat in it unless I felt fully assured it was going to be carefully driven.   We parted good friends, and Mr. Adamson invited me to dine with him a few days later.   When the company was formed Mr. Adamson himself asked me to become shareholders' auditor.   I was named, however, in the subsequent Bill simply as a promoter.   Other gentlemen shared my views and declined the directorate.

On the 15th December all arrangements were completed for depositing the Bill which provided for a capital of £8,000,000 in 800,000 shares of £10 each, and gave the company power to borrow £2,000,000.   The Provisional Directors were Messrs. D. Adamson, Jacob Bright, M.P., Henry Boddington, Wm. Fletcher, C. P. Henderson, Junr., Rd. Husband, Joseph Leigh, S. R. Platt, John Rylands, Marshall Stevens, John Walthew and James E. Platt.

The feeling in Liverpool was one of contempt for the Ship Canal.   Mr. Alderman Samuelson in the Council Chamber said :—

Liverpool could always hold its own against Manchester, and because there happened to be a dozen idiots in this world in regard to a Ship Canal, there was no reason why the Liverpool Corporation should constitute itself the thirteenth.   The whole thing was a bubble, and would, he believed, burst before another twelve months passed over their heads.

Correspondents in the Liverpool papers also advocated retaliation, and said :—

Let us be a manufacturing town.   Why should not cotton mills be built near Liverpool.

In Birkenhead some of the Councillors were favourable to the canal, and Mr. H. K. Aspinall, Chairman of the Woodside Ferry, gave it his blessing.

During the Lords' Parliamentary inquiry, Mr. Rodwell, Q.C., asked if Mr. Hugh Mason, M.P., did not represent Manchester on the Dock Board?   Through his answer Mr. Adamson got in conflict with that gentleman.   Mr. Adamson unfortunately stated that Mr. Mason was an investor in the Dock Board Trust, and a Director in several railway companies.   This gave Mr. Mason the opportunity of denial, and of saying that he had refused to back the Ship Canal, or give it encouragement, because he thought it would be a ruinous failure.   Whilst Mr. Mason did not hesitate to condemn the canal, many of the other Lancashire members were apathetic.   At a meeting in Ardwick an elector asked Mr. Houldsworth, M.P., "Whether he would assist (in Parliament) the promoters of the Ship Canal to obtain for the scheme the sanction of the legislature?"   This was to the point, and Mr. Houldsworth tried a diplomatic answer: "I think I am a disinterested witness, because I have not taken any active part in the promotion of this scheme. It is a business scheme, and, as you are aware, I have a great deal of business of my own, and I do not, as a rule, mix myself up with any scheme that does not come into connection, more or less, with my own business, if I can possibly help it," etc.   But his cruel heckler was not to be shaken off.   "I should like a more definite answer.   I am a Conservative, and I wish to know whether Mr. Houldsworth is prepared personally to support the Bill."   There was no escape, so Mr.

Houldsworth boldly faced the matter, and said, "Yes, I am". Henceforth he was committed to the canal scheme.

During 1883 many helpful contributions were made to Ship Canal literature. Amongst them the following series of articles :—

Our Masters of the Mersey, by "Verax" (Henry Dunckley).
The Liverpool Toll Bar, by Frank Hollins.
Past, Present and Future of the Manchester Ship Canal, by the *Manchester Guardian*.

The last step of the year was taken by Mr. Adamson, who, on 28th December, wrote to the Corporations of Manchester, Salford, Warrington and other surrounding boroughs, asking them to contribute one penny in the pound on their ratable value, in order to strengthen the subscription list, and to show the world the reality of the support accorded by Corporations and the public generally.

CHAPTER VIII.

1883.

SHIP CANAL BILL IN THE COMMONS—SPEECH OF MR. PEM
BER AND OTHER COUNSEL—SUMMARY OF EVIDENCE
BY THE SHIP CANAL COMPANY—BY THE OPPONENTS—
DECISION OF THE COMMITTEE—CLAUSES.

*The obstacles to the improvement of the Irwell between Warrington and Manchester
are so slight as to excite a smile in the engineer of the present time.—*BAINES' *History of
Lancashire and Cheshire.*

THE Ship Canal Bill having at last passed the Standing Orders, came before
a Committee of the House of Commons on the 1st May, 1883, consisting
of Sir Joseph Bailey, Mr. A. P. W. Vivian, Mr. Reginald Yorke and Mr.
Stafford Howard, the first-named being selected as Chairman, and certainly a gentleman better fitted for the position could not have been chosen. Many of the most
eminent barristers having been retained by the railway companies, it became
difficult for the promoters to obtain counsel equal to the occasion, but eventually
they selected Messrs. Pember, Michael, Balfour Browne, Pembroke Stephens and
C. A. Cripps. Arrayed against them were Messrs. Pope, Littler, Aspinall
Bidder, Saunders, Ledgard, Sutton, Moore, McIntyre, Noble, etc., representing
the Railways, Liverpool, the Dock Board, and numerous other bodies. In all
there were twenty-seven petitioners. Manchester entered a friendly opposition to
the Bill and was represented by Mr. Addison, Q.C.

Mr. Pember, Q.C., in an able speech opened the promoters' case. He stated
the proposed capital to be £7,500,000, £6,000,000 of which was to be subscribed
capital, for the rest he asked borrowing powers. The estimate for works was
£5,633,951. It was proposed to buy the Mersey and Irwell navigation from the
Bridgewater Trustees, and to make the Irwell navigable to Hunts Bank. The
worst railway gradient would be 1 in 114, and he pointed out there was already one

of 1 in 85 not far away.   He showed the serious disadvantages that trade endured at Manchester in consequence of hostile tariffs and costly transit, and that railways, having in one way or another got hold of many of the principal canals, had formed a Carriers' Trade Union.   To meet this, competition by sea, that could neither be crushed out nor bought off, was the only remedy.   Then he quoted the statement of Mr. Richard Moon (Chairman of the London and North-Western Railway), *viz.:* "That the sea and canals had done more to bring down railway rates than any competition among the companies themselves".   Again by their population and ratable value Mr. Pember showed the importance of Manchester and the surrounding towns, and gave the amount of their imports and exports, dwelling on the heavy cost of the carriage of cotton (raw and manufactured), and showing by a table of distances the saving that would be effected if Manchester became a distributing centre. Liverpool *must* be a dear port because she suffered from improvident finance in the past, and because she had spent £6,000,000 in useless docks at Birkenhead. Why should Manchester pay for this?   He went on to quote Mr. Findlay, manager of the London and North-Western Railway :—

I think that all has not been done that might, or ought to have been done at Liverpool for the economical and proper accommodation of the traffic. . . . There are no cranes or appliances, except what the ships possess.   If Liverpool had done what has been done in London, I have no doubt that the cost with which business could be carried on in the port of Liverpool would be very materially reduced, whilst the facilities would be enormously increased.

That the Liverpool dock accommodation was entirely inadequate, and that the port charges were extravagant was proved by the evidence of Mr. Findlay before a Select Committee in 1881-82, when he said that the charges on a ton of cotton landed at any of the following ports—from ship to truck—were respectively, Grimsby 3s. 2d., Garston 2s. 3d., Fleetwood 2s. 7d., Barrow 2s. 9d., Hull 3s. 8d., London 6s. 1d., Liverpool 8s. 1¼d.   After giving other statistics, Mr. Pember went on to argue that the traders of Lancashire could no longer bear the expense of Liverpool as a port, and that it was the intention of the promoters to create a cheap port at Manchester.   Nor could they submit to the excessive and unfair railway rates.   As an instance, iron sent by rail from Runcorn to Manchester (23 miles) was charged 6s. 3d., whilst from Middlesborough to Manchester (109 miles) it cost 8s. 4d. per ton.   Again, from Liverpool to Oldham the charge for cotton was 2·93d. per ton per mile, and from Liverpool to Bolton 4d. per ton per

E. H. PEMBER, K.C., LEADING COUNSEL, MANCHESTER SHIP CANAL.
After portrait by Holl.   *To face page* 142.

mile, whilst in France from Havre to Rouen, the freight was 1·18d. per ton per mile. He contended that it was necessary to reduce the carriage of produce to Manchester 50 per cent., and that it was impossible for the railways to do this, but that it might be done by water carriage, especially as regarded heavy traffic. Sir Edward Watkin (now an opponent) in a speech in 1883 said :—

It is absurd to suppose you can have any effective competition by water throughout the country unless you take the thing boldly in hand, and improve the whole navigation of the country. I do not think it will do the railway property permanently any injury at all.

Again :—

The way we looked at this scheme (speaking of the Ship Canal) was this, that the people of this great hive of industry (Manchester) have a right to carry it out with their own money, if they can. I repeat that I think a greatly improved waterway between Manchester and Liverpool is bound to be made, and that it would be a great blessing to Manchester and no damage to Liverpool.

As regarded vessels coming up to Manchester, the figures proved that 95 per cent. of the vessels afloat did not exceed 3,000 tons, and so could come up to Manchester. The tonnage he estimated at 2,000,000 tons per annum.

After Mr. Pember's opening address, the Chairman said it was useless going further into commercial or financial considerations till the Committee had been convinced that a connection between Manchester and the sea was possible.

Mr. Pember thereupon dealt with the various hostile petitions. The ground of the *Bridgewater and Mersey and Irwell Navigation Company's* petition was that shifting sands in the estuary would render it impossible to keep a reliable channel.

Of the Mersey Dock Board—That they had spent £1,500,000 on the Birkenhead estate and £16,500,000 at Liverpool. That they objected to the people of Manchester, with the assistance of the Corporation, making the canal. That they claimed the normal scour of the tide in the estuary, and opposed any interference. That in consequence of Runcorn Bridge, ships could not get up the canal; also that both the estimates and capital were insufficient.

Of the Trustees of the River Weaver—That there would be an interference with their claim to light and buoy the river.

Of the Corporation of Liverpool—That the Bill would depreciate the value of their docks, etc., and interfere with the welfare of Liverpool, which was dependent on its commerce, and so affect the security of the rates on which they had borrowed money.   That the estuary was in the shape of a bottle, and to reduce the area of the bulb would reduce the scour at the bar.   That the scheme was truncated and undefined, and the plans incomplete.   That the financial position of the promoters was weak.   That a Ship Canal could not compete with the existing railways and canals.   Lastly, that the course of the Vyrnwy pipe would be obstructed.

Of the Corporation of Birkenhead—That trade would be abstracted from their docks and the town ruined.   That the Ferry would be interfered with.   That new sand banks would be formed, and that the free course of sewage to the sea would be prevented.

Of the Owners of Land at Warrington—That their business rights would be injuriously affected.   That there would be an interference with a waterway hitherto free, by blocking up or diminishing the tidal flow.   That the river would silt up.   That they objected to dredging between Runcorn and the junction of the Mersey and Irwell.   Also to raising the water in the canal above the normal level of the river.

Of Sir Humphrey de Trafford—That his ancestors had resided on the Trafford estate for centuries.   That the canal would alter the natural boundary of his park, and bring polluted water close to his residence.   Also damage his Barton entrance and interfere with his drainage.

Of the London and North-Western Railway Company—That their gradients would be seriously affected in going under or over the river.   That damage might accrue to Runcorn Bridge, and that in consequence of traffic and circumstances having changed, the promoters had no right to avail themselves of existing clauses obliging railways to make swing bridges for a waterway.

Of the Cheshire Lines and Midland Company—That swing bridges ought not to be permitted.   That injury would be done to the working of traffic, especially in marshalling trains, and that the scheme could not be a success or commercially of advantage to the public.

Of Mr. Hargreaves, the Owner of Mill Bank Hall and the adjoining Paper Mills—That the canal would destroy the amenities and privacy of his house, cause smoke, dirt and steam, disturb his rookery, and possibly cause the subsidence of his buildings.   Also put an end to his ferry.

Mr. Pember then read a letter from Sir Richard Wyatt, Parliamentary agent for the Mersey Conservancy Commissioners, saying that body would raise no objection to deepening the low-water channel provided they could be satisfied the works were practicable, could be carried out without injury to the navigation or the dock approaches, and would not cause a silting up of the channel.

Mr. Pember then called the following sixty-seven witnesses :—

Engineers—

Mr. Edward Leader Williams, Manchester; Mr. Messent, Tyne Trust; Mr. Fowler, Tees Trust; Mr. Deas, Clyde Trust; Mr. Abernethy, C.E., London; Mr. J. F. Bateman, London and Manchester; Mr. Daniel Adamson, Hyde; Mr. R. Price Williams, London; Mr. James Brunlees, London; Mr. Alfred Giles, Southampton; Mr. Fred E. Dutton, London.

Shipowners and Builders—

Marshall Stevens, Garston; William Symons, Renfrew; Alexander Adamson, Glasgow; William McMillan, Glasgow; Thomas B. Seath, Glasgow; Alexander Neal, Glasgow and Manchester; Frederick Edwards, Cardiff; Peter Hutchinson, Glasgow; Henry H. Briggs, Hull.

Spinners, Manufacturers and Merchants—

Reuben Spencer, Manchester; C. T. Bradbury, Ashton-under-Lyne; Samuel Andrew, Oldham; Edward Walmsley, Stockport; Joseph Leigh, Stockport; G. B. Dewhurst, Lymm; Sam. Mendel, Manchester; Thomas Wilson (Wilson, Latham & Co.), Manchester; Gustav Behrens, Manchester; J. C. Fielden, Manchester.

Machinists and Metal Merchants—

George Little, Oldham; Henry McNeil, Manchester; R. B. Goldsworthy, Manchester.

Provision Trade—

Matthew Hudson, Manchester; J. T. W. Mitchell (Co-operative Wholesale Society), Manchester.

Oil Trade—

Patrick Moir Crane, Manchester.

Coal Trade—

Arnold Lupton, Leeds; Walter Rowley, Yorkshire; Joseph Mitchell, Barnsley; Horace Mayhew, Wigan.

Timber—

Rich. Lovett Stone, Hull.

Corn and Flour—

Frederick Moss, Salford.

Woollen Trade—

Louis J. Crossley, Halifax; T. Smith Scarborough, Halifax.

Carriers—

Alfred Hughes (W. Faulkner & Co.), Manchester; Bold Aldred, Manchester; William Clarke, Lock Keeper, Mill Bank.

Chemists and Scientists—

J. Carter Bell, Cheshire; Dr. C. A. Burghardt, Manchester; Dr. John Tatham, Manchester.

Nautical Men—

Captain Andrew Pearson, Liverpool; Commander Kingscote, Royal Navy.

Land Agents and Valuers—

John Dutton, Lymm; Hugh Cameron, Lymm; A. M. Dunlop, Manchester; J. S. Paterson, Manchester; William Raby, Manchester.

Corporation Representatives—

John Hopkinson, Mayor of Manchester; Richard Husband, Mayor of Salford; Samuel Warhurst, Mayor of Stalybridge; Thomas Isherwood, Mayor of Heywood; John Duckworth, Mayor of Bury; Joseph Davies, ex-Mayor of Warrington; F. H. Walmsley, ex-Mayor of Salford; Henry Whiley, Health Department, Manchester.

Members of Parliament—

Jacob Bright, Manchester; John Slagg, Manchester; Benjamin Armitage, Salford.

The Committee desired first to hear engineering evidence to satisfy them that the work could be carried out, and the engineer of the Ship Canal Company, Mr. Leader Williams, was the first witness. He explained the state of the Mersey in the past, the various Acts passed for the management of the river, and how seriously the trade of Liverpool was affected by the varying of channels in the upper reaches, also by the bar that had formed at the mouth of the Mersey. His proposition was to train a channel by half-tide retaining walls between Runcorn and Garston, of a width of 300 feet at the former and increasing to 600 feet at the latter. The navigation depth to be 24 feet at neap and 31 feet at spring tides, the depth at low-water neap tides 12 feet. The work was to be of a similar character to that which had been successfully carried out on the Tyne and the Tees. He estimated there would be 16,000,000 cubic yards of dredging at a cost of 9d. per yard = £600,000, and that he would use over one million cubic yards of rock in his retaining walls. He stated that under the Conservancy Act of 1842 there were powers to improve the estuary, and he intended to use those powers and in addition those of the Mersey and Irwell Navigation Act for the improvement of the river.

The witness was cross-examined by the opponents who urged that because of shifting sands it would be impossible to preserve a channel, and that outside the retaining walls the river would silt up and lessen the area of water that was necessary to keep down the silt on the Pluckington Bank. For the London and North-Western Railway Company it was claimed that the piers of the Runcorn Bridge would be endangered, and for the Shropshire Union that the entrance to Ellesmere Port would be silted up. Birkenhead was frightened of Pluckington Bank in-

SIR E. LEADER WILLIAMS, ENGINEER DURING CONSTRUCTION OF THE
MANCHESTER SHIP CANAL, SINCE CONSULTING ENGINEER.
Higginson Bowdon.   *To face page* 146.

creasing and impeding the Ferry; the Cheshire Lines Company seemed to dread the scheme being hung up half-way in case of financial failure. Mr. Leader Williams suggested that in the latter case it would become a public trust.

At this point Mr. Aspinall, Q.C., for the Corporation of Liverpool, put in a letter, dated 14th March, 1883, from Mr. Adamson to the Commissioners for the Conservancy of the river Mersey, which made clear the position of the various governing bodies on the Mersey.

TO THE COMMISSIONERS FOR THE CONSERVANCY OF THE RIVER MERSEY.

MY LORDS AND GENTLEMEN,

The promoters of the Ship Canal Bill consist of no less than 3,828 firms of merchants and other persons who have subscribed over £90,000 towards the expenses of obtaining an Act, and they venture to hope that now the long-looked for Manchester Ship Canal is before Parliament they will have the support of the Conservancy Commissioners. The necessity for the appointment of a Commission for the conservancy of the river Mersey appears to have been under the consideration of Mr. Huskisson in the years 1828 and 1829, and strong efforts were made by him to secure the appointment, but considerable delay took place, and the Commissioners were not appointed until the year 1842. In the meantime the officers of the Board of Trade had reported upon the matter, and their report was presented to the House of Commons on the 4th August, 1840. The report contains the following statement, namely:—

"It is not for Liverpool alone that a conservancy is wanting, nor for the navigation companies connected with the Mersey. It is of equal importance to Manchester and all the other manufacturing towns in Lancashire, Cheshire, Yorkshire and Staffordshire, and to the general commercial shipping interests of the kingdom. If the measure is properly carried into effect, it will be beneficial to the interests of the community at large."

Parliament accordingly constituted a Conservancy Commission, which represents not only the interests of Liverpool and the navigation companies, but also the interest of all parties concerned in the navigation of the Mersey and the trade of the district generally.

Further, by the "Upper Mersey Dues Act, 1860," Parliament enabled a body of trustees, including representatives of Manchester and Warrington, to purchase the town dues formerly levied by the Corporation of Liverpool, and subsequently by the Mersey Dock and Harbour Board, on all goods carried to or from any part of the Mersey above Garston, and these dues are now extinguished in the Upper Mersey.

Parliament has also granted powers by an Act passed in 1876 to a body of Commissioners (known as the Upper Mersey Commissioners) which include representatives elected by Manchester, Salford and Warrington to buoy and light the Upper Mersey and collect dues to defray the cost thereof. Further, these Commissioners had powers given them by a later Act, passed in the year 1879, to submit to the Conservancy Commissioners proposals for the removal of defined rocks, the object being as stated by the Acting Conservator, Vice-Admiral Spratt, C.B., F.R.S., in his annual report for the year 1880, to make a channel through a rocky barrier across the river Mersey, with a view to utilise the part of the estuary for deep-draughted ships, by obtaining a deeper and more permanent direction of the channel; and the Act provides that the Upper Mersey Commissioners shall proceed on the same conditions as the promoters proposed in the 26th clause of the Bill (being the clause having reference to the proposed low-water channel in this part of the Mersey), namely, with the consent in writing of the Mersey Commissioners, and in accordance with plans and sections to be previously submitted to and approved by them.

The valuable suggestions of the acting Conservator in his annual report for the year 1880, as to the character of the improvements necessary to allow ships of larger draught to navigate the Upper Mersey, have been fully considered by the promoters, and they are willing to adopt the recommendations therein contained.

The Bill as originally deposited in the Houses of Parliament was to make and maintain a low-water channel subject to the approval of the Conservancy Commissioners. The Examiner decided that a plan and section of the low-water channel should have been deposited, but the Standing Orders Committee of the House of Commons unanimously allowed the Bill to proceed, striking out the portion of the 26th clause relating to the low-water channel.

The promoters having originally proposed to leave themselves in the hands of the Conservancy Commissioners, are still willing and anxious to proceed with their Bill as now altered, and consider their position practically unchanged, as they have always contended that the Conservancy Commissioners have power under section 24 of their Act to authorise them to make such improvements in the low-water channel of the river as will enable vessels of deep draught to reach the entrance of the proposed canal.

The Mersey and Irwell Navigation from Warrington to Manchester is now almost disused, having passed into the hands of the Bridgewater Navigation Company, a majority of whose directors are directors of railway companies, which own lines between Liverpool and Manchester, and the Bridgewater Canal, which is in the same hands, is the only other navigation between those places. The result is there is no competition in the rates of carriage, which are considerably in excess of those charged by the same railway companies in other districts under the same circumstances where free navigation competes with them.

The powers of the proprietors of the Mersey and Irwell Navigation under their Acts of 7 Geo. I. c. 7 and 34 Geo. III. c. 37 extend from Liverpool to Manchester, and by the first-mentioned Act they were empowered to clear, scour, open, enlarge or straighten the rivers Mersey and Irwell, and from time to time, and at all times thereafter, to do all other matters or things necessary or convenient for making, maintaining, continuing and perfecting the navigable passage of the rivers Mersey and Irwell, or for the improvement or prosecution thereof, and by section 30 of the Mersey Conservancy Act, 5 & 6 Vict., the rights of the proprietors of the Mersey and Irwell Navigation are expressly reserved, and such rights have also been protected in all subsequent Acts relating to the rivers, or to railways, or roads crossing the rivers and are still in force.

The law officers of the Crown reported on these rights, 10th November, 1829, as follows :—

"We have seen the resolutions of the Mersey and Irwell Navigation Company, and the letter of their solicitor Mr. Eccles. They do not seem to us to oppose the grant of a Commission of Conservancy, but require only that none of the Parliamentary powers given to them by their Act of 7 Geo. I. should be infringed. We are of opinion that no Royal grant can impair these rights, which appear to us not to obstruct the general object of such Commission but to forward and promote it.

"The promoters by their Bill seek compulsory powers to purchase the Mersey and Irwell Navigation and their powers, and propose to convert the navigation into a Ship Canal between Runcorn and Manchester, and no fault has been found by the Examiners as to the plans for this portion of the project.

"The powers of the proprietors of the Mersey and Irwell Navigation if transferred to the promoters of the Ship Canal would enable them to improve the present low-water channel of the River Mersey from Runcorn to Garston, subject always to the approval of the Conservancy Commissioners.

"If the Bill be passed into a law without the portion of Clause 26, before referred to, the promoters would ask the Conservancy Commissioners to authorise them to straighten and deepen the existing low-water channel to a uniform depth of not less than 12 feet, which would enable vessels after crossing the bar to pass up to the entrance of the canal on the flood tide.

"The present low-water channel is in parts, of this, and even a greater depth.

"The promoters are willing to insert clauses in the Bill to enable the Conservancy Commissioners to call in professional advisers to report on the plans to be submitted to them for an improved channel and to hold public inquiries, and the promoters will agree to defray any costs incurred by the Conservators."

The Bill came before the Standing Orders Committee of the House of Lords on the 9th instant, and the further consideration was postponed to be re-heard before a larger Committee after Easter.

The opponents of the Bill then urged that the effect of the decision of the Standing Orders Committee of the House of Commons was to render the proposed canal useless, as the promoters had not the power left to execute the works in the tide-way necessary to afford them an access to the canal. The promoters, however, rely upon the powers of the Conservancy Commissioners and the rights they will acquire by the purchase of the Mersey and Irwell Navigation, but it is now a matter of importance to the promoters to know whether the Conservancy Commissioners will consider favourably the proposal for an improved low-water channel such as is now suggested.

That the project is desired by the public is shown by the fact that petitions have been presented praying for the dispensing with Standing Orders from the Association of Chambers of Commerce, 38 Municipal Corporations, 91 Local Boards, 31 Chambers of Commerce and 108 Companies, representing upwards of ten millions of capital, and by numerous landowners along the route of the canal.

The promoters have already incurred great expense in bringing the Bill forward to its present stage, and if it now fails to ensure a hearing upon its merits it will greatly retard if not altogether prevent another effort to secure an improved and independent waterway into the great manufacturing districts of Lancashire, Yorkshire, Cheshire and Staffordshire.

Under these circumstances the promoters trust the Conservancy Commissioners will be pleased to give this letter their early and favourable consideration   The promoters beg leave to enclose the statement they laid before the Standing Orders Committee of the House of Lords, which shows the reasons why they have relied on the control and powers of the Conservancy Commissioners.

I am, my lords and gentlemen,

Your obedient servant,

(*Signed*)   DANIEL ADAMSON,

Chairman of the Provisional Committee
formed for the promotion of the Bill.

The reply already alluded to from Sir R. H. Wyatt, on behalf of the Mersey Commissioners, was afterwards put in by Mr. Pember on behalf of the promoters.

28 PARLIAMENT STREET,
1st *May*, 1883.

MANCHESTER SHIP CANAL BILL.

SIR,

I am directed by the Commissioners for the Conservancy of the river Mersey to inform you that they have received and considered your letter of the 14th March in which you request, on behalf of the promoters, the favourable consideration by the Commissioners of the above scheme, particularly with reference to the proposed deepening of the low-water channel of the Mersey.

I am to state in reply that as at present advised the Commissioners see no objection in principle to the proposed deepening of the low-water channel, provided they are satisfied after full inquiry that the works are practicable and can be carried out without injury to the navigation or diminution of the area of anchorage and mooring grounds, and without interfering with the approaches to the docks or inducing a permanent silting up of that part of the channel.

It is in the opinion of the Commissioners essential for the protection of the interests which they represent that provisions be inserted in the Bill to secure the fulfilment of the above conditions, and I forward herewith a copy of a draft clause which may be taken as substantially embodying their requirements.

The Commissioners, however, reserve to themselves full discretion as to making any further objections or requisitions which may occur to them during the progress of the Bill and of taking all necessary steps to secure the adoption of their suggestions by the Committee to which the Bill is referred.

As there are no funds available for the purpose, I am requested to say the Commissioners must ask for an undertaking that the promoters of the Bill will pay all costs, charges and expenses to which the Commissioners have been or may be put in connection with this matter.

I am, Sir,
Yours obediently,
(*Signed*)   R. H. WYATT.

DANIEL ADAMSON, ESQ.

Resuming, Mr. Leader Williams showed that Birkenhead, whilst objecting to any reduction of tidal area, had herself abstracted 1,200 acres from the Mersey estuary when she made her docks.

The engineer's evidence was corroborated by several eminent men who had done similar work. *Mr. Messent*, engineer to the Tyne Trust, said that whereas at one time only a few ships over 500 tons could come up the Tyne, now nearly 5,000 ships over that size come up annually, a great many being over 2,000 tons; that his dredging had only cost $3\frac{1}{2}$d. to $6\frac{1}{2}$d. per ton, and that parallel walls would only cause silting if cross walls were also built.

Mr. Fowler, of the Tees, had failed with a narrow channel but succeeded with a broad one enclosed by 14 inch high half retaining walls. The Dee had silted up because the retaining walls had been specially designed to reclaim land.

Mr. Deas, of Glasgow, stated that in 1758 in the Clyde there was 1 foot 6 inches at low and 3 feet 8 inches at high water. Now at low there were 15 feet and at high water 26 feet. He had used low rubble walls and groynes. It was the cross and not longitudinal walls that caused mischief. His dredging cost $5\frac{1}{2}$d. per cubic yard.

Mr. Abernethy, consulting engineer, considered the powers of the Mersey Conservators sufficient, and that no fresh Parliamentary powers were necessary to build training walls. He estimated the cost of the canal at £5,634,000. If deepened to 24 feet, £291,000 more. He rejected the idea that a Conservancy Board, consisting of the First Lord of the Admiralty, the President of the Board of Trade, and the Chancellor of the Duchy of Lancaster, with Admiral Spratt as their adviser, existed simply to stop training walls from interfering with the tide. They had power to sanction improvements to the navigation.

Mr. J. F. Bateman, *C.E.*, gave evidence that the current in the Mersey was not due to anything done above Liverpool, but to the way in which Liverpool and Birkenhead had themselves altered the river by training walls opposite those towns.

After hearing counsel on both sides, the Committee retired to decide on the practicability of the engineering works, and whether there was power under the Mersey and Irwell Acts to carry out such works in the estuary.   On their return they consented to the Bill proceeding on the understanding that the work was divided into two sections, and they decided to restrain progress till plans had been deposited and approved of by the Mersey Commissioners, showing the depth to be given throughout the estuary, also till the necessary certificates had been given by the Board of Trade.

Thereupon the commercial witnesses were called, the first being *Daniel Adamson, J.P.*, Chairman of the scheme.   He stated that the costs of carriage to and from Manchester were excessive, and that it often cost more to send goods from Liverpool to Manchester than from London to Bombay.   In consequence some three years previously, when attending an iron and steel meeting, he consulted Mr. James Abernethy, C.E., as to the practicability of a Ship Canal.   As a result on 27th June, 1882, he called a meeting at his residence of leading citizens and engineers.   Afterwards he consulted Mr. Fulton, C.E., and Mr. Leader Williams, C.E., and arranged to form the Manchester Tidal Navigation Company. A Committee of that body, on the 26th September following, decided to instruct those engineers to make a joint report.   Not agreeing as to the plan to be adopted, they issued separate reports, and Mr. Abernethy was called upon to advise which was the best scheme, and he gave the preference to Mr. Leader Williams' Lock Canal.

Prior to 26th September, £14,000 had been subscribed for preliminary expenses, and afterwards a further sum of £64,000 for the Parliamentary Bill, which was applied for under the changed name of the Manchester Ship Canal Company. The Bill was approved of at a town's meeting, when 6,000 people were present. Mr. Adamson went on to say there were 1,561,000 people within a radius of 10 miles of Manchester, and that there had been an increase of 21 per cent. in the last decade.   That for timber and grain, taking into consideration mileage, the charge for railway rates was 50 per cent. more than in some other districts.   That for imported cotton the charges for dealing with it were Fleetwood 2s. 7d., Hull 3s. 8d., and Liverpool 8s. 1¼d., and this with insufficient accommodation at the latter port.   He put in tables to show that similar excessive charges existed more or less as regards railway rates and dock charges on all raw material coming to be manufactured, and for imported goods generally, also for exports.   He showed how water carriage had been practically destroyed by the railway interests, and how concerns like Sharp, Stewart & Co., Ormerod Grierson's, and others, had been driven out of the city to places where they could get cheap rates.   Also that the river Irwell being under the control of railway magnates had been utterly neglected, and had become little more than a huge sewer.   He quoted a resolution of the Associated Chambers of Commerce in 1882, petitioning that waterways should be emancipated from railway influence.   Also a speech of Sir Edward Watkin: "If it can be proved that the advantage of a navigation, which Parliament has given to the public, has been taken away, I think it would be quite reasonable on people coming to amalgamate, to say, 'You shall give up possession of this thing which is not properly used'".   Mr. Adamson went on to show that the Ship Canal would be a most useful outlet for the coal-fields, both of Lancashire and Yorkshire, and the readiest and cheapest means for the import of timber, and that it would accommodate all ocean-going goods steamers, and even some passenger steamers.   In cross-examination the opposing counsel attempted to elicit that capitalists had no confidence in the scheme, also that the Runcorn Bridge offered an insuperable difficulty, inasmuch as it would both be costly and cause delay if ships coming up had to telescope or fid the masts as was proposed.   Mr. Adamson denied the truth of these suggestions.

The next witness, *Mr. Marshall Stevens*, put in a list of ships in the Liverpool Docks, and showed that all goods steamers could come up the canal; also a comparative list of dock charges at Liverpool and Garston, to show that the average charge on goods was one-sixth at Garston of what it was in Liverpool.   He criticised the Liverpool system of appropriated berths, stevedores, master porterage, etc., and quoted Colonel Paris, a Liverpool Shipping Agent, who had passed severe strictures on the action of what he termed "The Family Party" *i.e.*, a body of shipowners who had too much power on the Dock Board.   In cross-examination Mr. Littler, Q.C., who is not the best-tempered counsel, allowed himself to be worsted by this witness.   He insisted on knowing the charge for cotton from Plymouth to Launceston (32 miles), to which Mr. Stevens replied, "There is no charge".   He repeated the question and got the same reply.   Then said the counsel, "Do you swear that?   Will you please answer my question?   If you are to be any use to the Committee, you must tell the Committee what you know.   You say there is no rate between Plymouth and Launceston for cotton.   Do you know the rate in the rate book is 11s. 8d.?"   "I may be in error," replied the witness, "in saying that there is no such rate, but I say there is no such traffic." Mr. Littler thought he had the witness in a trap, and charged him with being contumacious. He, counsel, had the rate book before him, which in triumph he handed to the Chairman, in order possibly to have the witness rebuked for being so obstinate in face of facts that seemed so clear.   However, when the book was handed back to Mr. Stevens, he turned the tables, and coolly pointed out that the counsel had not noticed in the rate book it was cotton goods, not cotton, that was quoted, and that there was no rate because it was an unheard of thing to carry cotton from Plymouth to Launceston.   The counsel being too confident thought he would rate the witness severely; instead Mr. Stevens quietly showed that his cross-examiner was totally ignorant of the difference between cotton and cotton goods.

Mr. Reuben Spencer came afterwards, and impressed the Committee by saying his firm employed 11,000 workpeople, and annually imported £250,000 worth of cotton.   Following him, Mr. J. C. Fielden quite astonished his hearers by the facility with which he dealt with figures as regarded the imports and exports of Manchester. He was listened to with marked attention, as also was Mr. J. T. W. Mitchell, Chairman of the Co-operative Wholesale Society.   At first his quaint appearance, loud voice and bluff manners puzzled the Committee, but when he told them he represented 500,000 co-operators, with a capital of between five and ten millions, and a turnover of three millions per annum, and that in his opinion the working classes whom he represented would save by the making of the canal as much money as would pay for it in twenty years, they seemed much surprised.   But the fun came when Mr. Pope took turn in cross-examining.   The two burly men evidently knew one another, and Mr. Mitchell was just as much at home as if he had met the advocate in the streets of Rochdale, and was going to have some verbal sparring with a familiar friend.

Q.—We will forget that we are not at a co-operative meeting.

A.—Then you ought to be. "Come with us and we will do you good."

Q.—Your society is a wholesale purchasing agency.

A.—That is so.

Q.—I thought I knew.

A.—You are very well informed.

Q.—Better than you think. You have also the business of shipowners.

A.—We have.

Q.—How long does your boat stop at Garston? The quicker the discharge the better.

A.—Hear, hear.

Q.—What do you buy at Liverpool?

A.—Sugar, fruits, etc. I need not describe them; it would only waste your time to describe them, because you know them so well.

Q.—That is a fine phrase, but it does not convey much meaning to my mind.

A.—It conveys a great deal of meaning because it is a splendid phrase.

Q.—I admire the phrase, but I cannot recognise the meaning.

A.—I am sorry for it.

The above questions and answers are typical of an examination that certainly amused the Committee. Passing over a number of witnesses, the next who specially gained the ear of the Committee was *Mr. Richard Husband*, Mayor of Salford, who said 1,100 acres of Salford were liable to be flooded, and he believed the canal would reduce, if not completely do away with, the evil.

Many coal owners gave evidence in favour of the Bill. *Mr. Arnold Lupton* proved that with the exception of Goole, Manchester would be the nearest seaport to the Yorkshire coal-fields, and the trade was badly in want of a port on the west side of England. In this he was supported by *Messrs. Walter Rowley* and *Joseph Mitchell*, the latter advocating the desirability of opening out the Barnsley coal-fields. Speaking for Lancashire, *Mr. Horace Mayhew* pointed out the Wigan coal-fields were only 9 miles from the Ship Canal, whilst they were 23 miles from Garston, and besides, the latter port was crowded out with coal and shipping, and to relieve the pressure they were working day and night.

Evidence was given for the timber trade, that in the west coast they could not get Baltic timber at a reasonable cost, whilst in the east the cost of carriage helped to preclude the use of American timber. It was estimated that 300,000 tons of timber would be used yearly in Manchester, and that a new port there would be a blessing to the timber trade.

Following came the evidence by representatives of the woollen trade, who considered it a disadvantage that the wool market should be in London whilst its main users were in the North of England. Then a number of local Mayors and afterwards local M.P.'s were in turn put in the box. The latter were certainly a disappointment, for whilst *Jacob Bright, M.P.*, gave a most hearty and loyal support, indeed capital evidence, *John Slagg, M.P.*, and *Benjamin Armitage, M.P.*, made it plain that they were half-hearted, and came to please their constituents rather than themselves. They were just the men that opposing counsel love to get hold of. It was elicited from Mr. Slagg that he was not sanguine of success, and that he had not himself subscribed to the preliminary fund; and from Mr. Armitage, that he personally did not understand the question, nor was he persuaded the venture was going to be a financial success.   He had come to speak on behalf of his working-class constituents in Salford, who were enthusiastically in favour of the canal, but he said the wealthy people were holding aloof.

Joseph Davies, J.P., of Warrington, was convinced the canal would prevent flooding in his town, and explained to the Committee that the conversion of the river into a Ship Canal had always been contemplated.   Hence the authorities had never consented to bridges being placed over the river, except on condition they should be replaced by swing bridges if required.   In order to show that many leading capitalists were supporting the scheme, *Councillor Goldsworthy* was put in the box, and the list of the names of leading supporters given with "Do you know So and So?"   Inadvertently counsel said to him, "Do you know Councillor Goldsworthy?" to which question he humorously replied, "I see him every morning in the looking-glass".

Various carriers proved that freightage on the river fell off from 1850, when the Bridgewater Canal Company bought the rival navigation, and it was shown that the draught of water had fallen from 5 feet to 3 feet, or to an extent that made the navigation practically useless.

After hearing several chemists and land agents, *Mr. Price Williams*, a specialist, stated that the worst new railway gradient was to be 1 in 114, whilst at present there existed gradients of 1 in 85 on the same line between Warrington and Newton; elsewhere 1 in 90, another 1 in 100, and on the Great Eastern 1 in 61.

A number of shipbuilders then gave evidence as regarded fidding or telescoping ships' masts to enable them to pass under bridges 75 feet above water-line.   They stated there would be no difficulty, and little cost or delay, and it was estimated that a 4,000 ton ship would cost at the outside £132.   *Captain Kingscote, R.N.*, thought the new canal would be less difficult to navigate than the Suez Canal, and *Mr. Edwards*, of Cardiff, expected that it would not take more than eight hours between Liverpool and Manchester, and that this small delay would not cause shipowners to make any extra charge.   Mr. Leader Williams being recalled, showed the connection that would exist by means of canals with all parts of England, explained the working of the Barton aqueduct, and showed there would be enough water to allow sixty-nine steamers to pass each day.

The case for the promoters closed on the 9th June, having occupied twenty-one days.

On the same day the opponents of the Bill began to call their witnesses, forty-three in number.

Captain Graham Hills, R.N., said that prior to 1872 the depth on the Liverpool bar was 11 feet at low water. Then a freshet brought down 5,800,000 cubic yards of soil, and the bar was reduced to 7 feet. The Pluckington Bank was also increased to 180 acres, but by flushing it had since been reduced to 134 acres. Even if it were washed away, the fear was that the sand would lie in the front of the other docks. Placing barriers on the Dee, and excluding the tide, had made usable 8,000 acres of land. In cross-examination Captain Hills admitted that in some estuaries tidal walls had been beneficial, but thought that in the Mersey they would cause a reduction of the tidal area. He admitted little had been attempted to improve the river, and said no dredging had been done for the last twenty-seven years. About forty-seven years ago, when Admiral Denham was in charge, he had made some experiments to improve the bar, and this was the only attempt that ever had been made.

Mr. George F. Lyster, engineer to the Dock Board, objected to the promoters' estimate for the new channel, and believed that instead of £600,000 it would cost £3,000,000. In cross-examination he did not believe railways round docks a necessity. In some cases they were useful. In docks made seventeen or eighteen years ago there were rails which were never used, and in the new docks they would not have them. He knew for a fact that shippers preferred to cart goods from the docks to the railway stations rather than send by rail from the ship's side. "To have rails within a shed would be simply ridiculous. It would be utterly absurd, and simply throwing away money, and obstructing the working to put rails through the sheds and then expect them to be used." When Mr. Findlay's evidence, given before the Railway Rates Commissioners, regarding insufficient railway accommodation at the docks, was quoted, the witness replied, "I heard of it, and I met him in the lobby a few minutes afterwards and said, 'What an extraordinary statement to make!' and, from my knowledge of it, it is inconsistent with the exact condition of the Mersey estate ".

Mr. Lyster then attacked the estimates, maintaining that instead of 700 to 800 acres of land being required he should say 3,000 to 4,000 would be wanted. However, he failed to make good his contention. He also stated that instead of the 9d. per cubic yard given by Mr. Leader Williams for excavations, the cost would be 1s. 8¼d. —this, in face of the evidence of Mr. Deas that the cost on the Clyde had been 5½d.

A curious episode occurred during the examination of this witness which caused some warmth. He had attacked the figures given by Mr. Messent, of the Tyne, for rock excavation, and thinking this gentleman had been under a misapprehension Mr. Lyster privately wired him for information. Mr. Messent sent a reply and also a copy of it to the promoters. Next day as the reply was unfavourable Mr. Lyster intended to lie low, but Mr. Pember would not let him off so easily. By degrees he pumped everything out, and induced the angry reply, "Do you mean to say that you have got my confidential reply?" Mr. Lyster ought not to have communicated with his opponents' witness without their knowledge, and could not expect a witness to keep correspondence from his principals.

Several engineers followed in support of Mr. Lyster, their main theme being there would be accretion behind the retaining walls, and that in time the area of the estuary would be reduced, and less water pass out to scour the bar.

Captain Kennedy, of the Inman Line, was of opinion there would be the greatest difficulty in passing up the canal, because of the fogs and insufficiency of water; that it was not safe to go to sea without ample masts; that ships would have to go up stern first, as he believed they did to Garston, and that a ship could not prepare to go under Runcorn Bridge without losing a tide.

The next witness called was *Mr. T. D. Hornby*, the very courteous and able Chairman of the Dock Board, who, however, seemed imbued with the idea that an attack was being made on the managerial capacity of the Dock Board, which it was his mission to repel. He stated that in 1852 the revenue of the Dock Board was £246,000 from 21,500 vessels, and the tonnage 3,912,000. In 1882 the revenue was £929,000 from 21,000 vessels, with a tonnage of 8,104,000 tons, the debt owing at the latter date being £16,373,000. In 1881 reductions in rates had been made of £120,000 per annum. He said there were twenty-four members of the Board elected by people who shipped through Liverpool, wherever they might reside, but as there was no proxy voting, everybody must come to Liverpool to vote. In addition there were four members appointed by the Conservancy Commissioners, of whom Mr. Hugh Mason, of Ashton, was one, and he considered if Manchester had had any special grievance, that gentleman would have brought it forward. He complained that Manchester, under the direction of Sir Joseph Heron, had compelled them to make a large expenditure on the Birkenhead Docks, and also to pay £1,500,000 for the town dues which had previously been appropriated for town improvements, building of churches, etc., in Liverpool, these same dues being partly provided by Manchester.

He defended the master porterage system, and believed it must be adopted in Manchester; also he admitted complaints about the insufficiency of dock accommodation. Mr. Pember in cross-examination drew the attention of the witness to the speech of Mr. Moon, Chairman of the London and North-Western Railway Company, when he said :—

Liverpool must have forgotten that it was the dearest port with which they traded; their wages were 23 per cent. more than anywhere else; they had no facilities for traffic; no room on the dock quays, and in the case of timber no place to stow it. If timber came to Fleetwood there was room for it to be stowed; and it was taken away in a week or a month, at the convenience of the consignee; but if it came to Liverpool, Mr. Hornby fined them if it lay on the quay twenty-four hours.

Again :—

Mr. Hornby knew that he could not undertake to do any cheap trade in Liverpool.

In reply, Mr. Hornby said this was a controversial speech by Mr. Moon, of which he would not admit the accuracy.

Mr. Hornby, in his evidence, having said the trade preferred carting, Mr. Pember read the opinion of the Liverpool Chamber of Commerce, "that the charges for housing and carting produce press heavily and seriously upon the trade of the port, and are largely the cause of diverting to out-ports trade which otherwise would flow to Liverpool". Again, "that there is a want of steam and mechanical appliances for discharging vessels and handling cargo on the quay, and that the large amount of manual labour now employed in those processes is a source of much of the extra expense incurred in the port of Liverpool, as compared with other ports".

In reply, Mr. Hornby would not admit the statements, and gave it as his experience that there were always complainers, and that it was impossible to satisfy everybody.

Mr. Ismay, of the White Star Line, considered Manchester would not be a safe port. Shipowners would require 5s. per ton more freight. Insurance would be higher, and he would not let his ships go up the canal.

Mr. Alfred Holt agreed entirely with Mr. Ismay, and said the proposed canal would be much worse to navigate than the Suez Canal. Cross-examined by Mr. Pember, he admitted the securing a cheaper system of inland conveyance had been his justification for introducing the Lancashire Plateway scheme, and that the cost of carriage by rail between Liverpool and Manchester had increased in the last fifty years.

Mr. Rodwell, Q.C., then addressed the Committee on behalf of the Dock Board, and got into an awkward fix about the Pluckington Bank, for the Chairman, Sir Joseph Bailey (than whom no one had more accurately watched the case), asked if he could explain the fact that his two chief witnesses, Mr. Lyster and Captain Graham Hills, had given contradictory evidence—one saying the Bank had increased and the other it had decreased. Mr. Rodwell leaned to the view of the Captain (that the length had decreased), and quoted his evidence: "I dare not touch the Pluckington Bank. I have told the Mersey Dock Board to leave it as it is. I do not know what danger would ensue if I were to disturb it." (In other words, they dare not do right for fear of doing wrong.)

Mr. Rodwell's peroration ran thus :—

To bring the sea to Manchester is, I believe, a thing which cannot be done. There may be plenty of Manchesters ; there can be but few Liverpools, and when my learned friend makes it a grievance that Manchester has been suffering under high dues which Liverpool has charged, I think he ought to have recollected that the prosperity of Manchester is due to the prosperity of Liverpool : they have flourished and grown together, one is the centre of commerce, the other is a great manufacturing centre; they have prospered together, and I believe it would be an evil day for Manchester if she estranges her relations with Liverpool, and becomes a rival instead of an ally.

He also portrayed the engineering difficulties, and said the sea was so capricious that it was impossible to make it a slave, as you could land. He concluded by saying :—

I ask you, in the interests of Liverpool and Birkenhead, not for a moment to imperil the grand estuary of the Mersey, the magnificent docks on its banks, and I ask you not to imperil those large and grand interests which are matters of national concern, in order to satisfy, it may be the jealousy or, perhaps, the cupidity of the Manchester people, who are promoting this Bill.

Mr. Clement Higgins, for Birkenhead, and *Mr. Dugdale*, for Ellesmere Port, then presented their respective cases.

Mr. Jebb, C.E., for the latter, spoke of the difficulty of keeping Ellesmere Port open, and said that they had 11 to 12 feet of water and could take vessels up to 200 tons. He admitted no money had been spent to keep the channel clear.

The Liverpool commercial case was next taken. *Mr. W. B. Forwood* being the first witness, said only 50,000 tons of cotton went to Manchester, whilst Oldham consumed 157,700 tons.   It was quite certain the freight would be 5 per cent. more to Manchester, and that this in addition to $\frac{1}{8}$ per cent. extra insurance would run away with the benefit obtained in other ways.   He twitted Manchester with never helping to get cheaper railway rates.   "I may say that I have been nearly seventeen years one of the leaders to agitate for cheaper railway rates, and I never heard a single complaint from Manchester, or received one single iota of assistance in our agitation."   He admitted saying that if railways charged the same to and from Liverpool as they did at some other ports, there would be a saving of £400,000 per year.   In his view a barge canal only was required between Liverpool and Manchester, and he believed one would be established within the next few years.

At this point Mr. Pember informed the Committee that the promoters had come to an arrangement with the Bridgewater Navigation Company to buy their concern, the price to be fixed by arbitration.

Mr. Aspinall, Q.C., then addressed the Committee on behalf of Liverpool, and called several witnesses.   His argument was, "that the canal would not create any fresh trade, but simply abstract it from Liverpool.   The latter would be poorer, and the former no richer because she would have to pay a high price for the trade. The risk was enormous, and the gain problematical."

Mr. Thomas Barham Foster, C.E., in giving evidence for the Trafford Estate and describing the Irwell and its weirs in 1868 to 1870, said :—

I have seen, I say, the whole river from weir to weir covered with dirty froth, so that you could not see any water at all, and through this froth countless bubbles rising.   Certainly one in every square foot of surface.   I went down professionally upon this river, in those years, and the stench was something incredible.   The river gets worse and worse till you get to Irlam, where the Mersey, which is a purer stream, joins it.

The next witness, *Mr. W. H. Watson*, agreed that while at Springfield, Lane, and Throstle Nest there was but a slight smell from the Irwell, at Irlam it was very bad.   He differed entirely with Mr. Bateman's opinion that the deeper the water the less the smell, and believed the increased depth would have a tendency to increase the putridity by depriving it of oxidation by the atmosphere.

Following the evidence of the chemists and the land valuers, came a number of engineers, and *Mr. George Findlay*, General Manager of the London and North-Western Railway Company, who stated that, in 1882, 104 passenger trains ran daily through Warrington, carrying 1,800,000 passengers in the year, and that in addition to 3,500,000 tons of goods and minerals, 25,000 waggons of live stock went through the same station.   He objected to the canal seriously increasing the gradients and destroying the utility of the pick-up water troughs.   Also to the tunnel under the Warrington and Garston line.   He considered the handling and warehousing accommodation at the Ship Canal docks quite insufficient, and he looked upon the idea of bringing Yorkshire coal to the Ship Canal as quite chimerical; this remark also applied to St. Helens and the Wigan coal-fields, for which Garston must remain the cheapest and best outlet.   He was quite sure the conferences of the shipowners who dominated the trade to India, China and America, and who not only pooled the freightage, but fixed what was to be sent from each port, would neutralise any effort to bring the shipping trade to Manchester, and benefit merchants by lower freights.   The chance of getting it there would be small.

Cross-examined by Mr. Pember, Mr. Findlay had to admit that the same kind of tunnel he objected to had successfully worked in London ; also that large quantities of South Yorkshire coal might, for various reasons, come to the canal.

The next witness, *Sir Frederick Bramwell*, was perhaps the best-known man in the Parliamentary Committee Rooms.   He stood over six feet high and was more than wide in proportion.   He had a large head with curly white hair, and a face that always beamed with humour.   He was a consulting engineer with a large experience, but it was said he had not constructed any works of importance in his life.   Hence counsel generally bantered him on the point.   His cross-examination ran thus :—

Q.—Mr. Pember.—Into the question of practical working I am not going to enter with you, because of the practical working of a railway I believe you have no experience whatever ?

A.—I do not quite know about that.   At all events, I may say that I have been consulting engineer to two railway companies, and I know something about engineering.   I am familiar with mechanical engineering, and I know as much about railway breaking as any man in England.

Q.—Now, Heaven help me, when I come to ask you about your figures !

A.—I trust it will.

(After a fight upon mechanical figures.)

The Chairman.—Both you learned gentlemen are thrashing away at this question.   I despair myself of understanding it, and I doubt if any one else in the room does.

At the conclusion of the London and North-Western case, *Mr. Pope* addressed the Committee.   He maintained the capital was insufficient.   That the project would bring injury and risk to many important undertakings without any commensurate benefit to Manchester.   That the damage to the railways was absolute and certain if they were compelled to carry out their Parliamentary obligations, and put swing bridges over the canal.   That the clauses, passed in 1846, when a Ship Canal for small ships and barges could only have been anticipated, and when there were few trains, should not be held to be operative now in the changed state of affairs.   He was quite sure shippers would always charge 5 per cent. more to Manchester than to Liverpool.   That coal-tips and sheds would be required and they had not been estimated for.   He concluded a magnificent speech by saying, "That sufficient unto the day is the mischief thereof".   That a Ship Canal was the clumsiest and most expensive way of reducing the cost of carriage.   "What would happen, think you, if the railways did find themselves pinched by the Ship Canal?   They would run it off the road in a week.   There is nothing for it but to reduce the railway rates, and what becomes of the Ship Canal?   I ask you without hesitation and with a clear conscience to reject it absolutely.   The kindest thing you can do for South Lancashire is to prick the bubble now."

Of all the opposing counsel no one commanded the respect of his adversaries more than Mr. Pope.   He had had his early business training in Manchester, he understood the honesty and bluntness of the Lancashire character, he liked to surprise witnesses by a few words in the native dialect, but he was candid and fair in all his examinations and transactions; he never descended to bullying witnesses, and he scorned to take a mean advantage of any one.

The case of the Midland Company was next brought before the Committee. Their General Manager, *Mr. Noble*, admitted they had a gradient of 1 in 90 on part of their system, against 1 in 114 on the proposed Bill; but he explained that the steep gradient had compelled his company, at a great cost, to make sidings at Heaton Mersey and Rowsley.   He always viewed the obligation as to swing bridges as not likely to be put in force.   It was a fact that the export and import trade of Liverpool grew rapidly from year to year and he thought before seven years were over they ought to quadruple their traffic, notwithstanding competition from other lines.

Mr. Pember, Q.C., then replied to the opponents' case.   He urged that the promoters having powers under the Mersey and Irwell Act to deal with the estuary, antecedent to the Conservancy Act of 1842, could not be debarred from making the proposed alterations in the estuary.   To his mind, the very fact that on an important main line like the Midland there were gradients of 1 in 90 and 1 in 100, killed the case for the Railway Companies, who now objected to 1 in 114. On the evidence of Mr. Adamson, which had been fully corroborated, there would be a saving of £440,000 per year to the cotton trade alone. The cost by canal in freight and charges to Newton Moor Mills would be 9s. 3d. against 18s. 4¾d. per ton by railway. Replying to Mr. Pope, who had called the scheme a gigantic bubble which he implored the Committee "to prick now," Mr. Pember addressing the Chairman said :—

Sir, Lancashire is not in the habit of blowing bubbles. Everybody else connected with the case admits that the Lancashire men mean it. Everybody admits the money will be found in Lancashire, except Mr. Slagg and Mr. Armitage. I say to them—the only persons who have thrown cold water upon the chances of the scheme—I say to them, " Pray, stay away," as I said to the Inman steamers. And just let me say when people talk about the stake of Liverpool in the Mersey, I should like to know whether all the towns in Lancashire have not got a stake in the Mersey. If we ruin the Mersey we ruin them. Surely the Mersey is more important to them than Liverpool which is only their emporium. The Mersey does not exist for Liverpool : it is equally for those towns which employ Liverpool. To suggest that those who have a primary interest in the Mersey want to ruin it, is to suggest suicide to them. No, sir, they want, not to ruin, but to save the Mersey, which Liverpool has neglected, and will neglect till the end of time.

Mr. Pember then went into the question of capital, and after giving a long list of subscribers and supporters of the canal, said :—

Are all these men mad? Are they ignorant? Are they visionary? All dreamers? Or do they represent a group of wild enthusiasts, about whom my friend, Mr. Pope, was so eloquent? Or are they really, as I said before, the representatives of the trade of this important country? talking about what they do know very well, and what they have considered.

Mr. Pember ended an eloquent and impassioned speech thus :—

In the days of one of the early Roman emperors a man invented a process for toughening pottery and glass. Any one who has ever been to Rome and seen the Monte Testaccio at Rome, a thing about the size of Primrose Hill, consisting entirely of potsherds, will be able to realise the waste involved in the fragility of earthen vessels in the days before iron came into household use. Well, sir, one day the inventor displayed his new wares before the emperor, and threw pots and pans on the marble floor of the Basilica. The emperor expressed his wonder and delight, and he said to the inventor: " Have you confided this process to any one ? " " No." " Have you committed the recipe to writing ? " " No." " Do you mean to say that it would die with you if you died to-day ? " " Yes," said the honest potter. " Then," said the emperor to the lictors, " take him out and strangle him," and they did so forthwith. Now the emperor no doubt thought the invention, beneficent as it was, might injure the potters as a class; so the inventor was strangled, his invention smothered, the potters were left to fatten, and the Monte Testaccio to grow.

Now my clients are as this inventor, their canal the toughened pottery, Liverpool and the London and North-Western Railway Company the potters, the bar of Liverpool the Monte Testaccio. I pray you, sir, not to let the Parliament of England—and I say it with all respect—in your person, and that of your colleagues, complete the parallel by playing the part of the reckless, the complacent and the short-sighted emperor.

The Committee room was then cleared. When the parties were called in the Chairman intimated they would allow the Bill to proceed provided the promoters accepted their clauses, eight in number, which must be inserted in the Bill, to be approved by the Committee before insertion.

They were as follows :—

I. The canal shall be divided into two sections, the lower section to include all works authorised in the Bill below and including Walton Lock ; the upper section to commence at Walton, and include all works above that point.

II. The company shall be restrained from proceeding with any of the works authorised by the Bill unless and until they shall have obtained the necessary powers to construct the estuary works.

III. The company shall apply to Parliament to sanction the details of any scheme on which they may have agreed with the Mersey Conservators for the construction of the works in the estuary.

IV. The company shall be restrained from proceeding with the upper section, that is to say, the works above Walton, until the estuary works have been so far completed as to show to the satisfaction of the Board of Trade whether or not they are likely to injure the estuary or the approaches thereto ; and until a certificate shall have been obtained from the Board of Trade, that in the opinion of that department and of the Mersey Conservators the works will not be injurious and can be permitted to remain.

V. The company shall bind themselves to continue, alter or remove, at their own cost, the said estuary works if directed to do so by the Mersey Conservators, and in such way as may be directed by them.

VI. The company shall give such security as the Board of Trade may from time to time direct, to ensure the observance of the foregoing stipulations.

VII. The company shall be restrained from making the railway deviations Nos. 1 and 2 until the aforementioned certificate as to the satisfactory nature of the works has been obtained from the Board of Trade.

VIII. The promoters must bring up a series of paragraphs to be added to the preamble giving an argued history of the case, as regards the estuary works and the bearing thereof on the other works.

The decision was given on the 4th July. The hearing occupied thirty-nine days.

On the following day (5th July, 1883) Mr. Pember produced his clauses, whereupon the Railway Companies, the Dock Board, Liverpool, and other chief opponents formally withdrew, and intimated they should oppose in the Upper House.

It was then decided to take the Warrington clauses, and the Chairman mentioned the views of his Committee as to the formula of the preamble to the Bill. He wished it to be made clear that whilst the promoters considered that under the Mersey and Irwell Act of 1721 and the Mersey Conservancy Act of 1842, they had power to do the estuary work, the Committee thought express Parliamentary sanction ought to be sought in the next session.

This Mr. Pember accepted, and the Committee considered the Warrington and the Race-course clauses. One clause, *viz.*, the right of the Warrington Corporation to contribute to the funds of, or hold shares in the Ship Canal Company, was the subject of much discussion. It was urged that the Corporations of Liverpool, Bristol, Preston and Boston had at one time or another all obtained sanction to assist undertakings of advantage to their towns, and that in 1872 there was a report in favour of public bodies promoting branch railways. The Committee, however, deleted the clause, it being thought that it would subsidise a competitor to railways.

On the 6th July Mr. Hargreaves, an opponent, intimated that he could not agree to the clause offered, and he should oppose in the Lords. Mr. Pember then took the somewhat unusual course of thanking the Committee for the great attention they had paid to the case.

During the Commons inquiry 15,594 questions were asked. The speeches and evidence occupy 1,702 pages of foolscap.

CHAPTER IX.

1883.

SHIP CANAL BILL IN THE LORDS—EFFORT TO WRECK IT ON STANDING ORDERS—LORD WINMARLEIGH RESCUES THE BILL—SPEECHES OF COUNSEL—EVIDENCE BY PROMOTERS AND OPPONENTS—UNFAVOURABLE DECISION.

Any improvement which will enable ocean-going vessels to discharge their cargoes in a commodious wet-dock in Manchester, would form an epoch of such magnitude in the history of Manchester as would quadruple her population, and render her the first, as well as the most enterprising, city of Europe.—Sir WILLIAM FAIRBAIRN *on the Improvement of the Irwell.*

WHEN the Ship Canal Bill came before the Standing Orders Committee of the Lords, its Chairman, Lord Redesdale, wished it to be thrown out, which would have ended the Bill for the session. However, to his great chagrin he was outvoted, the Standing Orders were suspended, and the Bill allowed to proceed. The usual course is for the Chairman to move the second reading in the Lords, but Lord Redesdale blocked the way by refusing point blank to do so, on the ground that it was against the custom of the Lords to proceed with any Bill that came from the Commons after the 21st June. In the dilemma Lord Winmarleigh very kindly undertook to perform the duty on 24th July, 1883.

On the second reading being moved, Lord Winmarleigh asked that the Standing Orders might be dispensed with, and explained the delays that had occurred to cause the Bill to be after date. The Earl of Redesdale vigorously opposed the suspension, and said the Bill came up at a time when it was impossible to make a good Committee on account of the want of attendance in the House. He complained that the Bill in its present shape was imperfect and gave no power to get ships from deep water into the canal. He said the completion of a scheme was one of the conditions on which private legislation was allowed to proceed, and that any other course would be introducing a most objectionable and flagrant principle. He objected to the imperfect character of the Bill as shown by the provisions laid down in it by the Committee of the House of Commons, which proposed to divide the work into two sections; the channel in the estuary and the canal itself. The company would first have to obtain the approval of the Mersey Commissioners to the estuary works before applying to Parliament to execute the same. It was avowedly an imperfect measure, and the Bill ought not to be passed. If it were it would create a precedent of a dangerous character.

Lord Winmarleigh in reply explained the enormous interests that were at stake, and said that the capital embarked in the cotton trade alone amounted to £100,000,000. He showed that the whole trade of the district was oppressed and strangled by the cost of the carriage of goods, and that whilst in 1863 the carriage of goods from Manchester to Liverpool was 6s. 6d. per ton, it had eventually arrived at a charge of 11s. per ton. The cost of timber from Liverpool to Manchester used to be 2s. 6d. per ton, now it was 7s. 6d. per ton. He urged that the preliminary expenses of this Bill were already £50,000, and that the House should not lightly stop a Bill on which so much money had been expended. He quoted Mr. Pember's story of the Roman emperor and the glass manufacturer which has been related in a previous chapter. Lord Winmarleigh hoped the House of Lords would not follow the example of the Roman Emperor, and so lose this wonderful canal.

The Earl of Derby sympathised with Lord Redesdale in his anxiety to stand up for the regularity of proceedings, but reminded him that the House of Lords had frequently suspended the Standing Orders, and allowed Bills to be read a second time. In this case the promoters were not responsible for the delay that had occurred. The inquiry in the House of Commons had lasted for thirty-eight days, a very unusual length of time to be given to any Parliamentary inquiry.

He might remind the House that the Committee of the House of Commons had sanctioned the canal, and reported, "that if the scheme could be carried out, it would afford valuable facilities to the trade of Lancashire". Also "that if the Bill pass, the company will be incorporated, who then can, and will, at once lay before the Mersey Commissioners plans and sections which the discussion and evidence in the House of Commons has rendered possible, with a view to obtaining their approval and the subsequent sanction of Parliament, whereas, if the Bill be rejected, a whole year, at least, will be lost, as the promoters will have no status to go to the Commissioners for their approval, and the whole contest will have to take place again in the House of Commons without the Bill ever having been investigated by the House of Lords". He added, on the highest authority, that the Mersey Commissioners were now perfectly satisfied with the clauses that had been inserted in the Bill, and that no opposition would come from them. If their lordships rejected the Bill, the promoters would think it very hard that, after bearing all the enormous expense, they should be thrown back, and have all their money wasted in consequence of a mere technical objection. The House then divided, with the result that 87 voted for and 24 against the suspension of the Standing Orders. This welcome result enabled the Bill to proceed, and it was remitted to a special Committee, consisting of the Earl of Camperdown, the Earl of Devon, Lord Methuen, the Earl of Aberdeen and the Duke of Bedford.

This Committee met on 30th July, 1883, the Earl of Camperdown in the chair. The promoting and opposing interests were generally represented by the same counsel that had appeared before the Committee of the House of Commons.

Mr. Pember again opened the promoters' case.

The Chairman called attention to the lower part of the estuary works being struck out of the Bill, when Mr. Pember said that the Bill was practically on all fours with the measure as presented to the House of Commons, except that there had been a modification as regarded the Bridgewater Navigation purchase, and that now it was not proposed to carry works into the deep-water channel of the Mersey. The former Bill took power to make a deep-water channel with retaining walls in the Mersey between Runcorn and Garston, and would have enabled the scheme to have been dealt with as a whole, but to this a technical objection in the Commons had been raised by the opponents of the Bill. They urged that the promoters had not deposited plans and sections of the deep-water channel between Runcorn and Garston. The reason for this was that it was not deemed necessary to do so, the promoters believing that they had powers, under existing Acts, to do the estuary work, provided they obtained the sanction of the Mersey Commissioners. The Bill passed by the House of Commons obliged the promoters to get the consent of the Mersey Commissioners, and to bring forward a Bill in the ensuing session to do the estuary works, at the same time depositing the necessary plans and sections. He might say the original Bill would have enabled the promoters to have begun their canal without the consent of the Mersey Commissioners, but they could not have begun the deep-water channel, and they would not have been such fools as to attempt it.   Mr. Pember then went into the early history of the Mersey and Irwell Navigation, and called attention to the ample powers given in the original Acts of Parliament, and specially to the provision that had been inserted in all Acts connected with the river, that, if ever the Navigation were converted into a Ship Canal, all authorised bridges should be converted into swing bridges. In the Mersey Docks and Harbour Board petition, it was urged that the Mersey ought not to be touched, and never ought to be touched, but it was ridiculous to suppose that the Mersey was the only estuary in Great Britain that could not be improved.   He urged that whilst practical men, like the engineers of the Tees, the Tyne and the Clyde, gentlemen who had studied that branch of their profession, saw no difficulty in the way, they were confronted by gentlemen who were not engineers, but critics of engineers, some of whom had never carried out a single engineering work, and had not made a special study of estuaries.   The idea of a Ship Canal was not a new one.   In 1824 there was an application for a Ship Canal from the Dee, and since then numerous schemes had been proposed to make a waterway from Manchester to Liverpool which should be available for sea-going ships.

Even recently, in order to secure cheap rates between Liverpool and the interior of Lancashire, a scheme had been brought forward by a number of Liverpool gentlemen, namely, the Lancashire Plateway scheme.   This showed the existence of a state of things which called for a remedy, and the Ship Canal would supply it.   Mr. Pember then dealt with the necessity of cheaper transit into Lancashire, and showed by figures how the various trades were suffering from the excessive rates charged in the district in comparison with the much smaller rates paid both at home and abroad.   He pointed out too that Liverpool was overcrowded, and could not accommodate a growing trade.   In consequence it was diverted abroad.   He dealt with the interest taken by the main Lancashire towns in the provision of a Ship Canal, and stated that not only had the various Town Councils petitioned in its favour, but that it had the monetary support of the mercantile classes from a large area round Manchester.   He gave many instances of dear rates, and then proceeded to deal with the petitions, commenting on the fact, that whilst there were twenty-eight petitions in the Commons there were now only twelve in the Lords.

There was considerable argument when the petition of the Shropshire Union and Canal Company was raised, Mr. Pember contending that the Ellesmere Port channel, about which they petitioned, was outside the purview of the present Committee, and that it was under lease to the London and North-Western Railway Company, who were already petitioners.

Mr. Dugdale, in reply, contended that Ellesmere Port would be included in the limits of the harbour and port of Manchester, that the water coming down the Mersey would affect his port, and that the Shropshire Union Canal Company's complaint was not included in the London and North-Western petition.   The Committee decided that the Shropshire Union Canal Company could not appear against the Bill, but might have a technical ground for appearing against the preamble.

Mr. Daniel Adamson, the first witness, said that so long ago as 1880 he had consulted Mr. Abernethy as to the possibilities of a Ship Canal.   As far back as 1712 Mr. Steers had proposed a Ship Canal between Manchester and the sea.   In 1720 the Mersey and Irwell Act was obtained to improve those rivers.   In 1824 an attempt was made to float a company for a waterway from the Dee to Manchester. In 1838 Sir John Rennie prepared a report for Warrington, in which he discussed an improved navigation between Liverpool and that town.   In 1843 Mr. Palmer, an engineer of that date, made a report as to a navigation suitable for large sea-going vessels, and afterwards Sir William Fairbairn, an engineer of great experience, spoke of the feasibility of the scheme.   He repeated the evidence he had given in the House of Commons, showing the vast increase there had been in the cotton industries of the district, and how they were oppressed by the toll that was levied on goods passing through Liverpool, and the subsequent heavy railway carriage.   At this point the Chairman of the Committee intervened and urged that, in consequence of the lateness of the session, the evidence should be abridged as far as possible.   Mr. Adamson on subsequent days gave the estimated saving to the cotton trade by a canal as £500,000 per annum.   He showed how industries had been driven from Manchester, especially locomotive making.   How the once active water navigation on the Irwell had died away.   How railways would benefit by the encouragement of industries which would give them an increased passenger traffic, very much more remunerative than goods traffic, instancing that a ton of third-class passengers produced 1s. 3d. per mile, whilst a ton of minerals did not bring a 1d. per ton per mile.

He stated that a deepened and widened navigation would prevent the floods which now at times devastated the district, and that a new opening would be afforded to the coal districts of Lancashire, the Midlands and South Yorkshire.

He concluded his evidence by saying that in Lancashire there had been nothing ever approaching the enthusiasm attached to this project since the Anti-Corn Law agitation, and that he felt sure the requisite capital would be found when it was required.

Cross-examined by *Mr. Rodwell*, he said that though they might not be able to bring 5,000-ton ships to Manchester, drawing 22 feet of water, he was quite convinced they could bring up 96 per cent. of the carrying trade of the country which was carried in ships not exceeding 2,500 tons burden. Also he was sure that ships in the docks would tranship into barges, and so, by means of the various canals, cheap distribution would be assured. Pressed as to the ability to find capital for purchasing the Bridgewater Canal, he said there would be no difficulty in finding money to buy a concern that already paid a handsome dividend.

Mr. Marshall Stevens said that the Liverpool docks were over-crowded, and the charges there heavy, also that there was no room for dock extension. It was a matter of history that though the Corporation of Liverpool had sold the town dues to the Dock Board, with the intention that they should be done away with directly the latter body had collected sufficient money to recoup the cost, yet, when the Dock Board had received all the money back, they still continued to collect the dues from the traders of Lancashire. Loaf sugar had to pay 7s. 1d. per ton (Liverpool charges), whilst the proposed charges in Manchester would be 2s. 6d. per ton. Comparing the Liverpool charges and railway freight with the canal rate and charges to Manchester, there would be a saving of 12s. 4d. per ton by the latter. It had been stated there would be an extra freight charged by shipowners for coming the 40 miles between Liverpool and Manchester. This would not be the case, because that distance would not enter into the calculation when thousands of miles had to be traversed. As a matter of fact his ships charged no more from Garston to Rouen than they did to Havre, though the former was 90 miles up a tortuous river. In all the schedules he had not reckoned ships' dues, which would allow a margin of about 1s. 8d. per ton (extra charge) if necessary. Liverpool always charged ships' dues. The port of Manchester would certainly influence and develop traffic coast-wise. The charge for Manchester goods to Plymouth varied from 45s. to 60s. per ton, whilst by sea they could be profitably carried for 12s. Garston is now a much cheaper port than Liverpool. Goods could be brought there, stored for a month free, and taken by rail to Liverpool cheaper than they could be landed in Liverpool direct.

When *Mr. Bidder* came to cross-examine this witness about master porterage and other Liverpool charges, the Chairman said he was utterly puzzled how a shipper could be his own master porter.   To which Mr. Pember replied that it seemed rather like a gentleman jumping down his own throat.   Eventually it was explained that in Liverpool a merchant could act as his own master porter, and charge himself for what he had done.

Following Mr. Marshall Stevens came several commercial witnesses.   *Mr. Sam. Mendel*, a retired Manchester merchant, said that in his time the rate to Liverpool had advanced from 5s. to 11s. per ton, and that railway competition had not stopped the increase.   He hoped for a saving on dock dues, charges and railway rates of from 6s. 6d. to 11s. per ton on Manchester goods.

Mr. J. C. Fielden, a manufacturing agent, estimated a saving by the canal to the cotton trade of at least £450,000 per year, and that there would be a great cheapening of the food supply to the million people round Manchester.   So severely was the cotton trade of Lancashire punished, that in four years, from 1876 to 1879 inclusive, there had been an aggregate loss of £24,000,000 sterling.   In fact he knew of one concern where there had been no dividend for seven years. He instanced the position of Todmorden, where there were now only four concerns left out of twenty-nine firms who were in existence ten years ago.   In the Bacup Valley there were twenty-three mills empty, which used to be engaged in the production of heavy cotton goods.   The turnover of the Lancashire cotton trade was enormous—£80,000,000 exports, and over £20,000,000 for the home trade.   Not only would the Ship Canal effect a great saving in carriage, but it would alter the brokerage system of Liverpool, which was a trades union on a gigantic scale, and levied £120,000 a year on the cotton that came through that town.   *Mr. Andrew*, Secretary to the Oldham Masters' Cotton Spinners' Association, gave the carriage of cotton as 6s. 6d. per ton in 1852 against 10s. which was paid to-day.   On being cross-examined by *Mr. Bidder*, he said that if the canal, estimated to cost over £5,000,000, should for any reason cost £10,000,000, it would abundantly pay the people of Oldham and Lancashire to make it.

Alderman Husband, Mayor of Salford, was of opinion that the canal would be of the greatest advantage in relieving his town from the disastrous floods that had afflicted it, because the water-level of the river would be reduced 10 feet.   He believed that if the canal were made, even at a cost of anything like £10,000,000, the traffic that would come would make it one of the busiest waters on the face of the earth.

Mr. L. J. Crossley, of Halifax, spoke of the large quantities of wool, hemp and jute that were consumed in Yorkshire, and said that the high charges in Liverpool had obliged them to import some of the latter through Hull. He could see no reason why the wool consumed by him should have to pass through London, when it added something like £3 a ton to the cost of the wool to get it from there. He estimated that almost one-half of the colonial wool sold in London was manufactured in the neighbourhood of Bradford. Yorkshire people paid 40s. per ton for the carriage of manufactured goods to London, whilst the manufacturers of Scotland could carry theirs double the distance for the same price. He believed that as Manchester would be the nearest port, it would effect a great saving in the carriage of the raw wool from London, and on the manufactured woollens exported. The wool coming annually from London was valued at from £8,000,000 to £10,000,000.

Mr. Mitchell, of the Co-operative Wholesale Society, Limited, said they had diverted a share of their trade from Liverpool to Goole, because the charges of the former port were 3s. 3d. to 3s. 9d. per ton against 6d. at the latter. If the canal were made, they would use Manchester and save the railway freight. They could now bring goods from New York to Liverpool at the rate of 9 miles for 1d., and to carry those same goods from Liverpool to Manchester (30 miles) cost 4d. per ton per mile. He believed they would carry to Manchester at the ocean rate, just as they carried from America to Rouen at the same rate as to Havre. If so, there would be a saving of 10s. 8d. per ton, which it now cost between Liverpool and Manchester. He estimated that the canal would save their firm £5,000 a year.

Mr. Henry McNiel estimated the saving on the iron that would come by the Ship Canal would amount from £100,000 to £150,000 a year.

Mr. Arnold Lupton repeated the evidence he gave in the House of Commons, and said that in the Barnsley coal-field alone they raised 30,000,000 tons per year, all of it a good steam coal of pure and hard quality, and well suited for bunker use or shipment on the Mersey at Partington. This would be by far their nearest port.

Mr. R. B. Goldsworthy, speaking of the support given to the project in the district, thought there was no doubt of the capital being raised, and expressed his intention to make further subscription in addition to the £1,000 he had already subscribed.

Mr. Joseph Davies, of Warrington, in answer to the question by *Mr. Pope*, "What will the canal do for Warrington?" said :—

At present we have a river that, comparatively speaking, does little or nothing from the sea upwards to Warrington. It is only at certain times of spring tides that we are enabled to have vessels come up to Warrington of about 120 tons burden, whereas, if we have this Ship Canal, we expect to have all the advantages that you have heard set forth for Manchester as they pass Warrington.

After several witnesses from Cardiff and Glasgow had given evidence as to the possibilities of the canal, the Chairman of the Committee made an appeal to the promoters to limit their evidence, in view of the risk there would be through the approaching end of the session, and no more commercial evidence was tendered.

Mr. Leader Williams commenced the engineering evidence by giving the history of the Mersey and Irwell, and Bridgewater Navigations. He then described the canal, and went into the question of estimates. An attempt had been made to discount the Ship Canal on the ground of the insufficiency and impurity of the water. He maintained the large body of water impounded would not be more impure than at present. The polluted water would remain a constant quantity, and the large amount impounded in flood time would diminish the pollution. The sluices when raised would allow the polluted water to pass away, and this would be replaced by a fresh supply in times of rain. The red sandstone surrounding the docks was full of water and could be drawn upon if necessary. When in times of drought the Bridgewater Canal required more water, he had sunk a well, put in a turbine of 80 horse-power, and utilised the fall of the river to pump a large quantity of good water. This he could do for the canal, in case of emergency, but the rainfall figures showed him that there would be an ample quantity to provide 69 large locks of water for full-sized steamers, or 1,141 small locks for barges per day. The Suez Canal paid its dividend upon ten or twelve large steamers per day. It should be borne in mind that a lock full of water was not lost on the passage of each vessel, inasmuch as the upper half of one lock would pass into the lower part of another. The pollution of the river Irwell arose largely from the dye stuffs, which, though they made it dark in colour, were not really injurious. Indeed, they helped to deodorise the water. He might say that the Local Government Board were forcing the neighbouring authorities to purify their sewage, and he believed that the river was improving every year. No doubt the design for passing the Bridgewater over the Ship Canal was a novelty. He had, however, had experience on the Weaver of raising vessels from a lower to a higher level, and this experiment had been successful.   He felt quite assured that what he proposed could be carried out without any risk.

Mr. Leader Williams' evidence was corroborated by *Mr. Abernethy* and other engineers, one of whom criticised a statement that timber ponds were a necessity.   At Millwall they had no timber ponds, and they were only wanted by merchants for a certain class of timber.

Mr. Richard Price Williams repeated his evidence before the House of Commons, and said it was quite possible to alter the junction near Warrington to suit the railway companies; also to have both a swing and a high-level bridge on the Warrington line if it were desired.

Mr. Alexander Adamson, of Glasgow, gave evidence that the cost of fidding a mast on a 4,000-ton ship to pass under the bridges would not exceed £132, and *Captain Pearson*, of Liverpool, an experienced navigator, saw no difficulty in ocean ships passing up the canal.

Dr. C. A. Burghardt spoke of the improvement of the Irwell, and predicted that in a few years it would be a fairly pure river.

Mr. Joseph Carter Bell, another eminent chemist, spoke of the general good health of the people living close to the river, and said that he himself lived near it, and had his garden going down to the edge, and would certainly have known if any injury had accrued to the public health.

Mr. Dunlop stated the land required for the canal proper as 714 acres, valued at £268,000; for the Warrington docks as 24 acres, and for the Manchester docks 142½ acres; total value, £175,500.   The railway deviations would cost £478,000.

Mr. Potter, Q.C., opened the case for the Corporation of Liverpool.   He objected to the extra cost caused by having to sink the Vyrnwy water pipes, for the supply of Liverpool, under the canal and the river Mersey.   Also that no plans, sections or estimates had been submitted for the estuarial work.   He stated that the canal would only serve a limited area outside Manchester, that there would not be sufficient headway for large ships to pass up, and that of every 150 ships engaged in the carrying trade, only twenty would be able to use the canal.   He characterised the canal as "a bit of a scheme" which would be absolutely useless unless supplemented by another Act of Parliament.   He asked their Lordships not to take a leap in the dark, inasmuch as the advantage to be gained was not commensurate with the risk involved.   He hoped the Committee would oblige the promoters to withdraw the Bill, and bring forward a complete scheme next year.

Sir William B. Forwood was the first Liverpool witness.   He at once attacked the evidence of Mr. Leigh, of Stockport, one of the promoters, and said :—

> I know Mr. Leigh's business intimately, and I can say that he could not possibly import anything except a very small portion of the cotton he uses.
> *Mr. Pember.*—Mr. Leigh has been here, and I am told distinctly he imports every ton.
> *Mr. Pope.*—This gentleman buys it for him in Liverpool, and Sir William Forwood knows, for he imports every ton.
> *The Witness.*—I buy in America.
> *The Chairman.*—If Sir William Forwood says that he buys it all for Mr. Leigh, that is a different thing.
> *Mr. Pope.*—He said he knows Mr. Leigh's business.
> *The Chairman.*—I decline to accept any person's knowledge of anybody's business.   If Sir William Forwood says, " I buy Mr. Leigh's cotton, and I know that Mr. Leigh could not buy it," that is a statement ; but if he says, " I know Mr. Leigh's business," that means nothing.   I infinitely prefer to hear Mr. Leigh about his own business.

On this Sir William Forwood explained that it was necessary for a cotton spinner to mix different descriptions of cotton before he could spin them, and that in his opinion a large spot cotton market was absolutely essential to secure the necessary selection of cotton.   His experience of the grain trade was that from 2s. 6d. to 5s. per ton more was charged to Rouen than to Havre.   He calculated that it would cost £101 19s. 6d. to bring 100 bales of cotton from New Orleans to Manchester, whilst the charge from New Orleans to the railway station in Liverpool would be £98 0s. 11d.   Of course when railway carriage was added, Manchester would have an advantage.   Modern ships bringing 9,000 to 10,000 bales of cotton could not go near the canal.   He produced a list to show that no towns in Lancashire would have cotton cheapened by the canal, except Manchester, and here the advantage would only be 10d. per ton.   He complained that whilst Liverpool had been for the last seventeen years struggling to get the railway rates into the interior reduced, Manchester had never given any assistance.   Indeed the various traders in Manchester, when appealed to, made no complaints worth bringing before a Committee of the House of Commons.   In his opinion, if the canal were made to-morrow, Manchester could not successfully compete with Liverpool, because if a port has a large miscellaneous freight to deal with, it can afford to take any special article at a comparatively low cost.   Recently, through improvements in marine engines and increased ocean competition, rates to Calcutta had been reduced from 60s. to 22s., to Bombay from 40s. and 50s. to 20s., and to New York from 60s. to 20s. per ton.   In order to reduce inland traffic rates a barge canal would be more economical than a ship canal.   Indeed, the latter was a very clumsy and expensive method, and like trying to make water run up a hill. Personally, he would never send a steamer to Manchester, but would barge or lighter every ton up to that town.   He feared there was no guarantee the Ship Canal would not be bought up by the railway.   It was a scheme entirely of engineers and promoters.   He had helped forward the Cheshire Lines Company to secure competition, but even before the new line was finished a joint purse agreement was made with the other companies running to Liverpool.   Lancashire would have to provide interest on a capital of from £10,000,000 to £15,000,000 for a Ship Canal, whilst a barge canal could be made at a cost of from £2,000,000 to £3,000,000.   Coasting vessels, on account of the extra time taken going up the canal, would have at least to pay double freights.   Indeed, looking into the matter as a man of business, the canal seemed to him the most visionary scheme, got up without due deliberation, that was ever brought before Parliament.

Cross-examined by *Mr. Pember*, he admitted there might be a chance of obtaining a back cargo from Manchester.   Also that Manchester had just cause to complain of the cost in obtaining raw material of all kinds, and he characterised the railway rates to that city as exorbitant.   He had since ascertained that there were high local charges at Rouen which might cause a higher freight than to Havre, but he still believed that Rouen was a much more convenient port than Manchester would be.   In his calculation he had put £3 13s. 6d. per ton as the cost of freight to Liverpool from New Orleans, and he considered it would cost an extra 5 per cent. on this, or 3s. 8d. per ton., to bring cotton up the river to Manchester.   When it was pointed out that 40 per cent. of the £3 13s. 6d. went in local charges in New Orleans, the only answer he could give was that it was a perfect fallacy to take it in that way, and the extra cost to Manchester would be 3s. 8d. per ton.

The witness believed he was the first person to propose a plateway which was afterwards taken up by Mr. Alfred Holt.   The object was to cheapen and improve the conveyance of goods and minerals..   The contemplated cost of the plateway was £800,000, but when the estimates came out at £8,000,000 he advised Mr. Holt not to pursue the matter further.   It did not strike him that a preliminary fund of £75,000 was out of proportion to an expenditure of £800,000. He had stated in the House that the Leeds and Liverpool and the Bridgewater Canals were under the influence of railway companies, and that though they were only permitted to charge 6s. per ton, under a penalty, they had for many years been charging at the rate of 10s. per ton. He was still of opinion there was a very close alliance between the different canals and the railways, and that the rates were too rigid and exorbitant. Indeed they were double what they ought to be, and he would be glad to see competition. Then there would be rivalry, and the rates ought to come down 30 to 40 per cent. The Leeds and Liverpool now pay 20 per cent., and the Bridgewater 8 per cent. dividend. If this Bill passed the House of Lords it would inflict serious and terrible injury to the Mersey, and therefore he was hostile to it. He understood that 20 feet would be the limit of the draught of water in the canal, and a 2,500-tons gross register cotton ship would draw close on 23 feet. In his evidence before a Select Committee of the House of Commons in March, 1881, he had stated that if Liverpool had the same advantages as other ports as regards interior carriage, there would be a saving of £400,000 per annum. In other words, they were overcharged that sum.

Though he would like to see the Bridgewater out of railway control, he would prefer a barge canal, made by merchants and manufacturers, to the proposed Ship Canal. He understood the latter was got up by promoters and engineers, such as Mr. Leader Williams, Mr. Adamson and Mr. Peacock, and it was his belief that the scheme had been engineered from beginning to end, and that the moneyed classes were not going to take it up.

Mr. Alexander Rendel, consulting engineer for the Indian State Railways, was of opinion that the estimates were insufficient, and the reason the Glasgow engine builders had beaten Manchester out of the market was, not because of the locality, but that the men did more work in the former place.

Mr. Francis Stevenson, engineer to the London and North-Western Railway Company, said that the increased gradients would put his company to the cost of an extra bank engine. He did not consider the Ship Canal Company had provided either sufficient sidings or quay space to work their traffic, and they had not included paving with square setts in their estimates. To his mind the dock area was not sufficient, neither were the estimates for excavations, cement, concrete, or pitching the slopes, and he put in tables to show what similar work had cost him elsewhere.

Mr. John Wolfe Barry did not think the Ship Canal work had a precedent in this country, and he was entirely opposed to making a tunnel under the canal near Warrington.

Mr. Elias Dorning said there was no person living who had had any experience in the buying of land for canals, because there had not been a canal constructed in this country for the last seventy years. He was of opinion that the estimates for the disposal of soil were not sufficient. It would be costly to provide places for tipping, and he added an extra £200,000 for their provision.

Mr. Frederick Leyland, shipowner of Liverpool, believed that vessels carrying Manchester goods would invariably bring mixed cargoes, and that would compel them to go to Liverpool even if they went up to Manchester.

Mr. George Findlay, general manager of the London and North-Western Railway Company, gave the annual value of the imports and exports of Lancashire as £195,000,000. His company carried 1,800,000 passengers over the lines proposed to be diverted at Warrington, and 104 passenger trains and 103 goods trains passed through Warrington daily. If the gradients were increased to 1 in 114, it would mean an extra pilot engine for all trains exceeding fourteen carriages, at a cost of a £1,000 a year for each engine. He preferred, however, this gradient to taking a tunnel at a gradient of 1 in 60, and he thought in proposing the latter the promoters were injuriously affecting them in the worst possible way. Cross-examined by *Mr. Pember*, the witness said he could not interpret what his Chairman, Mr. Moon, meant when he said that the sea and the canals did more to bring down rates than all the competition between the railways themselves. He admitted there were many gradients varying from 1 in 70 to 1 in 100 on their existing lines, and also that there was a Liverpool and Manchester conference, which comprised all the railway and canal companies, and that this conference fixed rates for traffic and the warehouse and carting charges, but he would not accept that this was a combination against the interests of the public.

Mr. James Grierson, of the Great Western Railway Company, objected to the lack of warehouse accommodation at the Ship Canal docks, and was of opinion that a railway tunnel at Latchford ought not to be allowed. His company had made the Severn tunnel, but it was 1 in 90 at one end and 1 in 100 at the other, with a flat length between, against the 1 in 60 proposed for the Latchford tunnel.

Mr. Pope, Q.C., then addressed the House on behalf of the London and North-Western Railway Company. He felt sure that no Committee ever sat before for ten days patiently listening to a Bill which, if passed, could have no possible operative effect unless another Bill were passed in another session by another Committee.   He maintained the engineering estimates were £2,000,000 too low, that instead of the 1,000 acres of land estimated for, an extra area of 1,600 acres would be required, and ridiculed what he called the ingenious answer of Mr. Dunlop, that they would not require to pay for the land not needed for works, and that the owner would be glad to let them have land for the deposit of soil, without payment.   By means of an agitation, clever, persistent and systematic as ever was brought to bear upon a political question, the promoters had succeeded in evoking in Manchester a certain amount of enthusiasm upon the subject.   He knew perfectly well they had started to raise a preliminary fund of £100,000, and had succeeded in raising £65,000, but still the moneyed classes of Manchester were not at the back of this scheme.   In the House of Commons Mr. Armitage, M.P., and Mr. Slagg, M.P., the district members, distinctly stated that neither they nor those they represented as the great commercial classes of Manchester had made up their minds that this scheme was anything but a delusion.   There had been plenty of enthusiasts like Mr. Mitchell and Mr. Fielden, but where were the leading men from Ashton and Bolton and Stockport, and other large towns? He believed Mr. Adamson was possessed with this scheme, and that he and his friends would do all in their power to support it.   But they must remember that Mr. Fielden had told their Lordships that within the last six years Lancashire had lost £24,000,000 of its capital, and had no money to spare.   He had never heard of a Bill in which all Parliamentary guarantees and checks had been so systematically disregarded.   To a limited extent Mr. Adamson and his friends knew their own business, but they were all of them mad upon the question of the Ship Canal, because, except in their own ranks, they would not find in England a calm and sensible merchant who would not at least hesitate as to the probabilities of this Ship Canal ever doing anything for the trade of Lancashire.   If cotton was at New Orleans or Galveston, the shipping agent would find vessels eager to go to Liverpool, but few would ever risk altering their masts for a single voyage, and then encountering the navigation to Manchester.   Besides, only a third of the vessels, and these the smaller ones, could go up to Manchester, and these would insist on a higher rate of freight.   It was difficult to divert traffic, every ounce of which had to be brought from some other port.   Glasgow was not a diversion but a development.   On the Thames ships were leaving the East India and West India Docks and going to Tilbury, lower down the river. The tendency, begotten of the necessities of the time, was to bring the docks to the sea-board, and give the railways the free access to them, so that they might be the distributive agencies over the country. It was not in the interests of the country for the railway interests to perish in order that the Ship Canal might survive. It was a cool proposition for the sake of this Ship Canal that the railway companies would be placed in the power of an arbitrator, who might put upon them the whole cost of making these deviations. He ridiculed the idea that the clauses passed in the early days of railways, in respect to substituting swing bridges, should now be put in force. At the time they were passed a ship canal of the present magnitude could not have been contemplated, and no idea could have been formed of the vast extension of railway traffic. He concluded by saying that the whole scheme had been shown to be so speculative and problematical in its benefit, that the Committee was not justified in inflicting certain and serious injury upon the interests of his clients. He invited them to say that the scheme was incomplete and ill-considered, and ought not to receive the sanction of Parliament, except as a whole, and after the full plans had been received and considered.

Mr. Francis Ellis, land agent, believed that the canal would render Trafford Hall, the seat of Sir Humphrey de Trafford, uninhabitable, and that gentleman would have to give up his home and leave the place. On cross-examination he admitted that the river was not now in a satisfactory condition, and that the canal would be farther from the house than the river was at present.

Mr. John Noble, of the Midland Railway Company, agreed there was an obligation on his company that they should provide a swing bridge in lieu of a fixed bridge if called upon to do so. He thought, however, the clause was inequitable and ought not to be enforced. There was no doubt there were existing gradients on the Midland lines worse than those proposed by the Ship Canal.

Mr. Charles Scotter, of the Manchester, Sheffield and Lincolnshire Railway Company, said that in his experience the working expenses of docks were from 40 to 50 per cent., and that, like the Amsterdam docks, they often cost double the estimates. He was quite sure that no South Yorkshire coal would be shipped at the canal docks, because there was no physical connection. The witness put in the Parliamentary notice when it was proposed in 1871 to sell the Mersey and Irwell Navigation and the Bridgewater Canal to the railway companies, and the circumstances seemed to cause considerable interest to their Lordships, who were

DEVIATION OF THE CHESHIRE LINES RAILWAY AT IRLAM.

To face page 182.

amazed at the attempt by the railway companies to possess themselves of these waterways.   They tried to get the fullest information from Mr. Scotter, when Mr. Littler, Q.C., intervened.

The Chairman (to Mr. Littler).—You are proceeding to answer questions put to this witness.   I should prefer he should answer my question before I allow you to put any question to him at all.   I said, " Do you know about this ? " and he replied, " Yes," and I said, " Do you know it officially ? " and he said, " Yes," and now he tells me that he does not know.
Mr. Littler.—Upon that particular point.
The Chairman.—He evidently knows nothing about the negotiations, or else he would tell me.   I prefer to get the answers out of his mouth instead of out of yours.   Then I may have any conversation you like with you presently.

The Chairman expressed regret that the witness had attempted an explanation on a subject about which he was evidently ignorant.

The witness went on to say that if all the tonnage carried by the Bridgewater Navigation Company on their canal was reduced 8d. a ton, their dividend would be gone, but their Lordships objected to this unsustained testimony.

In cross-examination by *Mr. Pember*, the question of the purchase of the Bridgewater interests by Sir Edward Watkin and Mr. Price, M.P., was again mentioned, when the Chairman said :—

I do not think it necessary to go further into this matter with this witness.   The reason I asked my question was at once to see whether he had any knowledge of this matter. Clearly he has not.   It is a mystery, that I am bound to say, such as I never heard of before, *viz.*, two Chairmen going and purchasing an undertaking and then re-issuing the capital to their own shareholders—the whole matter being perfectly independent of railways, and so on—and then when we have Mr. Price and Sir Edward Watkin, well-known gentlemen, we know that they do not usually act independently of railway interests.   Perhaps Mr. Littler will explain it—it is the strangest case in the world.   Just show me how it is that these two people, for the first time in their lives, so far as we are aware, took this trouble to negotiate with everybody, and then, finally, although their shareholders were not in any way concerned, they turn round to the shareholders and make the whole concern a matter amongst themselves.   Will you just tell me quite shortly ?

Mr. Littler replied that the two gentlemen named conceived the idea they would like to have the Bridgewater Navigation for their two companies.   They entered into negotiations, but in the year 1872 the Amalgamation Committee said that Parliament ought not to allow a canal to be bought up by a railway company. The Chairman thought they must have circumvented the Parliamentary Committee, but he would like to know how the Council for the opposition could maintain that

Mr. Price and Sir Edward Watkin acted independently of railway interests. "Why did these two gentlemen, who are notoriously railway men, act entirely independent of railway interests, and having done so, why did they issue the capital to their shareholders, who were in no way concerned?"

Mr. Littler in reply said that as the Bridgewater Trustees had sought Parliamentary power to sell their property to various railway and canal companies, Sir Edward Watkin and Mr. Price thought it a very good thing to buy it for their two companies, and they entered into an absolute contract to do so. When a Parliamentary Committee declined to sanction the sale of an independent navigation to a railway company, these two gentlemen offered the shares to their respective proprietors. The witness admitted that a shipping trade had been created at Grimsby, which had increased enormously, also he did not deny that at the Alexandra Docks, Newport, the working expenses were only 25 per cent. of their receipts.

Mr. T. B. Foster said that in consequence of the large area of water in the docks it would take 81 hours to flow away, whilst the present area of water only took $3\frac{1}{2}$ hours. Also that whilst it required only 14 hours rain to fill the present channel below Sir Humphrey de Trafford's property, it would require 264 hours rain to fill the enlarged area. Owing to the small velocity, the water would become corrupt and offensive. In cross-examination he admitted that the large quantity of comparatively pure compensation water would sensibly mitigate the evil.

Mr. W. H. Watson said that at present it was only oxidation of the water which prevented the Irwell being an intolerable nuisance to Sir Humphrey de Trafford, and that if the depth of the water was increased from 3 feet to 26 feet, and the velocity proportionately decreased, the water would undoubtedly putrify and cause a serious nuisance. Ships might disturb it, and the water, in consequence of the gases given off, would be most injurious to health. In cross-examination he said that at present the river reached Irlam before giving off its most putrid smell. Reduce the velocity and the smell would be produced much nearer the Trafford property.

Mr. Littler now addressed the Committee. He taunted the promoters that their leading capitalists, Messrs. Platt and Peacock, had not given evidence, and said he was not surprised at them giving each £1,000, because they had been influenced by a sort of wild terrorism, which was rampant in Lancashire. Any one who did not give would be morally boycotted. The agitation had been fomented by publications of every kind which were seasoned with a good spice of hatred for Liverpool. Mr. Adamson had been lured by the glowing accounts of the 18 per cent. earned on the Suez Canal, and his mouth watered for a similar dividend in Manchester. Mr. Hamilton Fulton, who was the original engineer of the scheme, whose experience was as ten to one compared with that of Mr. Leader Williams, had been thrown overboard. He characterised the scheme as having been hastily brought forward, and said he proposed to show that it was Manchester against all England. "Manchester may be great, Manchester may be powerful, and Manchester may be rich, but it is not only that, my lord, that is understating it, because it is a section of Manchester, and that not the most moneyed, influential, or even the most intelligent portion of Manchester against the more intelligent and more moneyed people of England. That is the issue." He said there was the strongest reason why swivel bridges should not be authorised. The condition of things in 1845 was so different that obsolete provisions then made should not be put into operation if they affected the interest of the public who travel on railways. He ridiculed the idea of the promoters doing the proposed work under the provisions of the Mersey and Irwell Navigation Act, or of the Act passed in 1842, and said Parliament ought not now to set a precedent by passing an incomplete Bill. On his complaining that an undue outside influence had been used by the friends of the promoters to interfere with the Chairman's judgment, the Earl of Aberdeen, one of the Committee, said he too had had a similar communication on the other side. Mr. Littler went on to say that no time had been fixed for the completion of the works, nothing had been estimated for spoil banks, and he predicted that the rates and tolls contemplated were so inadequate that in a few years the promoters would be asking Parliament to increase them, as had been the case with the Regent's Canal. He said the scheme was rotten, and asked the Committee to throw out a Bill which was intended to make profits for a private company, and which would interfere with the large vested interests of the Mersey Docks and Harbour Board, Sir Humphrey de Trafford, and many others.

At the instance of *Mr. Rodwell, Q.C.*, Mr. Adamson was recalled to explain his evidence that Mr. Hugh Mason was a director of two or three railways, and had considerable investments in the Mersey Dock Board. Mr. Adamson admitted that he had been somewhat in error, and ought to have said that Mr. Mason was a member of the Liverpool Dock Board, and also a director of the Bridgewater

Navigation Company and of the Midland Railway Company, in both of which latter concerns he was an investor.   Mr. Mason's letter to the *Manchester Guardian*, correcting the mistake, was read, in which he said: "I could not join the Canal Company because I thought it would prove a ruinous failure".

Mr. Rodwell, Q.C., then, on behalf of the Mersey Dock Board, recalled Captain Graham Hills.   He was of opinion that shutting off the flow of the tide at Bank Quay would deprive the lower part of the river of the accustomed flow of tidal water, and do away with the scour.   That particular part of the river would become a settling-point for silt, and the tidal area would be diminished.

Mr. J. S. Swire, a London shipowner, was sure an extra freight would be charged to Manchester, and that all vessels would have to call at both Liverpool and Manchester, because the latter could not provide outward cargoes.

Mr. T. B. Hornby, Chairman of the Mersey Dock Board, admitted Mr. Hugh Mason did not represent Manchester in an official capacity, but said he afforded a medium for all grievances, and that he had never yet made any complaints.   Since the commencement of the Ship Canal agitation the Liverpool charges had been reduced £112,000 per annum, and he pleaded that the scheme should be deferred and considered as a whole.   In cross-examination he stated that the Dock Board paid the Corporation £1,500,000 in order to do away with the town dues.   He considered they were only town dues in name, because they were applied for trade purposes, though he admitted they were bringing in about £250,000 a year.

Mr. Rodwell, Q.C., then addressed the Committee.   He asked them to reject the Bill as an innovation on the principles of legislation which had hitherto prevailed, inasmuch as it was an incomplete Bill which threatened the estuary of the Mersey.   It was too great a power to put into the hands of the officials of the Board of Trade to deal with the interests of Liverpool and the Mersey, and to say if the Bill ought, or ought not, to go before Parliament.   It was obvious the Chairman in the other House felt the difficulty when he said that, "If we should hereafter pass the Bill, and works were executed above the estuary, the works of the canal proper, costing, say, £6,000,000, very great pressure would be put upon the Commissioners of the Mersey".   Appealing to the Committee to pass the Bill, Mr. Pember would be sure to say, "If you don't pass it, all this will be thrown away and wasted".   The Bill and estimates were crude.   They provided for no conveniences or accommodation at Manchester.   No railway communications, no sidings, and no warehouses were included.   If the moneyed people of Manchester were so liberal and bountiful, why did these great Manchester merchants borrow the shillings of the working classes for preliminary expenses?   Why did not these gentlemen put down their names for £20,000? instead of saying, "Oh! we do not care what it costs".   The scheme was brought forward by promoters and engineers, and as Sir William Forwood said, "it had been engineered from beginning to end at Manchester," where they had attempted to carry the thing through by agitating the popular mind, and not the moneyed classes.   He intimated that if the Bill had been for simply purchasing the Bridgewater Canal and Mersey and Irwell Navigation there could have been no possible objection, and he concluded by asking for the rejection of the Bill.

Mr. Pember, in his final reply, said he had pleaded, and would plead to his dying day, that the powers contained in the Mersey and Irwell Acts, and in the Act of 1842 constituting the Mersey Commissioners, were amply sufficient to enable the promoters to carry out all estuarial works, if the officials, representing the Mersey Commissioners, reported favourably on the plans and estimates of the promoters, and would give their certificate that they entailed no damage to the Mersey, nor to the interests of the towns thereon, nor to the shipowners using it. But when the Chairman of the House of Commons Committee heard the lawyers upon both sides argue it, he said with perfect frankness :—

I am a layman, these arguments have gone to a point I am utterly unable to cope with.   All we shall consider is, whether those grounds are such at this moment as to stop the Bill.

Finally the Commons Committee reported :—

And whereas it seemed doubtful whether the said Acts were not in excess of the powers conferred by the said Acts, and might lead to much contest in the Courts of Law.

Further :—

And whereas there are divers interests requiring careful protection, and it seemed to the Committee that works such as those proposed ought not to be undertaken except under the express sanction of Parliament.

It was this decision that obliged the promoters to come before the Lords Committee with their present Bill, which had been characterised as incomplete. They were not asked to pass an incomplete Bill and to take a step in the dark, as had been represented.   It was true that their estimates would be exceeded, but that was because instead of buying merely the Mersey and Irwell Navigation, they were now going to buy the Bridgewater Canal as well. But it must be borne in mind that the latter concern was paying 8 per cent. dividend, and was a concern on which money could be raised. Even supposing they were from £200,000 to £400,000 short, it was an easy matter to come to Parliament again, when they came next year for their estuarial powers. It had been said that this was a vicious scheme, because it must be consented to by the Mersey Commissioners. But this meant that such eminent men as Admiral Spratt or Sir John Coode must have reported favourably. As regards the taunt of shilling subscriptions, he would say that only £830 had been received in subscriptions of under £1. So far as the bubble Mr. Pope spoke about, and which he intended to prick, he might say there were dozens of Bills which, like this, had suspensory clauses. There was a precedent in the Holyhead and Chester line in which the Admiralty had a veto. Mr. Pope had no need to distress himself about the sword of Damocles hanging over the trade of Lancashire. The people were not afraid of the dire misfortune he predicted. If the London and North-Western Railway Company had to provide three bank engines at a cost of £3,000 per year, what was this to the benefit they got by getting rid of the swing bridge clauses? It simply meant one seven-hundredth part of their income. After dealing with the immense savings which would be effected, and would help the trades of Lancashire (now very seriously handicapped), and the industries which would certainly settle down on the banks of the canal, Mr. Pember pointed out the advantages of a progressive policy to Glasgow and Goole, and urged that they must keep pace with the cheapening of transit abroad. He pleaded that a frustration of the promoters' efforts would mean disappointment, rejection and despondency, from which it would be difficult for even the resolute and courageous Lancastrian projectors of this Bill to rise. After alluding to the importance of the industries supporting the Bill, and the subscriptions from large employers of labour, he asked that the Bill might be allowed to pass.

The Committee then cleared the room. After some time, when the public were readmitted, the Chairman gave the decision, *viz.* :—

The Committee have arrived at the decision that it is not expedient to proceed with this Bill in the present session of Parliament. I am instructed to say that in the opinion of the Committee the promoters are very much indebted to their counsel for the very great ability and energy with which the case has been presented to the Committee.

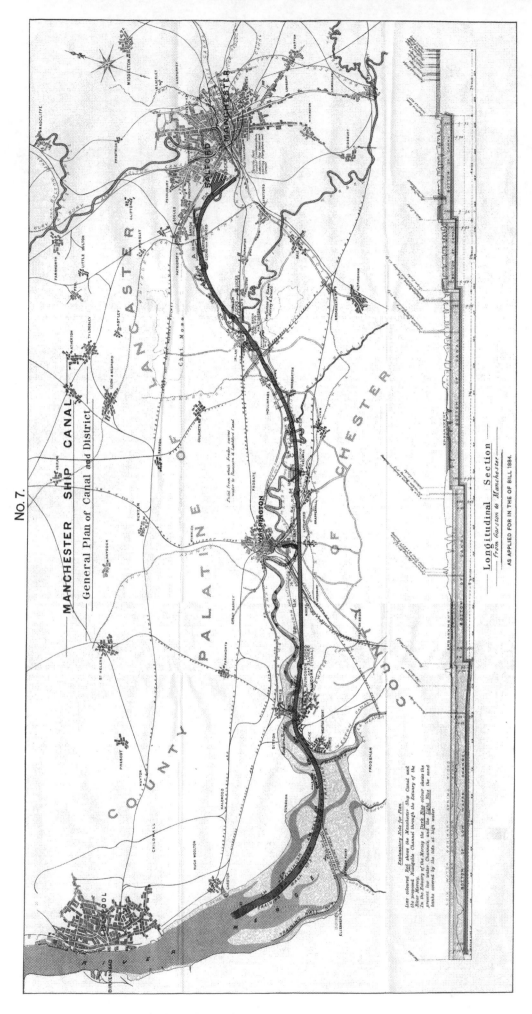

The material originally positioned here is too large for reproduction in this reissue. A PDF can be downloaded from the web address given on page iv of this book, by clicking on 'Resources Available'.

CHAPTER X.

1884.

MANCHESTER THOROUGHLY ROUSED — EVENTS OF THE YEAR—ACTION OF THE CITY COUNCIL—OF THE CHAMBER OF COMMERCE—PUBLIC MEETINGS—ODD INCIDENTS CONNECTED WITH THE PARLIAMENTARY FIGHT — DEMONSTRATION AT POMONA GARDENS — OPINIONS OF THE PRESS — LIVERPOOL UNEASY — BILL REJECTED — CORPORATION AID—DETERMINATION TO APPLY NEXT YEAR.

> Bid harbours open, public ways extend;
> Bid temples, worthier of God, ascend;
> Bid the broad arch the dangerous flood contain,
> The mole projected break the roaring main;
> Back to his bounds their subject sea command,
> And roll obedient rivers through the land.
> These honours, peace to happy Britain brings;
> These are imperial works, and worthy kings.
> —POPE.

THE Ship Canal Bill having been deposited,[1] the first business of 1884 was to prepare and hand in the estimates, the amounts being :—

Dockworks at Manchester and Warrington	£1,121,263
Branch and deviation railways	456,172
Canal works	3,920,172
Estuary works	1,390,419
New roads	16,161
	£6,904,187

[1] See Plan 7.

Strange to say both in 1883 and 1884 the Bill bore precisely the same number, "57". This was unique in the history of private bill legislation. Mr. Adamson started the campaign at Ashton, where he had a most enthusiastic reception; his prophecies, however, astonished even his friends.

The 15th January being the last day for making the Parliamentary deposit, arrangements were made to pay early into the Bank the £280,729 required, and Mr. Adamson, with two other directors, went to London for the purpose. They found memorials against Standing Orders had been presented by Liverpool and the Dock Board, who pleaded want of compliance, because certain plans of deviations had not been deposited. The various railway companies did not this year attempt to oppose the Bill on Standing Orders.

On the 24th January the promoters gained their first success. Though the Dock Board offered a strenuous opposition on the ground that the plans were not sufficiently explicit, the Examiner on Standing Orders decided, "That for all practical purposes the promoters had complied with the Standing Orders".

Meetings now became an every night occurrence. At one held at Withington Mr. J. A. Beith said :—

There was, in his opinion, a most cruel monopoly of carriage between Liverpool and Manchester in the combination of three railways and one canal. That monopoly was prejudicial in every way to Manchester and to Lancashire generally. Besides that monopoly there was in Liverpool a Dock Board which was antediluvian in its management and expensive beyond all necessity. There was an immense future for a canal between Liverpool and Manchester. The heavy trades were being killed out of Manchester for want of a canal. If the canal could be made for £6,000,000 he thought it would pay, but he did not think it would pay at first.

The irony of fate afterwards placed Mr. Beith on the very Dock Board he so derided, and it is a singular fact that it was his casting vote that ensured the dredging away of the Liverpool bar that had been a stumbling-block to the shipping trade of Liverpool for numberless years. Mr. Beith was very proud that the voice of Manchester enabled such a success to be achieved. His own words to me were :—

I reached the Dock Board meeting late, and found them in the midst of a discussion about the bar. The progressives were urging an attempt should be made to dredge away the Bar ; the fossils were declaring Liverpool would be ruined if any failure occurred and pleaded for it to be left alone—the parties were just equal. My casting my vote with the progressives sealed the fate of the bar, and caused a work to be carried out that might well have been done years before.

Mr. Jacob Bright made a most stirring speech at Oldham :—

My belief is that the commercial prizes of the future will be as great, if not greater, than those of the past.   But they won't be won by communities that are guided by timid councils.   They won't fall into the hands of men who dare risk nothing, and who cannot lift their thoughts above the routine occupations in which their lives have been passed.   I say the commercial prizes of the future are greater than those you yet have seen.   What will our trade with the United States be when that country instead of having 50,000,000 people will have 100,000,000 people ?—and the time is not remote when there will be 100,000,000 people there—and when probably duties will be moderate, or when she may even approach to the free trade system which we ourselves adopt.

Following this meeting the Oldham Chamber of Commerce passed strong resolutions in favour of the Bill, and many of the Limited Spinning Companies volunteered contributions of 10s. to 20s. per thousand spindles.

The attitude of the Manchester Chamber of Commerce, with Messrs. Lord and Steinthal at its head, was termed in the Press as one of "masterly inactivity" and came in for severe criticism.   One correspondent, Mr. Frank Hollins, held that the merchants and packers to the East were of opinion that their trade would be entrenched upon and inwardly felt,

> Let spinning, weaving, trade and commerce die
> But leave us still our packing industry.

As a result there was a semi-revolt, and the intention was openly expressed of altering the composition of the Board of the Chamber of Commerce at the coming election by infusing some new blood.   When the day of meeting arrived, the attendance was so large that an adjournment took place to the Mayor's parlour, in the Town Hall.   The President, Mr. George Lord, announced his intention not to seek re-election.   Mr. Adamson's name was included in the new Board, and a series of resolutions were passed in favour of the Ship Canal, the only dis-sentient being Mr. Joseph Spencer.   Mr. Jacob Bright argued that the Chamber should not be "too fastidiously scrupulous," and it was evident that Messrs. Slagg and Houldsworth were being impelled forward by public opinion rather than of their own volition.   Eventually a compromise was effected, and Mr. Adamson and some of his friends were placed on the Board, with Mr. Hutton as Chairman. Shortly after a meeting took place between the new Board of Directors and a deputation from the Ship Canal Committee, when Mr. Leader Williams attended with his plans, and fully explained the scheme.

One of the burning questions of the day was Mr. Adamson's suggestion that the Corporations of Manchester and Salford should contribute from the rates towards the fund for promoting the Bill.   The *Guardian* in a leading article said :—

There ought not to be a moment's hesitation on the part of the City Council in giving a negative reply, and that even if a vast majority of the ratepayers would sanction the appropriation of the money, the principle involved would be not one whit the less pernicious and dangerous.

And when the Salford General Purposes Committee decided to recommend financial assistance, the same paper said it was illegal, and it was unlikely such a clause would receive the sanction of Parliament.   But the unkindest cut of all came from Mr. Bleckley, of Warrington (once a supporter of the canal), who said the Ship Canal Committee had to send the hat round to get civic support, and ridiculed the idea of vested railway interests being affected.

If the Lancashire and Yorkshire Company did nothing until the capital for the Ship Canal was got they would wait a long time.

The *Liverpool Daily Post* thought people would hesitate before they threw good money after bad, and it was not an exhilarating process in bad times to pay money which would never bear a shilling of interest.   Liverpool was not afraid of the canal.

We do not believe that the money to make the canal will ever be found, or that it would pay if it were made.   No canal that could be dug would tempt our great steamers to undertake a difficult and dangerous inland navigation.

The article ended by urging a barge canal.   Mr. Ismay, the broad-minded Liverpool shipowner, would fight the canal in another way.   " Sweep away dock and town dues," he urged, "and substitute port charges according to a tonnage scale.   The costly, unremunerative and dangerous Ship Canal owes its origin to a desire to reduce the charges on raw material and manufactured goods."   The Dock Board should adopt a scientific and economical system of taxation, more conducive to the public interest, than its present cumbersome, capricious and vexatious mode of collecting its dues.   Strong words to come from Liverpool! Early in February it was announced that Lord Redesdale, who the previous year refused to touch the Bill, would this session move the second reading, and that this year a Committee of the House of Lords would first consider the Bill.   During the month large and enthusiastic meetings in its support were held in all parts of the county, and on the 29th May the Manchester Corporation, in view of the advantages to the city, decided to contribute £10,000 in case the Bill passed. The Directors of the Chamber of Commerce also selected four influential members to give evidence in its favour.

The bitterness entertained in some quarters to the canal was exemplified in a letter to a daily contemporary signed "Truth," the writer of which declared the promoters were the instigators of the agitation among the unemployed of Manchester and Salford. The writer, a Bank Manager, was traced, confessed his guilt, and apologised. The proposition that the Ship Canal should pay interest out of capital during construction was also the subject of severe criticism in the Press. It was urged that in the Hull and Barnsley case this had been prevented by an injunction filed by a Manchester gentleman, but on the other side it was shown that Parliament had afterwards legalised such payments in the case of the Regent's Canal Company.

From the 24th April to the 23rd May the Ship Canal Bill was under inquiry by a Lords Committee, who, strange to say, sat for longer hours each day than did a Committee of the Commons. Whilst the Lords were sitting and discussing the possibility of ships passing under the bridges, an incident bearing on this part of the case was pointed out by the leading counsel for the promoters. A shrill whistle was heard, and attention was at once directed to the river, which flows past the Houses of Parliament. A large screw collier of 1,000 tons, plying between London and the Tyne, was seen steaming slowly down the river. Her masts were lying flat on the decks, and but for her size one would have thought she was an ordinary river boat. Her passage was witnessed with evident interest by all the noble Lords as she glided under Westminster Bridge. Besides lowering her masts she also lowered her funnel, thus showing how easy it is for vessels of large tonnage, specially designed to navigate inland waters, to pass under even ordinary roadway bridges. Another incident of the inquiry was the appearance of Mr. A. J. Hunter, of Messrs. Wm. Graham & Co., Manchester. It is said he went to London about six weeks previously to give evidence in favour of the Ship Canal Bill, but somehow the opponents got hold of him and he became a witness for the Corporation of Liverpool, but in cross-examination he really gave strong evidence in favour of the Bill. This inconsistency was much commented upon in the Press of the day.

One argument of Sir William Forwood's was that though Glasgow was a port, and could get cheap cotton, yet the spinning trade had almost died out there. It was proved, however, that a powerful organisation, which dictated its own terms to the employers, had helped to stamp out the trade. This organisation refused to permit (without their sanction) even the sons of a master to learn the practical working of the business, and when they gave their permission, it was clogged with irritating and foolish stipulations. It was shown this was one of the reasons why the Glasgow families once connected with spinning had separated themselves from it, and that the cotton trade was driven away mainly by the arbitrary action of trades unions.

As the inquiry in the Lords neared its close, speculation became rife. Friends were very anxious because the Duke of Richmond from time to time made it plain he was not captivated by the Bill, and the unusual course of having a private sitting, lasting three hours on Ascension Day, showed there were differences on the Committee. It was evident the opposition had impressed the Committee with the feeling that moneyed men were not backing the canal, and that the scheme had been got up by speculators and engineers. Counsel urged the promoters to strengthen their case as regarded capital. In this pass they telegraphed me on the Thursday to put myself in communication at once with Mr. Samson, a member of the promoters' legal firm, who was in Manchester, to try and beat up some capital, and then go up to London and give evidence not later than the following Wednesday week. Mr. Samson and I started on the Saturday afternoon by calling on Mr. Rylands, at Longford Hall. He was alone, and received us very kindly; we explained the necessity of the case, and pointed out to him the signal service he would render the Ship Canal, in which he had always been deeply interested, by taking £50,000 in shares. He received the suggestion more favourably than we had expected, promised he would help, and arranged for us to call on him at his Manchester warehouse on the following Monday morning at 8.30 A.M. for a reply.

On the Sunday I dined with Mr. Samson at the Conservative Club, and then went to see Mr. Hilton Greaves, at Oldham, and got from him a conditional promise of £15,000 in shares. On the Monday morning when we called on Mr. Rylands we were met by Mrs. Rylands, who absolutely refused to let us see him. She did not know us, and had got the idea that we had been pressing her husband unduly, and that as an old man he ought to be protected. We explained that Mr. Rylands had already contributed largely to the preliminary expenses, and this would be lost if the Bill were thrown out. All explanations and entreaties were useless, she would not allow her husband to be disturbed.   This was a dreadful blow to us, and we retired to take counsel with a mutual friend, who suggested we should ask to see Mr. Rylands in the presence of his wife (under the promise not to push business), and simply ask him to talk the matter over with two gentlemen who were his confidential advisers.   Our request was acceded to, and the next morning I was overjoyed to receive a letter, asking me to call, when Mr. Rylands in my presence signed a promise to take £50,000 in shares.   For some days I was on the track of Mr. George Benton, between his house and the hotel where he did his business.   At last I went to his house and waited till he got up about noon.   We then had a long chat, and I induced him to sign a promise to take £50,000 in shares.   He told me the curious fact that he laid his plans and made his calculations for the big works he had on hand whilst he lay in bed in the morning.

The hunt for money ended on the Tuesday night, when I went to London with promises for £130,000 in shares, all obtained in five days.

When Mr. Pember started to make his final speech, the Lords had sat thirty-eight days of five hours each, equal to forty-seven days of a Commons Committee. They had heard 25,367 questions asked from 151 witnesses, occupying 1,861 folio sheets of printed matter.   Mr. Pember's speech was a masterpiece of eloquence. Feeling the Committee were impressed by the opponents' statement that the capital could not be raised, and that the works might be left in an unfinished condition, he met it by volunteering that £5,000,000 should be subscribed before they were commenced.   The prediction of damage, nay ruin, to the estuary, he met by showing that it had actually improved, notwithstanding Liverpool herself had made large abstractions and he argued she ought not to prohibit other people doing what she herself had done.   He twitted the Dock Board on the disgraceful state of the bar, and said they did not dare to do right from the pusillanimous fear that they might do wrong.   The excitement in the corridor whilst the Committee were deliberating was such as rarely has been experienced.   For over two hours hundreds of people blockaded the door of the Committee Room, and when it was opened the excited partisans rushed in pell-mell regardless of consequences.   When the Duke of Richmond, in a few crisp sentences, declared the preamble proved, all the rules of the House were set at defiance, and the supporters burst into a hearty cheer, which, of course, was at once suppressed.

The news quickly passed to Manchester and Oldham, and in both places the liveliest feelings of satisfaction were expressed.   Outside the Market Street Committee Rooms the street was blocked by people anxious to read the telegrams ; on the Manchester Exchange there was much cheering, and peals were rung on the Cathedral bells till late in the evening.

On Mr. Adamson's return he met with a triumphant reception, both on his way home and at his works, where his workmen presented him with an address of congratulation. From the station a band escorted him home, playing "See the conquering hero comes," and the populace insisted on taking the horses out of the carriage and drawing him home.

The Press of the country outside Liverpool joined in a pæan of praise on the courage of Manchester in attacking a vast monopoly, and congratulated them on their success. Even the *Manchester Guardian* had a good word to say on the perseverance and determination that had been displayed, but they guarded themselves in a warning article, and said :—

Engineers may be wrong, and although the preamble of the Bill has passed the searching ordeal of the Lords, the duty of guarding, in every reasonable way, against a disastrous failure, remains not the less imperative.

The *Liverpool Mercury* wrote :—

Nothing can be further from the truth than the idea that Liverpool is afraid of Manchester competition. We repudiate the idea, as we have a right to do, that our sea-going steamers will trust themselves to the inland navigation, or that any mere volition on the part of Manchester manufacturers, however strong, can change the course of the world's trade. Natural forces are too strong for the most obstinate wills, and all the world over natural forces are taking trade from inland ports down to the sea, but not in the opposite direction.

Another Liverpool paper professed to know the preamble was only passed by a majority of one, and that whilst Lords Norton, Dunraven and Lovat voted for it, the Duke of Richmond and Lord Barrington were in opposition.

A third paper said Manchester had succeeded in giving Liverpool a back-handed slap in the face, but it was her own fault.

It is useless to disguise the fact that the Dock Board and some of their officials are in no small degree responsible for the initiation of the Ship Canal scheme, and the pertinacity with which it has been fought. Excessive rates and snobbish sauciness angered many influential Manchester men, who felt it would be unsafe to allow the Dock Board to have its hand perpetually at their throat. Hence the zeal (almost vindictiveness) with which the canal scheme has been fought. No doubt Manchester will now have a big crow. They are welcome to their victory. We sincerely hope that the scheme may improve the trade of Manchester, whilst we are not amongst those who think that it will do a vast deal of harm to the port of Liverpool and the great interests connected with it.

Sir William Forwood wrote to the press protesting against the decision, and repeating a previous statement that, as only 2,400,000 tons of produce were consumed in Manchester and 15 miles round—and this was now divided between four railways and a canal—the Ship Canal could really have little to carry, and that if only 5 per cent. increased ocean freight were charged there could be no advantage by the canal.

Does any sane man suppose that the railway companies will stand idly by and allow the canal to take their traffic?   A reduction of 3s. per ton would still leave them a paying trade, but would effectually starve out the canal and reduce it to bankruptcy.

This letter provoked a long correspondence, and on it the *Liverpool Daily Post* wrote :—

Sir William Forwood does not anticipate that anything he can say will prevent those who have the money from being imposed upon by those who have the brains : he places before the public a compact array of uncontrovertible facts, which should induce intending investors to hesitate ere pitching their money into Mr. Leader Williams' ditch. . . . The promoters set their faces against the probability of higher freights to Manchester than to Liverpool.   They labour under the delusion that if there is a canal there must be trade, and that the great ships and steamers that bridge the ocean could not resist the temptation to struggle between the reefs leading to the seaport of Manchester.   They appear to have no doubt whatever that all the additional expense involved, and risks incurred, will be cheerfully borne by shipowners as tributes to Manchester ambition.   Except so far as the training walls imperil the navigation of the Mersey, Liverpool cares nothing for the advance of civilisation indicated by this wild enterprise for enriching lawyers, engineers and contractors.

A demonstration, under the auspices of the Manchester and Salford Trades Council, took place on Saturday, the 24th June, in the large Agricultural Hall at Pomona Gardens, which was filled to overflowing.   There were about 50,000 people present.   The procession started from Albert Square, and had five bands with them to enliven the proceedings.   The object was to strengthen the hands of the Provisional Committee in the coming struggle in the House of Commons.   Mr. H. Slatter was in the chair.   Alderman J. J. Harwood moved the first resolution :—

That the thanks of this assembly of the industrial classes be given to the Chairman and members of the Provisional Committee for their untiring energy and their successful efforts in establishing before a Committee of the House of Lords the necessity and advantage of connecting Manchester, Salford and the district with the sea by means of a Ship Canal. And this meeting earnestly desires that the Provisional Committee will continue to press forward the Ship Canal scheme to its final completion, and they pledge themselves to render all possible assistance to accomplish that object, believing the result will be the establishing of new industries and the further development of those already in existence.

This was carried unanimously.   The meeting was a magnificent success, and showed the extraordinary popularity of the Ship Canal.

From the Lords the Bill was remitted to the Commons, and it became evident that if it were to go through no time must be lost, both from the natural effluxion of the session, and because Government was in a perilous position as regarded the Franchise Bill, which, if rejected, might lead to a dissolution.   It was at one time hoped to secure Mr. W. E. Forster as Chairman, but neither he nor Mr. Corry was available, so Mr. Sclater Booth was ultimately made Chairman.

Liverpool was now thoroughly aroused.   Hitherto the Corporation, the Dock Board and the Chamber of Commerce had not always worked in unison : they had even been disposed to blame one another and the railway companies for Liverpool being a dear port—really the cause of the Ship Canal movement; now they were banded together against the common foe, resolved to strenuously oppose, and if possible upset the Bill.   As the *Liverpool Courier* put it :—

The question at issue is not whether Parliament will allow Manchester to compete with Liverpool for maritime supremacy, but whether Manchester shall be allowed to utterly destroy this port in futile efforts to make itself into a seaport.   If the Mersey navigation be materially injured, Liverpool trade will be crippled, if not annihilated, and Manchester trade will share in the catastrophe.

Up to this time Liverpool had laughed at the idea of the capital being raised, but now Sir William Forwood told them they were labouring under a delusion. He was convinced Manchester could and would raise the money.   The same gentleman, by repeating his extraordinary statements, induced a special meeting of the Liverpool Chamber of Commerce (the presence of only about twenty members showed their interest in the matter) to send a strong petition against the Bill.

A leading Liverpool paper, commenting on this meeting, wrote :—

There is a fear that many people will be deluded into the belief that it is really a practicable project, and thus £5,000,000 may be raised to fill up the Mersey with sand so that where great ships now ride at anchor grass may be grown.

At a subsequent monthly meeting of the Liverpool Chamber of Commerce, it was decided, in consequence of a letter from Mr. Samuel Smith, M.P., to take, what Sir Richard Cross warned them would be an unusual course, *viz.*, to promote opposition to the second reading of the Ship Canal Bill.   In the course of the debate surprise was expressed that neither Lord Derby nor Lord Sefton opposed the Bill, and Colonel Paris cautioned his colleagues to look before they leaped, and recalled to their memories the prophecy of an eminent engineer, that if the Leasowe embankment were built, the port of Liverpool would be blocked up; and also that of another engineer, that if training walls were not built for a distance of $1\frac{1}{2}$ miles in a straight line from New Brighton, they would lose the entrance over the bar.

Mr. Aspinall, of Birkenhead, went further.   He wrote:—

The costly and really foolish opposition by our local bodies to the Ship Canal is unworthy of this great seaport.   I write as a practical man, for twenty years in our Local Board, and Chairman of the Woodside Ferry Committee, and thoroughly acquainted with our river from Runcorn to the North-west Light Ship.   Born and bred here on the banks of the river, spending many years in steamers and boats, my knowledge of the tides, in fact, the entire area of its estuaries, is fairly accurate, and I confess my surprise at the reports of the local evidence brought before the Lords Committee by the opponents of the Bill, most of which was easy of refutation.

On the 26th June, when the Ship Canal came up for the second reading, Mr. Samuel Smith, M.P., told the House that in the opinion of the highest authorities the works which it was proposed to construct in the upper part of the river would have the effect, in time, of closing up the Mersey, and reducing to beggary the 800,000 people who resided within 5 miles of the Liverpool Exchange.   To many of them it meant a sentence of death, and though he did not intend to oppose now, he asked that the Bill be referred to a very strong Committee, reserving full liberty to oppose it at a later stage.   This extraordinary speech brought up Mr. Jacob Bright. He said people had come to the conclusion the Mersey was not safe in the hands of Liverpool alone, that they had utterly neglected it and the bar, and that the whole system of shipping at Liverpool was antiquated and costly; indeed, that Liverpool was the dearest port in the country outside London.   He was supported by Mr. Houldsworth, who said that 60 per cent. of the whole cost of conveying goods to Calcutta was incurred before the ship left Liverpool.   Though Messrs. Rathbone, Whitley and McIver followed Mr. Smith in his opposition, the House became impatient at this unusual attempt to damage a Bill that had already passed the Lords, and it was read a second time.   Sir John Mowbray promised on behalf of the Selection Committee that the Bill should be relegated to a strong Committee.

The discussion was unfortunate for some who took part in it.   Poor Mr. Smith got small thanks for his action.   He was told by the Liverpool Press that he had rushed into the fray unnecessarily, and without previous consultation and instruction; indeed, he had done more harm than good, that his alarmist speech had come to a lame conclusion and that it showed a gross want of tact.

Mr. Houldsworth, M.P., wrote to the *Manchester Guardian* that he was reported as having said in his speech, "I believe myself in the success of this scheme," and he was not aware of having said so. This provoked some very plain letter-writing. One correspondent asked if he intended making a sphinx of himself, and wanted an official interpretation of the speech. Another wrote, "Let him at once tell us if he has written this letter to damn the scheme by having it used by the enemy, as Mr. Hugh Mason's letter formerly was, or tell us what he does mean—that is, if he knows himself."

Warrington lying half-way between Manchester and Liverpool had hitherto held aloof, but at a town's meeting held at the end of June, and addressed by Mr. Adamson, strong resolutions were passed in favour of the Bill.

On 27th June a deputation from the Mersey Dock Board and the Liverpool Chamber of Commerce, accompanied by the Mayors of Bootle and Birkenhead and several others, waited on Lord Northbrook and Admiral Spratt, two of the Conservators of the Mersey, who listened to them and promised due consideration. A few days later, at a meeting of the Dock Board, the Chairman, Mr. Hornby, made the important statement :—

If they (the promoters) were to drop the estuary works, and bring up a scheme which would not touch the estuary, and would be free from the objections arising on that point— if they were to bring such a scheme, the language we have used and the line we have taken would preclude us offering opposition against what would simply be a communication between Manchester and the sea. We might still have considered it an injudicious proceeding, and a dreadful waste of money, but as regarded our opposition, the point we wished to lay stress upon would be gone.

These words ought to be well remembered, inasmuch as they turned out to be a complete will-o'-the-wisp to the promoters.

Meanwhile petitions against the Bill were asked for and secured from all the Liverpool trading bodies who had been practically apathetic before, and the Liverpool Press was teeming with antagonistic articles. One paper wrote :—

Imagine shipowners spending hundreds of pounds on adapting the masts of their vessels to the necessities of a bridge canal, and incurring the extra expense of lowering and hoisting the top masts—and all to gratify Manchester vanity ! and fancy underwriters accepting risks on the same terms as for vessels which complete their voyages at Liverpool or Birkenhead !

These sarcasms are amusing in the light of after events.

On the 7th July the Committee of the House of Commons commenced their labours, an account of which is given in another chapter. From the first it was a fight against time.   The promoters knew full well that under most favourable circumstances there was a risk of Parliament dissolving before the Bill could get through, and then all their labours would be lost; but they were buoyed with the hope that the searching and prolonged inquiry in the Lords would be read by their successors in the Commons, and have an influence upon them.   Most reluctantly they decided to cut down their evidence and counsels' speeches to a narrow limit. The opponents, on the other hand, were fully alive that this was their last chance; they must bring all the evidence and all the influence in their power to scotch the Bill, and if they could only spin out the hearing, the effluxion of time would give them the victory.   From the beginning they therefore opposed or parried every suggestion of the Chairman to quicken the pace.   The general feeling in Liverpool was that the die was cast and that the passing of the Bill was a foregone conclusion. Their only chance to avert this was to talk the Bill out, or divert the attention of the Committee by illusive promises.   "Reject the estuary," said they, "and we will not oppose you going on its borders."   As the *Daily Post* said:—

If Manchester is pining for the smell of salt water, her olfactory nerves can surely be gratified in a less perilous way.   Let her cut a canal to the sea in any other direction she pleases. . . . Let them cut it through by land to Garston, where they will find deep water, and where their pet project will be unable to work much mischief.

Lured by these promises made over and over again both inside and outside the House, the Committee (never a strong one) on the 1st August came to the unexpected conclusion to reject the Bill.   It was a staggering blow to the promoters. They reeled under it, but neither lost their heads nor their courage, and at once started to repair the breach.   It did seem hard that, after an inquiry by a Committee of five in the Lords, who had spent over ten weeks in going into the case most fully and passing the Bill, their decision should be reversed by four Commoners who had hurried through the inquiry in under four weeks.   On the fatal day I had gone with the City Council over the waterworks at Longdendale, all the time burning to know how things were going in London.   My friends were very confident, but somehow I had my misgivings, for throughout the promoters' case had been unduly hurried by the Committee.   On returning to London Road Station there was a depressing silence, as if some mishap had taken place.   When we got outside I heard a cabman in furious tones denouncing the Houses of Parliament, and sending them all to perdition.   A cold sweat came over me, and I said to my friends, "Sure enough they have thrown us out," and so it was.   The cabman in his vexation was relieving his feelings by strong language. For a second time our hopes were wrecked, and all our time, trouble and money were lost. Every one was asking, "what shall we do next?" It is impossible to gauge the minds of a Committee, but the general feeling was that the additional weight of engineering evidence by the opponents, backed by the statement they had made, that a canal less costly and as useful, could be made without even running the risk of damaging the estuary, turned the scale in their favour. The greatest sympathy was expressed for Mr. Adamson. If he had not secured success, it was felt that by his hard work and his dogged determination he had earned it, and that he deserved the gratitude of all well-wishers of the canal.

It took some days for the Provisional Committee to recover from the crushing blow. However, on the 5th of August they met and unanimously resolved to persevere in their efforts to make Manchester a port, though it was admitted it might be necessary to change the route and modify the scheme. At the meeting Mr. Adamson called attention to the pledge given by Sir William Forwood, and ratified by Mr. Aspinall, Q.C., counsel for the Dock Board, that if these changes were made Liverpool would desist from further opposition. On the motion of Mr. John Rylands, seconded by Mr. Hilton Greaves, of Oldham, and supported by Mr. James E. Platt, the Mayor of Salford, Dr. Mackie, of Warrington, and others, it was resolved :—

That this Committee notwithstanding the rejection of the Bill lately by the House of Commons, is of opinion that the movement to connect Manchester with the sea by a Ship Canal should be continued and prosecuted to a successful issue, and further, that the concessions made by the opposition warrant the consideration of an alternative scheme avoiding the estuary, which will secure a satisfactory canal. This Committee is also of opinion that a meeting of subscribers, as well as a public meeting should be summoned at an early date to receive the report of the Provisional Committee, and to test the feeling of the district as to further operations.

On the 7th August, the Chairman, Mr. Adamson, with a deputation from the Provisional Committee had an interview with the Corporation Special Committee "Re Ship Canal". They represented that the £10,000 already voted could not be claimed now that the Bill was rejected, and that they feared being able to continue the struggle unless Manchester and other municipalities came to their aid. The Chairman suggested a contribution of 2d. in the pound on the rates towards the Parliamentary expenses, and stated that now they were adopting a route suggested by Liverpool, the expenses of a Bill in the coming session could not be heavy.

Subsequently a proposition was made to the Council by the Special Committee that on an indemnity being given against any further liability, the sum of £18,000, or 2d. in the pound, on the city rate be contributed towards the proportion of the Bill. This was approved of by the Council subject to its receiving the sanction of the ratepayers at a Borough Funds Meeting.

One service done by the rejection of the Bill was to arouse an overwhelming feeling in its favour. From all sides, from corporations, local boards, trading bodies, limited companies, trades unions, etc., came letters of sympathy, words of encouragement and offers of help—even working men came forward to promise a day's wages to the fund. When a deputation waited on the Manchester City Council with a petition signed by 2,272 ratepayers, the Mayor at once promised to call a town's meeting to consider it.

This meeting was held in the Free Trade Hall on the 15th August. Earlier in the day the subscribers had been called together. At the subscribers' meeting the auditor stated the balance in hand was £12,500, but there were still some liabilities. The Chairman, Mr. Adamson, said that the solicitor, engineer and others, in view of the defeat, would make considerable deductions in their charges, and were prepared to throw their energies into another fight, when he was sure they would win; he hoped with an altered route, sanctioned by Liverpool, and a scheme improved by experience, that the Corporations of Manchester, Salford, Oldham, Warrington and other towns would render still greater assistance than they had done in the past; indeed, he trusted they would subscribe to the great undertaking, and become co-promoters of the Bill. A resolution, authorising the Committee to secure a new route and go for a new Bill was moved by Alderman Mark, seconded by Alderman Walton Smith, and supported by Mr. Houldsworth, M.P., who liked the spirit of the promoters not to cry after spilled milk, and had great faith that heaven helped those who helped themselves. The public meeting the same night was even more enthusiastic and determined. It was addressed by Mr. Adamson, the members for Manchester and other influential citizens, and by resolution it pledged itself as to the necessity of the canal, and expressed a hope that the Corporations of Manchester and the neighbourhood would give financial support. Mr. Adamson was exceedingly severe on the action of the Bridgewater Navigation Company. "The greatest enemy the promoters of the canal had in the House of Commons was the solicitor for the Bridgewater Canal. He trusted that gentleman would get his deserts." He fell foul, too, of the Liverpool opponents, and said they had made a promise not to oppose if the new scheme kept clear of the estuary, and they would be kept to it, though from the day's papers it would be seen that Liverpool had commenced the first dance in the double shuffle.   He asked, "Was it right that the Liverpool Corporation should oppose the scheme out of public funds, if the friends of the project have not the right to ask Manchester to step in and help them?"   Then Mr. Houldsworth, M.P., in a careful, non-committal speech said :—

> The scheme in order to be successful should commend itself not merely to people who dream but to commercial men, who were asked to put their money in it.

He roused a strong feeling of indignation, and a man in the body of the hall shouted, "He's on the Liverpool side"

Mr. Jacob Bright carried the meeting with him by an able, courageous and cheering speech.

> He did not understand the reasoning of people who could applaud a Town Council for spending money in pictures and books, and who, in the next moment, could condemn the Town Council if it took part in opening a great highway to the sea, by which every man, woman and child in this city would be benefited, and by which the city of Manchester would be placed on equal terms commercially with the foremost communities in the world.
>
> Great things had been done by Lancashire men in the past.   There were ways of accomplishing the object they had in view, and the enterprising, courageous and hard-working people of Manchester would find out a successful way.

The splendid spirit shown by the citizens of Manchester and the subscribers, together with the offers of assistance, left the Provisional Committee no option, and they determined to lose no time in inquiring into the suggestions for avoiding the estuary made by Liverpool in the Parliamentary Committee.   A consultation was held in Liverpool with some broadminded shipowners there who had a kindly feeling towards the canal.   Both the Lancashire and Cheshire shores were visited by the Committee to seek for a good entrance.   I well remember the autumn day when, accompanied by the Chairman, Mr. Hilton Greaves, Mr. William Johnson, the well-known Liverpool shipowner, and others, we walked several miles along the Lancashire side from Garston.   How fagged we all were, and how welcome a cup of tea was to our parched throats when we reached the hotel!   We saw enough to convince us that the Lancashire side would not do, and that the entrance to the canal must be from the Cheshire side.   Mr. Leader Williams, the engineer, also being of this opinion, he was instructed to get out plans and estimates, and submit them to the Committee.

Meanwhile, the papers were full of a controversy on the propriety of the Corporation giving financial help to the promoters.   A few correspondents, headed by Mr. E. H. Fuller, opposed vigorously, and that gentleman compared the undertaking with the South Sea scheme, which became the South Sea Bubble.   He prophesied the same fate would attach to the Ship Canal scheme, and he objected to any addition to the rates.   On the other side, it was shown that most successful ports had, at some time or other, been helped from the rates, and with great advantage.   Without such assistance Liverpool, Glasgow, Newcastle, Bristol and other places would have been scotched in their infancy.   The public was reminded of the dictum of Sir Edward Watkin, himself a railway magnate :—

> I repeat that I think a greatly improved navigation between Manchester and Liverpool is bound to be made, and that it would be a great blessing to Manchester and no damage to Liverpool.

The position of the Weaver Navigation was also pointed out, where the rates of Cheshire were assisted by that waterway.

In order to prevent Corporation assistance being given, a ratepayers' memorial to the Board of Trade and the Home Secretary was got up, headed by such well-known citizens as Mr. Roby, M.P. ; J. F. Hutton & Co., Dale Street ; Schunck, Souchay & Co. ; Fogg, Braddock & Co. ; Railton & Son, East Street ; Earle, Sons & Co., Brown Street ; J. Clapham, J.P., King Street, and others.   Their fear was that if the Corporation once embarked as promoters of the scheme, it would be an easy step to levy a further and permanent canal rate, and they regarded the canal as a speculative scheme and a private venture.   It was also urged that to fight railway and vested interests owning £190,000,000 capital was hopeless.   On the other side it was ably argued, especially by Mr. Reuben Spencer, Mr. J. W. Harvey, Mr. J. M. Fletcher and others, that it was a case of "nothing venture, nothing win," and that if the Corporations of the leading commercial cities, including Glasgow, Liverpool, Bristol and Hull, had laid the foundation of their success by giving monetary assistance to dock enterprise, Manchester must either follow suit or be content to take a back seat.   Further, that it was good policy for the city to levy a rate of 2d. in the pound to aid the Bill, and identify itself with a valuable scheme.   If the railways, because of their unity and wealth, were allowed to dominate the trade of the country, then good-bye to its success.

We should have been a nice pigmy race of slaves to the Romans and others had our forefathers had no more pluck than some of this degenerate race who bow their knee to the power of the railway interest.

The *Manchester Guardian*, dealing with the position in an article on 19th September, 1884, said :—

We are not concerned at this moment to decide whether all this is economically sound or not ; what we have to point out is, that we are in the presence of a movement, observable in a greater or less degree in all the leading countries of the world, for embodying in a practical form the idea that it is in the interests of the public to provide traders with ample and cheap facilities for transport ; even if this is to be done only by the help of the taxpayer.

The *City News* thought :—

The reasoning of the objectors might have had weight a score of years ago ; it has none now.   Parliament has long recognised the expediency of allowing corporate authorities with the sanction of a majority of the ratepayers and the Local Government Board to levy rates for promoting in Parliament bills that will benefit inhabitants, . . . and it is the cheapest mode of carriage which Lancashire wants to fit her to maintain her industries.

A further reason for Manchester contributing was to prevent so great a scheme now or hereafter falling entirely into the hands of private individuals.   Dr. Pankhurst had a tilt with Mr. Henry H. Howorth, of Salford, over the question, the latter protesting that—

If it be right and just that our Aldermen and Councillors should call upon us to devote our money to constructing the Ship Canal, I cannot see how they can refuse to do the same for any other venture which is steered in the same masterly manner, and we shall presently be found embarked in a scheme for the extraction of good palm oil from worn-out paving stones, a process which, if successful, must of course be of immense benefit to Manchester.

On the 6th October, 1884, the Borough Funds Meeting was held.   The Mayor (Councillor Goldschmidt), whilst objecting to the Corporation being in any way concerned in the execution of the works, thought a Ship Canal would be of the greatest benefit to the town, and moved that £18,000, equal to 2d. in the pound on the rates, be voted towards assisting the promoters to secure the Bill.   This was seconded by Alderman Harwood ; he estimated the canal would effect a saving of £508,035 to the trades of the city, and remove an anomaly he was told now existed, *viz.*, that the working men of Liverpool could live 15 per cent. cheaper than the working men of Manchester.   The contribution meant 2d. in the pound on the rates for one year only.   There was nothing to lose and everything to gain by this project.

The resolution was put to the meeting and the vast audience held up both hands.   There were only about a score of dissentients.   The result was received with much cheering.   Several ratepayers rose to demand a poll, among the rest Mr. Lynde, solicitor to the London and North-Western Railway Company, who announced that in demanding a poll he was acting professionally for that company.   At once there were cries of "who is to pay the cost?"   A reply of "the London and North-Western" caused an irate ratepayer to exclaim "stop their traffic".   Sir Joseph Heron said there was nearly always room for repentance, and if Mr. Lynde should receive those instructions which he hoped might come from headquarters, he would still be in a position to withdraw his request for a poll.

There were loud and frequent complaints about the course adopted by Mr. Lynde.   It was said it was hard lines to make the city spend £1,500 on a poll.

Next day Mr. Lynde wrote :—

I am instructed to inform you that having entered a protest against public funds being used for private enterprises, the London and North-Western Railway Company will not put the city to the expense of a poll, and therefore I withdraw the demand I made yesterday.

Had this step not been taken there is no doubt a great many tradespeople would have shown their displeasure by withdrawing their traffic.   On the other hand, the action of Mr. Lynde had prevented other ratepayers demanding a poll, and many angry letters appeared in the Press, one correspondent saying that a private arrangement between Mr. Lynde and Sir Joseph Heron had practically superseded the vote of the citizens of Manchester.   This Mr. Lynde indignantly repudiated.   His graphic description of the feeling of the meeting when he demanded a poll is worth recording :—

A poll was demanded in the midst of such yells and execrations as rarely fall to the lot of a murderer.

Several bitter letters afterwards appeared signed by Mr. W. A. Lynde, but he subsequently withdrew them, saying they had been published without his authority.

At the Warrington Borough Funds Meeting the resolution to subscribe 2d. in the pound was passed with only two dissentients, and these demanded a poll. When, however, the opponents called a meeting they were defeated, and a resolution in favour of the rate was passed.

The legality of the Manchester Corporation assisting the Ship Canal was fully discussed in the Press.   The result of the town's meeting fomented the previous opposition.   Mr. Roby, M.P., wrote to the *Guardian* :—

I question the expediency, if not the right, of the Council to force ratepayers to contribute to it.

Committees were formed both in Manchester and Salford, and signatures obtained to petitions to the Local Government Board.   In Manchester Mr. Charles Pierson and Mr. F. A. Woodcock were the secretaries; the latter was vigorously attacked as being a solicitor, and the nominee of the London and North-Western Railway Company, and without doubt the statements made and the carelessness exhibited in getting signatures were very reprehensible.   Many men disowned their signatures attached to the petition.   Still it cannot be denied that many leading citizens signed it; amongst others, Barbour Brothers, William Graham & Co., George Frazer, Son & Co., Carver Brothers, G. H. Gaddum, Robert Barclay & Co., J. H. Agnew & Brothers, A. & S. Henry, Chamberlain, Donner & Co., D. Matheson & Co., and others.   The legal questions, both as regarded the poll and the power of the Corporation to contribute, were ably argued in the Press by Mr. W. H. S. Watts for the Corporation, and Mr. F. A. Woodcock for the memorialists.

On the 21st November the Local Government Board wrote to say, "the matters referred to in the Bill (and memorial) will not be such as come within their jurisdiction," and on the 3rd December Mr. John T. Hibbert, M.P., forwarded the decision of the Home Secretary (Sir William Harcourt) giving his assent to the proposal of the City Council to contribute 2d. in the pound out of the rates.   The following is a copy of his letter :—

WHITEHALL, *3rd December*, 1884.

SIR,

I am directed by the Secretary of State to transmit to you herewith, to be laid before the Town Council of Manchester, with reference to your letter of the 5th ultimo, an instrument under his hand signifying his approval, pursuant to the Act 35 & 36 Vict., cap. 91, of a resolution passed by them on the 29th October last, to the effect that it is expedient for the Corporation of Manchester to contribute to promote in the first session of Parliament a Bill for the purpose of incorporating a company, and empowering such company to construct a canal from the river Mersey to the city of Manchester, navigable for ocean steamers and ships of large burden, and for that purpose to apply the funds under the control of the Town Council, or such portion thereof as may be necessary, to the payment of the costs attending the same, not exceeding a sum produced by a rate of twopence in the pound upon the annual rateable value of the city.

I am, Sir,
Your obedient servant,
JOHN T. HIBBERT.

During the sessions 1883-84 the Ship Canal promoters had silenced the opposition of the Bridgewater Navigation Company by offering to buy their property, but stimulated no doubt by the many suggestions of a barge canal, the Directors of that undertaking at an Extraordinary General Meeting held in November, 1884, suggested that instead of selling their concern to the Ship Canal, they should spend £324,000 in improving the river Mersey, and thus make it very remunerative. The Chairman, Mr. Cropper, took much pains to prove that the project was but the revival of an old scheme, but he admitted that it must be in opposition to the Ship Canal, as both could not be carried out. That the Navigation Company were working in collusion with Mr. Woodcock and the memorialists is evident, for the latter pleaded the Bridgewater Navigation Company's action would make the canal unnecessary.

The plans and books of reference for the 1885 Bill were deposited in the middle of November, 1884. They differed from those of the previous year inasmuch as training walls in the river were dispensed with; the channel commenced at Eastham, and was carried through land skirting the Cheshire shore. The docks designed to be made on the race-course were given up, and docks for coasting ships placed in Manchester on the site of the Pomona Gardens. A list of advantages put forth included the carrying of main sewers alongside the canal, and the widening and deepening of the river to prevent floods.

The reply of the Home Secretary to the memorialists of Manchester, and the fact that the result of the Warrington poll showed 6,355 in favour of a twopenny rate and only 515 against, choked off the Salford memorial against contributing, which contained the names of some very influential men, among the rest Messrs. Charles Heywood, Elkanah Armitage, George W. Agnew, Andrew Knowles, Nathaniel Shelmerdine and others. A counter-memorial in favour of the rate was presented by Mr. J. G. Groves, but it was felt the decision at Manchester disposed of the case.

The action of the Bridgewater Navigation Company stirred up the Liverpool Press. The *Daily Post* thought :—

Manchester people are at length awakening to the fact that the Ship Canal scheme is far too colossal a project to have any practical chance of being carried into effect. They had better look at the state of the river nearest home before finding fault with Liverpool, for they had said very unkind things about Liverpool not removing the bar, and not improving the condition of the river.

Sir Edward Watkin had recently written a scathing article to the *Times* saying that on his return from America he was kept hours outside Liverpool because there were only 8 feet of water on the bar; also that "Liverpool is a place where the dogma of absolute perfection is accepted as a religion," but, that if a change did not take place, Antwerp and other ports would run away with our trade. Mr. Gibbons, of Liverpool, in another local paper said:—

Talk about the so-called conservatism of the House of Lords! Why they are en-lightened liberal reformers compared with the studied conservatism in some important matters in the administration of our Dock Board.

Mr. Falk, of Liverpool, also wrote:—

Where there is no will there is no way. In view of my own experience, I should advise any sanguine inventors who enter the dock offices to bear in mind Dante's instruction over another place, "He who enters here must leave all hope behind him".

Criticisms and suggestions from Mr. Russel Aitken, Mr. Harold Littledale and others, following on Captain Eads' evidence before the Parliamentary Com-mittee, no doubt created a public feeling that ended in the removal of the bar. Admiral Spratt, the acting Conservator of the Mersey, in his yearly report gave it as his opinion that the deepening of the bar would be a comparatively economical work, and said he could see no reason to oppose the alternative plan of taking the Ship Canal from Runcorn close to the Cheshire side of the estuary, and he believed it would be both feasible and less costly in the end than to take it through the estuary. Nevertheless, directly the particulars of the new route became known, instead of mollifying the Liverpool opponents, they seemed to become more bitter in their opposition, and Sir William Forwood, who had been most definite in his promises, appeared glad to find an excuse for not keeping them.

The Ship Canal Bill for the coming session was duly deposited on 16th December, 1884.

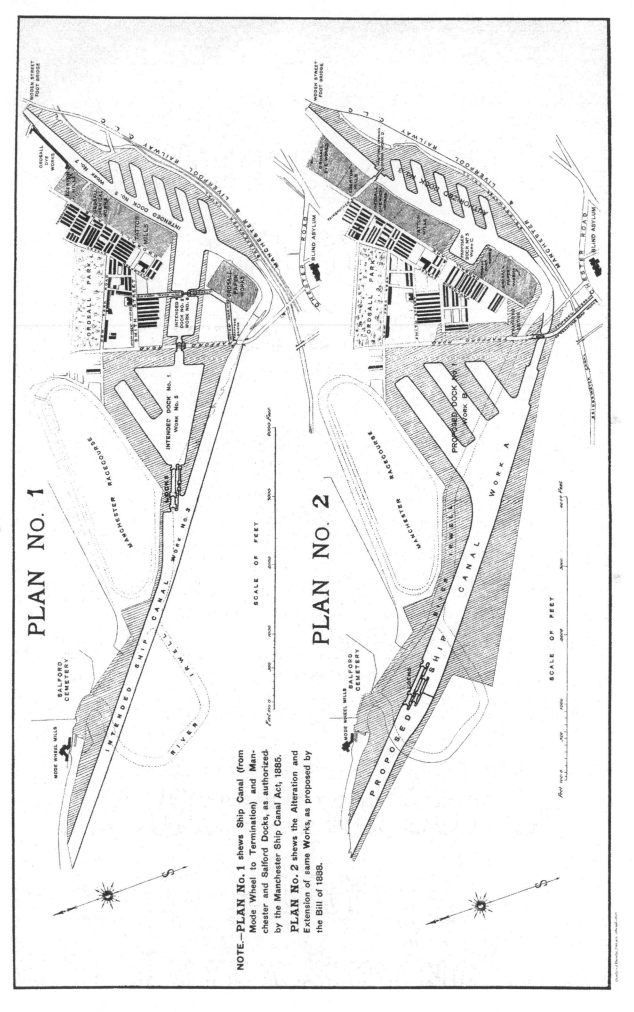

NO. 9.

PLAN No. 1

PLAN No. 2

NOTE.—PLAN No. 1 shews Ship Canal (from Mode Wheel to Termination) and Manchester and Salford Docks, as authorized by the Manchester Ship Canal Act, 1885.

PLAN No. 2 shews the Alteration and Extension of same Works, as proposed by the Bill of 1888.

CHAPTER XI.

1884.

HISTORY OF THE PARLIAMENTARY BILL IN THE LORDS— COUNSEL'S SPEECHES—EVIDENCE PRO AND CON—CAPTAIN EADS OF NEW ORLEANS—THE DUKE OF RICHMOND NO FRIEND—INCIDENTS IN THE HOUSE—BILL TO PROCEED.

There can be no question, of course, of the great advantage of cheap carriage in these times. It is almost as essential as quick transit. If the canal were made, merchants would be able to put their goods on board at their own quays with much less risk of damage, and greatly decreased cost of carriage.—*Daily News*, 19th October, 1882.

THE Ship Canal Bill was introduced to the Committee of the House of Lords on 11th March, 1884. The Committee consisted of the Duke of Richmond (Chairman), Lords Norton, Shute, Lovat and Dunraven.

Counsel for the promoters—Messrs. Pember, Michael, Balfour Browne, Cripps, Nash and Sington. With five counsel engaged, there were times during the 1883 session when not one of them was at liberty to conduct the case. This year the Chairman made up his mind not to be in the same fix, so he determined to take up two local barristers who would always be present, watching the case. One of them, Mr. Tom Nash, who had been very helpful at Manchester, was an effective speaker, and had delighted the Chairman at one of the meetings by an onslaught he made on the capitalists, who were then severely holding aloof, as the speaker said, "to see how the cat jumped". He characterised them as a cowardly lot.

In London the juniors were unused to the work, and one of them made a great mistake by asking a witness, "Do you know any reason why Manchester should not be a port of call?" Every one was aghast at the question, and Mr. Pember turning round was very cross and used strong language.

Inasmuch as the Bill was to a great extent on similar lines to the one of the previous year, it is intended here as far as possible to avoid the duplication of evidence and of counsels' speeches, new evidence and new matter only being dealt with.

There were twenty-six petitioners against the Bill, eight of whom appeared by counsel at the opening of the case.

Mr. Pember, Q.C., said except for trivial matters the scheme was the same as the one presented in 1883.   Last year the Bill had been in jeopardy, because plans and details of the estuarial works had not been presented.   On the present plans the works were shown in blue.   Last year the promoters believed the consent of the Mersey Commissioners was sufficient, now it was proposed to take Parliamentary powers.   After describing the main features of the canal, he stated that the Ship Canal was the same depth as the Suez Canal, and nearly double the bottom width. Also that it would accommodate all the cotton ships afloat, and all others except a few passenger steamers.   He then explained how floods would be prevented, and mentioned that it was proposed this year to carry the Warrington and Stockport line over the canal instead of in a tunnel as previously arranged.   There was an alteration too in the estimates ; they now came to £10,000,000, of which £6,904,186 was for works.   The former sum included interest during construction, and the purchase of the Bridgewater Navigation properties.

They proposed with the Mersey to follow the example of the great industrial rivers of Great Britain, the Tyne, the Clyde and the Tees.   Ten miles only out of a length of 40 miles, and 700 feet, out of a breadth of 3 miles, were affected in the estuary, and only one-fiftieth part of the cubical capacity of the estuary would be dealt with.   It was a disgrace to a community like that of Lancashire that the estuary should remain in its present neglected condition.   Liverpool was too timid to take the matter up—she was secure in her monopoly ; and the ports above her had to be content with tortuous, dangerous and shifting channels.

It was time commercial England stirred up, and utilised and extended her waterways.   Continental nations were going ahead and profiting by cheap carriage, and if the Dutch could spend £4,000,000 on canals, etc., at both Rotterdam and Amsterdam, surely Manchester and Lancashire ought to spend £8,000,000 or £10,000,000 to improve their main water avenue.

Within 45 miles of Manchester there was a population of 7,000,000 ; and whilst the population of England and Wales in the last decade had increased 14 per cent. that of Lancashire had increased 22 per cent.

Mr. Pember then showed numerous cases of Manchester freightage rates being dearer than elsewhere in England, and put in tables to prove this. He also gave many instances of continental rates. For example, cotton from Rotterdam to Mulhouse, 432 miles, cost 25s. 6d. per ton, whilst for 40 miles out from Liverpool the cost was 18s. 4d. Hamburg to Elberfeld, 236 miles, cost 14s. for goods, whilst 40 miles out from Liverpool it was 18s. "Can you wonder, then," said he, "that we are beaten by competition?" Cotton from Liverpool to Oldham used to be 6s. 6d. per ton, now it is 11s. The enormous increase in the capital account of railways, and the increase of expenditure, made it impossible for them adequately to reduce their rates and give the necessary relief to trade, and it was only cheap water carriage that could give it. The tonnage of ships entering Liverpool had increased 77 per cent. in twenty years. The port was crowded out now, and if a proportionate increase went on, a new port on the Mersey was an absolute necessity. There was no fear of the capacity of the canal; it was broader and as deep as the Suez Canal. "If they can take a ship of 5,000 tons, so can we." The tonnage of goods in and out of Liverpool and Birkenhead, was at least 15,000,000 tons, and he claimed that one-third of it must naturally adopt Manchester as its most convenient port. Mr. Pember then dealt with the petitions against the Bill.

First that of the *Mersey Docks and Harbour Board*, who claimed they had expended over £16,000,000 on their estate, and that if the canal were made the passage of ocean-going steamers and ships of large burden would be impracticable, through the silting up of the channel. They also asserted that damage would be done to the Vyrnwy line of water-pipes, that the estimates were insufficient, and they objected to the payment of interest out of capital during construction. Mr. Pember denied any intention to damage the dock estate; he said the scheme was intended to protect the trade investments of Manchester and district, amounting to at least £100,000,000. If £16,000,000 had been spent at Liverpool, Manchester could surely find the capital for the canal. The opposition was really prompted by commercial fear and jealousy, and the idea that if Manchester became a rival port the trade would go up the river. As regards a trust the clause was intended to prevent railways ever getting possession of the canal, but the public must have gone to sleep if ever they allowed this to take place.

Mr. Pember then referred to the petition of the *London and North-Western Railway Company*, who in addition to the objections of last year said: "The scheme would wholly fail in affecting the object of making Manchester a port". They also denied "that the scheme could ever succeed as a financial speculation, nor did they believe that even, if authorised, funds would be forthcoming for its execution ".

As regarded the petition of *Sir Humphrey de Trafford*, who complained of the large area of polluted water, Mr. Pember pointed out that the present river was nearer to him than the proposed canal would be, and that he had the Salford sewage works close by.

Mr. Daniel Adamson, the first witness, said that since the rejection of the Bill in 1883, there had been a large and enthusiastic meeting, in favour of the renewed Bill, that overflowed the large room in the Free Trade Hall. He estimated 7,000 people came to support the Bill, and in all that number only one hand was held up against it. The Chamber of Commerce, too, in the largest meeting of that body ever held, was similarly unanimous. They had also appointed four of their members to give evidence before Parliamentary Committees. The Manchester Corporation had voted £10,000 and the Salford Corporation 1d. in the pound, equal to £3,000, in aid of the cost of a Parliamentary Bill. The Manchester Guardian Society for the Protection of Trade had voted £150, and the London Cotton Brokers' Association had passed a resolution in favour of the Bill. Defeat had only aroused the enthusiasm of the people, and a very great number of meetings had been held, not only in Manchester but all over the county, and they were in the heartiest sympathy with the movement. In his opinion the many millions wasted at Birkenhead, and the £1,500,000 paid to Liverpool for the dock dues, must tend to make Liverpool a dear port. He then put in a large number of statistical tables showing amongst other facts that for short distances the Liverpool railway rates were dearer than the average rates to other English towns—on grain to the extent of 47.95 per cent. and on timber 55.73 per cent. That whilst the dock dues on cotton at Liverpool were 3s. 6d. per ton, at Bristol they were 2s. and at Hull 6d. That the charges from the ship to the railway truck were 8s. at Liverpool, 3s. 4d. at Glasgow, and 2s. 6d. at Hull. That whilst the railway rates for cotton from Liverpool to Lancashire towns were 3.1d. per ton per mile, the charge in other parts of the kingdom quoted by railway companies averaged 1.67d. per ton. Also that at the same time that 2.53d. per ton per mile was the rate from Liverpool, the average rate in other parts of the kingdom was 1.17d. per ton. The witness was quite sure that about 95 per cent. of the vessels coming to Liverpool could safely use the Ship Canal. He believed the increase of population and trade going on would soon block the present means of transit to Liverpool, and that the canal would relieve the over-pressed lines by carrying the heavy goods, which were the least remunerative to railways. Since the Ship Canal came to the front the Dock Board had reduced the Liverpool charges on cotton 14 per cent., a saving to the trade of £18,693 per annum.

In cross-examination, *Mr. Aspinall, Q.C.*, took occasion to question the witness about what he called his "popular oratory," when at a meeting he criticised one of the Parliamentary Chairmen of the preceding year as "a most fractious, disturbing and interrupting Chairman," and alluded to his attack on Mr. Pope, referred to in Chapter VII., about pricking the bubble which would eventually burst and overwhelm him. As to the first charge, Mr. Adamson replied that he believed he was speaking the truth, and about the Q.C. that it was retribution for the counsel's treatment when cross-examining him; indeed, it was the only sentence he wrote out when he made the speech.

Mr. Reuben Spencer (of Messrs. Rylands & Co., Limited) complained that whilst the present freight from Liverpool to Manchester was 3d. per ton per mile, the cost of bringing the same goods from Hull to Leeds was 1¾d. per ton. *Mr. Edward Walmsley*, of Stockport, said the carrying ring was so oppressive that they (the railways and canals) settled the rates between them, and though the canal flowed by his door, cotton was actually taken past to the terminal warehouse, and he had to pay cartage.

Following on many commercial witnesses came *Alderman Hopkinson*, ex-Mayor of Manchester, who hoped to see a main sewer along the bank, constructed *pari passu* with the canal, which should convey the sewage of Manchester to tidal waters. This gentleman created much amusement by declaring that 11,000 was the number of water-closets *put into his mouth*, and this was accentuated by the witness being unconscious of having said anything out of the way—he meant the figures had been supplied to him.

The next witness, *Sir Joseph Heron*, Town-Clerk of Manchester, had so often crossed swords with Liverpool before about the management of the docks, that it was expected his well-known ability as a witness would be put to the test. He claimed that the anxiety of the leading merchants of Liverpool to get a plateway to Manchester at the cost of £7,000,000 in order to cheapen carriage, was the admission of a grievance that required rectifying. His answer to the suggestion that the Ship Canal would soon join the band of monopolists was, that the reduced rates fixed by the Bill would secure the public from loss; besides, there was a clause in reference to a public trust.   From public reports it appeared there was a constant crying out for more dock accommodation in Liverpool, and he was told the Dock Board's debt (in 1857, £6,000,000), had now become nearly £18,000,000.   He considered the dock dues now levied were an unnecessary and unjustifiable burden on trade.   Of old the Liverpool Corporation had levied town dues on every article that went in or out of the port, which in 1857 amounted to £180,000 per year, of which not a sixpence was spent on the port, but applied for the benefit of the borough.   He had in hand the carrying through of the 1857 Act which dealt with this abuse.   The Dock Board were to pay the Corporation £1,500,000 in lieu of the town dues, and it was understood as soon as the Dock Board had repaid themselves the town dues were to cease, as they had done at Garston under similar circumstances.   Instead, they still remained, and a sum of £300,000 per year was now improperly collected from the goods passing through Liverpool.   He repudiated the suggestion that Manchester had compelled the Dock Board to buy the Birkenhead Docks, or were in any way responsible for £6,000,000 wastefully spent there, and he quoted the speech of Mr. Moon (Chairman of the London and North-Western Railway) to the Dock Board: "Gentlemen, do not forget that you are the representatives of the dearest docks in Great Britain, with the exception of London".   He objected strongly to the system of election to the Dock Board, and when he considered the wealth of the district, he felt sure the capital for the Ship Canal would be subscribed.

In cross-examination, Mr. Aspinall pressed the witness hard as to the probability of the capital being found, and went so far as to put the pertinent but inconvenient question, "Have you any intention yourself of putting any money into this?"   Without the least perturbation, and in the coolest manner possible, Sir Joseph Heron replied by another question, "Did you ever know a Town Clerk who had money to invest?"   This clever reply caused a burst of laughter, in which the Committee joined, and the examining counsel dropped the question with the remark, "Pointing at one Town Clerk, I should anticipate it very strongly".   Counsel tried hard to justify the continuance of town dues, and to prove the Birkenhead Docks had been thrust on Liverpool by the action of Manchester, but Sir Joseph could not be shaken.

After hearing the Corporation witnesses, one of them (*Alderman Thompson*), speaking of the burdens on local trade, said:—

The people have been something like the old man of the sea, they have a burden upon their shoulders that they cannot shake off, and it has been in vain for them to appeal to the railway companies and others.

Mr. Joseph Leigh, of Stockport, denied the statement that he bought the bulk of his cotton through the firm of which Sir William Forwood was a member.

Then followed the commercial witnesses.

Mr. Wilson (Wilson, Latham & Co.) said it cost as much to carry goods from Manchester to Liverpool as from Liverpool to Bombay, *via* the Suez Canal. The freight from Manchester to Bombay, including 10 per cent. primage, being 19s. 3d. per ton.

Mr. Gustav Behrens estimated that 77 per cent. of the whole through rate from Manchester to Bombay was swallowed up by carriage to the port and port charges; he bore witness, too, that no extra charge was made by long-distance ships. The freights to Hong Kong and Shanghai were the same, though it took four days longer to reach the latter place. The rates were the same to Genoa, Leghorn, Naples and Palermo, though the distances vary considerably. He had no doubt in time the freights to Liverpool and Manchester would be identical. By having a man to look to his own shipping charges he hoped to save £500 per year, and there would be less damage done to the packages.

Mr. Joseph Lieben (of Nathan & Sons) estimated the Ship Canal would save his firm £1,000 per year.

Sir Joseph Lee (of Tootal, Broadhurst & Lee), after stating that in 1877 the Chamber of Commerce passed a resolution in favour of a Ship Canal, said that the saving at their Bolton Mills would be 4s. per ton, and that they were now paying 25s. per ton railway freight to London on goods not requiring speed, whilst a coasting steamer could well carry them at 10s. per ton. The witness confessed in cross-examination that up to that date he had given no monetary support to the scheme.

Mr. George Woods, representing the Warrington wire trade, pointed out that Germany was fast beating England out of the foreign markets in consequence of the burdens and difficulties they had to contend with. Australia was a great market for wire. Germany could deliver wire in London for shipment at 10s. 6d. per ton, whilst the cost from Warrington to London was 17s. 6d. The position now was that, unless a special brand be specified, England had no chance against Germany in the Australian market. There was no extra freight charged to Natal over Cape Town, though there was a difference of some hundreds of miles.   To make the canal would be a great godsend to Lancashire, and to the wire interest in particular.   He believed that the wool market would eventually come to Manchester ; he understood wool was handicapped to the extent of £3 per ton in coming 200 miles from London, whilst the average distance from Manchester of the wool-using towns was 38 miles.   The land that grows the wool is fenced with wire, and there ought to be an advantageous interchange.

Mr. Bowes (of Barningham & Co., Pendleton) said it cost 7s. 6d. per ton to take railway tie bars between 30 and 40 miles to Birkenhead, and 12s. 6d. thence to Calcutta, 8,500 miles, and the burdens on the iron trade had driven it from the inland towns to the seaports.   By the canal scale of charges it would cost 6s. 11d. per ton from Ardrossan to Manchester, against the present charge of 12s. 1d. by railway.   Belgian manufacturers could send iron to London at a lower rate than his firm was charged from Salford to London.   A great deal of iron work was now exported from Antwerp to the East Indies and Australia through the cheapness of continental rates.   There the railways largely belong to the Government, who foster the various trades.   This enables the Belgian iron girders to undersell British productions.   He estimated his firm would save £1,350 per year by the canal.

Other iron merchants confirmed the above statements.   One of them, *Mr. Thomas Ashbury*, of Openshaw, gave an instance of a German firm carrying off an order of £50,000 simply through their cheaper carriage.   He calculated the Ship Canal would save their firm close on £5,000 per year.

Mr. Moir Crane, an oil merchant, complained that the charge on his goods to London was 25s. per ton, whilst his rival at Liverpool could send for 16s. 8d. because there was competition by water.

Mr. Samuel Ogden, an overseer of Manchester, spoke of the terrible depression that had come over the city.   The assessments of one township increased on an average £6,000 per year up to 1879.   Thence up to 1882 they decreased on an average £900 per year.   He put in a long list of factories, workshops and warehouses that had been closed during the last few years.

Mr. Thomas Garnett, of Bradford, said within the last ten years property in his town had depreciated 40 per cent., and that if wool could be brought through Manchester there would be relief to the extent of £1 per ton.   That city was 31 miles nearer to Bradford than Liverpool.

Mr. Max Baerlein said the cost of exporting waste at Manchester would be 6s. 7d. against 13s. now charged at Liverpool. Besides avoiding damage, the charge for exporting machinery would be 5s. 7d. in place of 12s. 8d. now paid at Liverpool.

Mr. Fred. Render, a corn miller, estimated the canal would save him (if going full time) £7,000 per year.

Mr. William Warburton, also a corn merchant, considered the present railway charge from Liverpool to Manchester exorbitant; it came to 8s. 11d. per ton for 32 miles, whilst at Dublin the cost was 8s. for 140 miles. It cost for freight and charges 39s. per ton to bring wheat from Bombay to Manchester. A saving of 10s. per ton would be made if the ship came direct to Manchester. He believed the saving in corn and grain brought to Manchester by water would be £90,000 per annum.

Mr. James Platt (Platt Brothers, Oldham) stated that the canal would effect a saving to his firm of £7,000 to £8,000 per year on his machinery exports and timber imports at Liverpool alone. In addition they would make a considerable saving on goods passing through eastern and other ports. The Oldham Council were quite willing to support the canal up to 1d. per £1 on their rental.

Mr. W. A. Nicholls complained of existing burdens; it cost as much to bring apples from Faversham in Kent (230 miles) as it did from New York to Liverpool (3,000 miles). The cost of fruit from Liverpool to Manchester was 17s. 1d. per ton. By the canal (if no extra sea carriage) it would be 8s. 2d., or a saving of 53 per cent. Coming through Liverpool, there were ten different handlings of the goods, causing much damage. If sent by Ship Canal there would be four handlings. Potatoes from the Channel Islands, *via* Southampton, cost 40s. per ton; they could come by canal for 24s., or a saving of 40 per cent. If from Cornwall it would amount to 50 per cent.

Mr. Harrison, another fruit merchant, said 70,000 tons of potatoes came into Manchester from Scotland and Ireland, and a few from Yorkshire. If these came by canal there ought to be a saving of £10,000 per year.

Mr. Capper, to show existing inequalities, said that fish coming from Ireland cost £4 10s. for carriage to Liverpool, whilst to Manchester the cost would be £7 per ton.

Mr. George Brown, shipowner, Glasgow, had no doubt of the freights from abroad being the same to Liverpool and Manchester. They were the same to

Glasgow as to Greenock, and ships from San Francisco sent to Falmouth for orders charged no more to Leith than to Liverpool.

In re-examination, Mr. Balfour Browne asked witness, who spoke with a strong Scotch accent, what was the distance between Glasgow and Ardrossan "as the crow flies?" This puzzled him, but at last he replied, "but ships dinna gang that way". When the witness came out of the box he asked why the people had laughed? He had not realised what a quaint reply he had made.

Mr. Wharton (of the Salford Cattle Market) estimated a saving by coming direct to Manchester of 2s. per head in cattle, and 7½d. in sheep; besides, they would be in a better condition.

Mr. Marshall Stevens repeated his previous evidence, and was sanguine of the canal carrying 5,000,000 tons per year, or one-third of Liverpool's present business.

Several gentlemen spoke of the benefit to the Lancashire, Yorkshire and Midland coal-fields, and of the advantage to them of a new and convenient outlet for coal, such as would be provided by coal-tips at Partington.

Mr. Leader Williams, after giving a full description of the canal works, went on to explain the changes on the previous year's scheme. The Warrington Corporation were now friends instead of opponents. Their difficulty as to saline water had been overcome; the river was cut off by a dam so that it stopped at Latchford Locks, and no tidal water could get into the river. By a change of position better gradients were given to the railways, and where the depth of the canal was 24 feet it had been increased to 26 feet. Many of the slopes had been flattened, and this year they had provided ample land for spoil. In reply to the Chairman, witness said he understood Manchester contemplated a culvert for sewage alongside the canal, and either taking it to the tidal waters or else pumping it on to Carrington Moss.

The engineers of the Clyde, Tyne and Tees, as well as Mr. Abernethy, repeated their evidence of last year, approving the scheme, and declaring emphatically there would be neither reclamation of land nor loss of capacity in the estuary. Following came a whole bevy of shipowners from Glasgow, Cardiff and London, who were quite sure the same freight and insurance would be charged to Manchester as to Liverpool, and that when ships came with cargo to Falmouth and other ports for orders, Manchester would not be excepted from the charter party.

Mr. Croft (of the Cork Steamship Company) said he had been sending ships up the Amsterdam, Rotterdam and Ghent Canals, of sizes varying from 750 to 3,500 tons without any difficulty.

Mr. Adam Stott, J.P., Flixton, estimated the damage done by floods to the valley of the Mersey in the year 1872 at £80,000, and floods of less magnitude were of constant recurrence. He believed the effect of the canal would be to do away with them, and explained that at the locks there was a fall of nearly 18 feet, and that this drop would prevent backing up, especially at Irlam Locks, close to which the Irwell and Mersey joined. To his mind they would never in future be afraid of a flood.

Mr. Alfred Hughes stated that when railways came to the front, there was fierce competition with carriers by water, and in 1849 the rate to Liverpool by water came down to 2s. 6d. per ton, but when the railways and water carriers entered into a conference, the rate went up to 8s. 4d. by water and 10s. by rail. Shortly afterwards the rate became uniform at 10s.

Mr. Charles Ross, yarn agent, felt sure there would be sufficient cargo to fill daily one 800-ton steamer either way, to and from Glasgow, and where they were now paying 25s. per ton by rail, and 16s. 8d. *via* Liverpool, he believed the charge would not exceed 12s. per ton by Ship Canal.

Mr. Lionel Wells, engineer, speaking of the Runcorn Bridge, showed that the piers of the bridge were driven over 40 feet deep in the rock to provide for deepening the channel at some future time.

Mr. Jacobs, borough surveyor of Salford, believed the canal would prevent a recurrence of flooding both in his borough and from Throstle Nest downwards, and that deeper water would lessen the liability to smell. Various chemists were called to prove there would be a reduction of existing nuisances, and that there would be no danger from secondary decomposition.

The Writer of this History was the next witness. The reasons for his being called, and the episodes connected therewith, are related in a previous chapter. He deposed to the enthusiastic support given to the scheme; he had promises to take up shares to the extent of about £130,000, producing letters in support of this statement. When the first was handed in, *viz.*, one from Mr. John Rylands, for £50,000, objection was at once taken on the ground that he should have come in person. Mr. Pember replied that old age prevented him, and the witness was his accredited agent. Upon this the Committee clustered round the Chairman, and there was an animated and almost angry conversation for five minutes. The Chairman was determined to stop the evidence, but his colleagues out-voted him. The witness heard one of the noble Lords say :—

But they have brought this upon themselves; all along they have been saying that no men of capital are willing to support the scheme, and now when we have a person who can give us definite information we ought to hear what he has to say.

The Chairman in no pleasant tone replied :—

Against my judgment my colleagues feel rather inclined to admit it.   I thought myself it was too irregular.   I do not think it so irregular as not to admit it, and my colleagues think it is better possibly to hear something about it.

The Chairman :—

Q.—(To the witness.)   Before we allow you to read the letter, I must ask you, is it in the shape of an authoritative document to you?

A.—It is.   I was coming to London, and I wished to be fortified with some authority for the statements I made, and I got two letters from two gentlemen who——

Q.—Does this bind this gentleman, so far as he is capable of being bound, to subscribe whatever he may specify in this letter?

A.—It expresses his intention.

Q.—Let me look at the letter.   Did you see him sign it?

A.—Yes.

Q.—Did you write the body of the letter?

A.—No.   His clerk wrote it for him.

Q.—And he signed it?

A.—Yes.

You may put it in.   Perhaps you had better read it.   The witness then read it.

NEW HIGH STREET,
MANCHESTER, *March* 29, 1884.

DEAR SIR,

I authorise you to state to the Lords Committee on the Manchester Ship Canal Bill, that I am willing, in case the Bill is passed and the company formed, to subscribe for at least £50,000 of shares.

I am,

Yours truly,

John Rylands

On an attempt being made to put in a letter from Mr. George Benton, also promising £50,000, the Chairman declined to receive it on the ground that Mr. Benton could have come himself, but the witness succeeded in making this and other promises known to the Committee.   Asked by the opposing counsel :—

How long will it take the canal to get fairly to work. Ten years?

The reply was :—

It will pay in a much shorter time, but traders through cheapened carriage will benefit at once.

Q.—When you say, as a gentleman in business and a promoter, that you are going to take shares, do you mean that you will do it from your enthusiasm for the scheme, or as an investment?

A.—Jointly, on both scores.

Q.—You have thought about it a great deal?

A.—I believe, first of all for the benefit of the district, and secondly for investment. The benefit to the district will come immediately, and the investment will come shortly afterwards.

Q.—Your first reason is patriotism, and your second profit?

The witness replied that his patriotism would lead at once to remuneration, because business men will at once reap a substantial advantage by the canal being made.

Mr. Marshall Stevens originally estimated Manchester would get 5,000,000 tons of the Liverpool traffic, but thought with additions from collateral sources and the increase out of succeeding years, there was a probability of 9,000,000 tons. He put in tables showing a traffic of 9,650,850 tons, a revenue of £1,491,505, and a reserve revenue of £187,500, if and when ship dues were charged as at Liverpool.

Mr. George Hicks, insurance agent, was sure that, as on the Suez Canal, no extra premium would be charged. He was one of the people who first brought the question of a Ship Canal before the Chamber of Commerce in 1877, when favourable resolutions were passed. He believed in Mr. Marshall Stevens' figures on the ground that two tons per head of the population represented the foreign trade of the country. When last year's Committee reported, " If the scheme could be carried out with due regard to existing interests, the Manchester Ship Canal would afford valuable facilities to the trade of Lancashire, and ought to be sanctioned," the Chairman remarked, "That is the strongest preamble I ever heard put into a Bill."

Mr. Jacob Bright, M.P., put the position in a nutshell : "We do not seek in the slightest degree to injure any other interest, to injure Liverpool, or to injure the railways. We seek to benefit ourselves; if in benefiting ourselves we injure these other interests, we regret it, but we cannot help it." He believed the list of supporters of the canal which he put in was such a list as probably had never been seen in Manchester in connection with one single object. Mr. Pope, in cross-examination, tried hard to make capital out of Mr. Armitage's qualified evidence of last year, but failed, the witness replying, "I am happy to say the doubts in Mr. Armitage's mind have been removed; the moneyed classes of Manchester have got riper on this question, and Mr. Armitage is one of them". The witness then put in the list of subscribers of over £50 to the Parliamentary Expenses Fund, *viz.* :—

31 subscribers of £2,000 and not less than £500 each.

28	,,	,,	500	,,	,,	250	,,
154	,,	,,	250	,,	,,	100	,,
176	,,	,,	100	,,	,,	50	,,

Mr. John Slagg, M.P., having sat on a Parliamentary Committee to inquire into the question of canals, said England did not keep pace with the Continent or America. France alone had voted £40,000,000 sterling for the improvement of inland navigation. He had always spoken warmly in support of the scheme, and since last year had given a subscription as a manifestation of his interest. He believed if the Committee passed the Bill, the capital would be found. He put in tables to show that as regards cotton, machinery, timber and wool, the average cost of carriage in Lancashire was more than twice as much as that on the Continent.

Mr. Arnold, M.P., advocated the scheme as a prevention of flooding in Salford. He also said the great local wants of Manchester were cheap food and cheap transit.

Mr. A. M. Dunlop, land agent, then put in his revised land estimates, including the race-course, the total being £1,168,003, and these were confirmed by other valuers.

Mr. Aspinall then addressed the Committee on behalf of the Dock Board. He gave a history of the Liverpool Docks and of the struggle about dock dues, claiming they had been kept in existence because they were non-ratable, whereas, if revenue was raised by dock rates, such would have to contribute to the city rates to the extent of £20,000 per year. He claimed that no competitive scheme should be passed that would affect the bondholders of the Mersey Docks and Harbour Board, who held securities to the extent of about £16,000,000, which, he said, existed by the authority of Parliament, asked for by Manchester. He contended that city had forced Liverpool to buy the Birkenhead Docks and made it a dear port, and now they wanted to be relieved from it. In respect to master porterage, he said the system was a safeguard and convenience to shipowners and buyers. He admitted the railway freights were unjust and excessive, but said Manchester had never exerted herself to get them reduced. He denied the contemplated saving could be made, even if ships did come up the canal, and pointed out the length of time that must elapse before it could be remunerative. He ridiculed Mr. Marshall Stevens' idea that rice and many other similar articles on his list would ever find a market in Manchester. He doubted the financial power of the promoters to carry out the work, and pictured the disaster if they broke down in the middle of it. Speaking of the Lancashire people, he said they would be lunatics to find these millions of money, unless they did it by way of public subscription and for the sake of patriotism.

He then dealt with the engineering difficulties, and attacked the estimates on the lines of previous inquiries. He called *Mr. A. T. Squarey*, solicitor to the Mersey Dock Board, who, after giving a history of the creation and working of the Dock Trust, stated the total debt was now £16,322,000. In addition to the ordinary dock receipts, there was a conservancy fund for lighthouses, etc., to which all vessels using the Mersey contributed. Out of £46,609 collected last year, £3,306 came from ships using the upper Mersey ports. In 1880 the Board made reductions in their dues amounting to £124,240 per annum, and in 1884 a further remission on cotton of £29,707 a year. Some time back the Board of Trade fixed the rates for master porterage, but on the representation of the steam shipping trade that they were too low, 10 per cent. was added. The majority of the master porters were themselves mercantile men and shipowners. In cross-examination, the witness admitted the liability of town dues to rates was an undecided question, and that the receipts at Birkenhead were £89,000 on an expenditure of £6,000,000. Further, that the opposition to town dues was led by Liverpool merchants, who also opposed the system of master porterage. He did not deny that in 1880 before a Committee in the Lords he gave evidence as follows :—

I know over and over again we have had at the Board instances where the master porterage of a vessel has been sold by the man who was entitled to it to another man, on the consideration of the payment of money—that means, there is a profit in it.

Dock and town dues were collected for the maintenance and improvement of the port, as expressly stated in the Act of Parliament in the words—

To remove any article or thing being in their judgment an obstruction or impediment to such use or navigation in any dock, sea channel or elsewhere.

Witness was asked :—

Q.—Is that a clause which you say is compulsory?

A.—Upon the opinion of Justices Quain and Day it is compulsory, and not optional.

Q.—Have you expended any money whatever in improving the estuary of the Mersey?

A.—No. I do not think any money has been spent upon other improvements of the upper estuary.

Q.—Or upon any work whatever in order to improve the estuary itself?

A.—No; and that has been advisedly because it was thought that the estuary was best allowed to remain in its present condition.

Q.—Has the desirability of works for the improvement of the estuary been discussed at the Board?

A.—It has been thought Pluckington Bank might be removed, but the opinion was that if you removed it no one could tell where it would go, and it was better to leave it alone.

Q.—Have any reports been received as to the necessity and possibility of improving the upper estuary?

A.—None. No application has been made to any engineer, nor report received, till quite recently from Mr. Eads.

Q.—Has any action been taken as to the desirability of removing the bar?

A.—None. No report received from any engineer, or action taken, except from Mr. Eads.

Q.—Did the Board consult any eminent engineer with respect to the bar or any works rendered necessary.

A.—I do not think they consulted any one but their own officers.

This evidence shows that though Liverpool had been receiving dues applicable to the removal of the bar she practically had done nothing, the bar remaining all the time a hindrance to the harbour and a peril to shipping.

Captain Eads, on the representation of counsel that he must return to America, was allowed to interpolate his evidence. This witness got perhaps the largest retaining fee ever paid, the Liverpool Dock Board giving him £4,000 to come to England. He said he was an American, but had been fourteen years a member of the London Institute of Civil Engineers. He had been identified with the Mississippi for forty years, and during that time till he took it in hand the river was obstructed by a bar at its mouth, which gradually grew worse, and was a great detriment to shipping. By means of the jetty system he had dealt with one of the mouths, and increased the depth from 8 feet to 30 feet, and this was maintained. There was considerable analogy between that river and the Mersey. The Mississippi brought down large quantities of heavy silt, and spreading out its waters at the mouth like the veins of a fan lost its force, and, in consequence, the heavy matter settled and formed a bar a distance out to sea. Across the bar were three places deeper than the rest, through which ships passed at high water. He was allowed to operate on one of these. By means of wire-sunken grids with weights he caught and utilised the detritus, consisting of tree-roots, branches, etc., thus forming piers about 2 miles long on either side of the opening. The water rushing through, by its velocity forced a deepened channel across the bar. Thus, by using the forces of nature, he had made the delta of the Mississippi navigable at a very moderate cost, and restored the trade of New Orleans. In one season's floods 16 feet of deposit had been captured by the grids. He maintained the proposed tidal channel in the Mersey would stop the fretting away of the banks so necessary to maintain the capacity of the estuary, that in consequence there would be less water to carry away detritus to sea, and that it would settle on Pluckington Bank and the bar, and probably silt up the docks.

Cross-examined by *Mr. Pember*, the witness admitted the drainage area of the Mersey was a pigmy compared with the Mississippi, and no comparison between the two could in any respect hold good. Also that the character of the silt in suspension at ebb and flood tide, as demonstrated by the examinations of Dr. Burghardt, varied from that of the Mississippi, and if correct would upset his theory. He did not know of any immediate effects of the frets of the Mersey on the Liverpool bar, and when reminded that Captain Graham Hills, a Liverpool witness, had stated that the fret of 1872 had carried down 5,800,000 yards of matter, and that afterwards the bar was shoaled from 11 feet to 7 feet, the witness expressed his doubts, and said he was not prepared to admit Captain Graham Hills was right in his assumption.

Captain Eads after parrying the question a long time agreed the habits of the two rivers were entirely different, the tidal rise of the Mersey being 30 to 31 feet against 1 foot 2 inches in the American river. Mr. Pember closed his examination thus : "A man who comes here to say the Mersey is comparable with the Mississippi will say or admit anything, and I have no further questions to ask him".

Captain Graham Hills, marine surveyor to the Dock Board, described the estuary of the Mersey, and generally repeated his evidence of the 1883 inquiry. He was in the box the greatest part of three days. The question whether training walls at the mouth of the Seine, which had admittedly benefited Rouen, had, or had not damaged Havre, was a bone of contention for a whole day. The engineer of the Seine works calculated it would take 21,000 years to silt up the estuary; the witness maintained that in twenty-five years serious damage had been done.   He explained that when last year he said the great fret of 1872 in the river Mersey was coincident with the shoaling of the bar to 7 feet, he had intended to convey the idea that the process had been gradual.   It was quite true that in last year's evidence he had said the shoaling of the bar followed upon a fret in the upper reaches of the river, but that answer was not correct, because he had already said it was coincident with it.   He maintained Dr. Burghardt's conclusions on the amount of silt in the ebbing and flowing tides were based on a fallacy, because he had not calculated the volume of water in the river when he took the samples.   Speaking of ships crossing the bar, and getting into the canal on one tide, he ventured the opinion that vessels of 20 feet would be exceptional in the canal, and if any of that size came up, they would have to wait for high tide.   In cross-examination, he admitted that though he pictured all kinds of accidents through ships getting on training walls and banks, yet the Tyne, the Clyde and the Tees, with similar walls, were navigated in the dark.   Also that though he was now objecting to small encroachments, yet in the past Birkenhead and Liverpool had taken 1,000 acres of the tidal area without having any bad effect on the bar.   Further, that the Dock Board were tipping yearly an average of 500,000 cubic yards of dock dredgings in the Narrows between Prince's Wharf and Seacombe.   Now, they only tipped it there in bad weather. In reply to the Chairman, the witness said the Board had power to dredge, but the bar had not been dredged in past years, because even at a 10 feet neap tide they had 30 feet of water, and that was as much as there was in New York and other ports.   He opposed the scheme because it meant the loss of tidal water and tidal power.   The accretion caused by training walls would exclude fully half the tidal water now flowing up the river, and that meant shoaling the channel up to, and over the bar.   In reply to the noble Lords, witness said that from 1835 to 1838 Admiral Denham did not dredge but tried to harrow or rake the bar, so as to disturb the silt, but his successor gave up the attempt.   Personally, his opinion was that dredging would be a very difficult operation, and that it was decidedly better that everything to do with the bar should be left entirely to natural causes.

Mr. Thomas Stevenson came to support the evidence of Captain Graham Hills, but he was the unfortunate writer of a work on estuaries, and when he was cross-examined by Mr. Michael his book was quoted against him, and he replied, "I put in many things I am not responsible for".   Also as regards Captain Graham

Hills' tidal velocity, "I do not place the least value on that, he merely guessed it". When asked by Mr. Bidder, "I want the consequences which will follow—which you say always do follow—where the space is limited and the silt is present," witness replied: "There can be no doubt whatever in my mind that the whole of the estuary of the Mersey would be silted up and covered with grass. I cannot tell you at what date." Further asked by Mr. Aspinall what would be the effect of these walls? witness replied: "I have no doubt whatever upon the subject, that the whole of that area will be in the end accreted up to high-water spring tides".

Mr. R. N. Dale believed extra insurance would be charged for the canal, but admitted no difference was made on the Bristol Channel ports.

Mr. L. F. Vernon-Harcourt and *Mr. John Wolfe-Barry*, both said the Mersey was a most exceptional river, dependent on the tidal capacity of the estuary to maintain a channel, and prevent the bar shoaling up. They considered training walls dangerous unless carried to deep water. The latter ventured to remark that if Manchester wanted a canal, it might be made in a less objectionable way.

Mr. George F. Lyster, engineer of the Dock Board, after describing the docks and stating the amount of money spent thereon, said his objection to the Ship Canal arose from a belief that the interference with the river would involve physical results, which would go far to destroy the working conditions of the dock estate, and injure the approaches to the Mersey. It meant ruination to the docks and river. No indemnity would cover it. Manchester could not pay for the damage that would be done. The proposed slopes would not stand; they should not be laid at a less angle than $2\frac{1}{2}$ to 1. He thought the promoters were entirely wrong as to the character of the rock they expected to find. On cross-examination, he admitted that 1,400 acres of the estuary had been enclosed by Liverpool and Birkenhead, and that civil engineers in 1852 informed the Liverpool authorities that if they constructed large docks and made such encroachments upon the river the port of Liverpool would be ruined. Also that a vessel drawing 24 feet of water could not enter the new north docks in a 10 feet neap. He did not agree with Mr. Stevenson, a witness on his own side, when he said there was nothing to say against Mr. Leader Williams' design. When asked what steps had been taken to get rid of the bar, witness replied "none". Some attempts were made many years ago by raking the bar, but it was a failure—the work of one day was destroyed by the sea the next. They had not tried to blast away the bar, because it was all sand—live sand. When asked if he had tried to dredge away the Pluckington Bank, he replied:

"Never, except at the tail end, and as the sills of the adjacent docks had been fixed to suit the bank, it really did not do much harm; even sluicing it away might involve serious consequences. He estimated the cost of channel work at £2,870,675 against the estimate of £1,390,417 by the promoters.

Sir Frederick Bramwell was of opinion that training walls did no damage on the Tyne and Tees because they were carried out to the bar, nor did they damage Havre at the mouth of the Seine because there was no bar. In the latter case training walls caused accretions in the Estuary, and they would do so in the Mersey estuary, and allow less water space for flushing the bar. In cross-examination, he admitted cross walls in the Seine had helped accretion, and that such were not to be placed in the Mersey. He did not see any possibility of dredging the Liverpool bar, and he thought if Manchester wanted a canal she should take it to deep water.

Mr. T. D. Hornby, Chairman of the Liverpool Dock Board, repeated his defence of the policy of the Board, and claimed that in 1880 and 1884 they had made reductions in their charges to the amount of £154,000 per annum. In cross-examination, it was elicited that between 1855 and 1879 the rates and dues had been raised 50 per cent., and that the Liverpool Chamber of Commerce was bitterly complaining still about master porterage and other charges. When it was pointed out that the charges on sugar were at Liverpool 2s. 1d. against 10d. at Greenock and 1s. 3d. at Glasgow, the witness said it was absurd to compare such towns. He admitted that in Liverpool goods transhipped in the river paid dock dues, though they never used the docks, also that in December, 1883, Mr. Harrison, chief of the Works Committee, said "we have more trade than we can accommodate". Further, that the *City of Brussels* was run down by collision when waiting for water outside the bar, but he maintained the latter was through injudicious anchoring. He agreed with Captain Graham Hills that the bar must be left to natural causes; nothing could be done to improve it except at an enormous cost.

Admiral Grant had constantly known vessels get athwart of the Suez Canal. He admitted, however, that it was narrower at the bottom and had worse bends than any on the Ship Canal.

Mr. Dugdale then addressed the Committee on behalf of the Shropshire Union Railway and Canal Companies, and called their engineer, *Mr. G. R. Jebb*, who said that ships of 300 to 400 tons now came up the Mersey to Ellesmere Port, and that he feared any interference with the channel caused by the river Gowey.

Mr. Thomas Hales, traffic manager, said the usual size of ships trading to Ellesmere Port was 150 to 250 tons; the largest he had seen was 340 tons, but a vessel of 500 tons had discharged in the tidal basin at that port.

Mr. Thomas H. Jackson, Chairman of the Liverpool Steamship Owners' Association, thought ships could not well anchor, and that it would not be safe for vessels of any considerable size to navigate the canal, and if they did, it would require two tides to reach Manchester, and in winter the ice would be liable to damage propellers. Rather than let one of his ships with a consignment deliver at Manchester, he would pay the freight by railway from Liverpool to Manchester.

Mr. John Laird, of Birkenhead, objected to an increased number of hopper barges which might obstruct shipping on the river. By claiming the Mersey as a semi-private domain of Liverpool, he caused some astonishment; but he admitted that though Manchester might secure some trade, the general effect of the canal would be to make the Mersey a more important emporium than at present. Mr. Pember twitted witness about the attack his father had made on the management of the docks, about the heavy dues and charges, and the lack of railway accommodation; also as regards the delay of a week or ten days that sometimes occurred before large ships had a sufficiency of water to enter the Liverpool Docks. The witness's reply was: "You must leave my father's sayings to speak for themselves".

Mr. Alexander J. Hunter, partner in the firm of William Graham & Co. (the only Manchester shipper who ventured in Parliament to oppose the canal), said it would effect a saving of about 3s. 9d. per ton on goods for Bombay, equal to ¼ per cent. on the value of the goods exported, but he did not think this would have the slightest effect in altering the course of the export trade as at present constituted. His experience was that when the 5 per cent. *ad valorem* duty on Indian goods was repealed in 1882, the people engaged in the trade did not benefit, and were disappointed that the change neither developed trade nor made it more remunerative. Manchester goods would never provide a whole outward cargo, and he could not see how Manchester could provide miscellaneous goods to fill up or afford the facility of despatch equal to Liverpool and Birkenhead. It was a matter of opinion, but he did not believe the required £8,000,000 would ever be raised.

In cross-examination, witness admitted his firm shipped 25,000 to 30,000 tons of Manchester goods, which would give a saving of £5,625 per annum at 3s. 9d. per ton either to his own firm or somebody else, and this would be an important matter; but he maintained the abolition of the 5 per cent. *ad valorem* duty was no benefit whatever. "If the abolition had done good it would show itself in the increase of the exports from England of Manchester goods. It has not done so." Three years ago freights to India came down from 50s. to 22s. 6d., yet no benefit came into the pocket of the shipper, it went to the ultimate consumer in India. He felt sure more insurance and freight would be charged to Manchester than to Liverpool. In reply to a noble Lord, he said it now cost nearly as much to send goods from Manchester to Liverpool (31 miles) as from Liverpool to Bombay (6,500 miles).

Mr. Frederick Massey (George Warren & Co.), Liverpool, maintained it would take £1,200 to alter a 4,000-ton ship so as to pass under the Ship Canal bridges, and that it would need five weeks to do the work. In his opinion it would cost 30 to 50 per cent. more to do the coasting trade to Manchester than to Liverpool, and the former port could not compete with his rate of discharge, *viz.*, about 73 tons per hour. In cross-examination, he said his firm had an appropriated berth, and admitted having made many complaints to the Dock Board about the costs of cartage and other inconveniences at the Liverpool Docks.

Mr. Hoult, of Liverpool, estimated that it would take a day and a half to come to Manchester and the same to return, and that it would not be safe for ships to navigate the canal during the night.

Sir William B. Forwood was under examination and cross-examination for the greater part of two days, and as he had been a principal witness in each previous inquiry, only a brief *résumé* of his examination will be necessary. He denied any feeling of rivalry in Liverpool; there was a feeling of indifference as regarded competition, and in the scheme they failed to recognise either thinking or moneyed Manchester. Glasgow spinners had failed to create a cotton market, and so would Manchester. Liverpool had had to succumb to London in the wool trade. The cost of lowering the masts, and the dangers and difficulties of the canal must cause such a heavy freightage as to be prohibitive. Sailing vessels never could go up. He had always believed Liverpool had suffered from heavy railway rates to and from the interior, and he had taken a prominent part in getting the Railway Commission appointed, but they could get *no* assistance from Manchester and neighbouring towns, and now they wanted a most clumsy and costly method of reducing railway charges, devised by ambitious lawyers and engineers who were raising a spurious agitation to support it.

Mr. Pember in cross-examination obtained an admission that a difference of a day or two in a long journey would not necessarily increase freightage to Manchester. It was not the depressed condition of trade that induced the Ship Canal agitation. "It is chronic for Manchester people to complain of the condition of their trade—I never knew it otherwise. They do not reckon their profits till they have put away a large sum for depreciation and all sorts of things; but notwithstanding their complaints they wax rich and grow fat." Speaking of heavy brokerage charges, the witness admitted Messrs. Paton & Co., of Liverpool, were fined £50 for selling to the Oldham Cotton Buying Company without charging double brokerage, the latter firm not being members of the Liverpool Association. He attributed provisions being 5 per cent. dearer in Manchester than Liverpool to the fact that Liverpool had a better market, and that it cost more to get to Manchester.

Mr. Pember then questioned Sir William Forwood on statements made in 1881 before the Committee on Railways, when he said that Liverpool had suffered great injuries from unfair and excessive railway rates, and that thereby the Calcutta trade had been driven to London, and the grain trade to Fleetwood, Hull, etc. He had to admit the heavy charge of 43s. per ton. ocean freight was a main factor, and that when they reduced to 22s. 6d. the trade came back again. During his examination witness stated that Mr. Adamson's saving of £158,000 on goods within carting distance was incorrect, and ought to be £73,562, and that it would cost 8d. per ton per mile by cart to Oldham and Bolton. He was now confronted by his own evidence in 1881, that the maximum charge from Liverpool to Manchester by canal was 2d. per ton per mile, and as to carriage by road for a large quantity, that a teamster had offered to carry at 3d. per ton per mile. Again witness corroborated Mr. Hunter when he said 5s. per ton, or $\frac{1}{4}$ per cent., was a trifle on the value of a commodity, whilst in 1881 he had given in evidence, "a difference of 5s. upon manufactured goods is sufficient to turn the trade". Mr. Pember also reminded Sir William of his statement the same year, that if the railways carried as cheaply to and from Liverpool as they did from other ports, there would be a saving to certain trades only of £400,000 per year, thus justifying the application of Manchester to cheapen her carriage. And further, that he had said that Liverpool was no better off with railways than she was before they were constructed. Certain it was that in the skilful hands of Mr. Pember, Sir William Forwood's previous evidence placed him in a singularly embarrassing position.

After *Mr. Francis Stevenson*, engineer of the London and North-Western

Railway, had criticised the estimates, Mr. Pember read to the Committee the terms for purchase by arbitration of the Bridgewater Navigation property.

Then followed the examination of several engineers, one of whom, *Mr. A. M. Rendel*, suggested that instead of following the Mersey channel they should bring the canal round the edge of the estuary from Runcorn and avoid the channel altogether.

Mr. George Findlay, manager of the London and North-Western Railway, repeated his previous evidence that outside a 12 mile limit of Manchester no benefit could be conferred by the Ship Canal, and justified the charge of 9s. 2d. on bale traffic from Manchester to Liverpool, of which he said 4s. 8d. was the terminal cost for handling, etc., 2s. 9d. was the cost of delivering and collecting, and only 1s. 9d. went to the railway company for haulage.   In cross-examination, he admitted that Lancashire imported largely through east coast ports like Hull, Grimsby, etc., but did not think that avenue of trade would be displaced.   He did not deny that it cost more to carry iron girders from the ship side at Liverpool to Manchester than it took to carry the whole way from Antwerp to Manchester, *via* the east coast ports, or that Californian wheat came to Manchester *via* Hull.   He admitted cotton goods paid 25s. per ton to London if exported, and 40s. per ton if used there, and that it was almost as cheap to carry from Manchester to Bombay as from Manchester to London; also that goods could be sold as cheaply in Calcutta as in the Metropolis. He doubted the statement that the export rate of 25s. by railway would be affected by the proposed 10s. canal rate to London, but he admitted that in time past there had been a pooling of receipts by shippers and carriers, a certain portion of goods being attributable to London and another portion to Liverpool.   Re-examined by Mr. Pope :—

Q.—I daresay you know that Mr. Leader Williams described his system of sluices and locks and hydraulic machinery for the working of them, everything to be as perfect as it can be in the great Ship Canal : it will not be worked or maintained for nothing?

A.—No ; and granting they were absolutely correct with regard to their own estimate of the amount of traffic they were likely to get, and they were actually to get the tolls they have estimated, the cost of working would be so much greater than they have calculated that it could not be less than 50 per cent. I am of opinion that there is not a sufficient basis to enable them to go to the public and say, now, this is an undertaking which is likely to be profitable and advantageous; they would not be able to raise a shilling on such a prospect of traffic estimate and working expenses.

If they were successful in getting their Bill, the whole thing would have to become a question of dealing with financial agents, and giving preferences to one part of the work over the other, and it would end in nothing but ruin and confusion to all the people associated with it.

Mr. Francis Ellis, speaking of Trafford Hall, said the Trafford family had lived there since before the Conquest, and would have continued to live there as long as they were allowed to do so.   He opposed the Bill because of the nuisance that was sure to arise from a huge body of polluted water.

At the conclusion of Mr. Findlay's evidence, *Mr. Pope* addressed the Committee on behalf of Liverpool and the London and North-Western Railway Company. He assured the Committee there was no jealousy on the part of Liverpool, and that but for the estuary works, that Corporation ought not, could not, and would not have raised a single word upon the question.   Speaking of the obligations by various railways to place swing instead of fixed bridges in case of a Ship Canal being made, he said they were utterly dead, because no President of the Board of Trade under the changed circumstances would dream of giving his certificate for such a crossing that would inflict permanent injury on the railways and the public.   As to raising the capital, he asked :—

Does anybody believe that if your Lordships were to grant this Bill the capital would be found for the Ship Canal?   What have they got to raise? they talk about millions as if they were threepenny bits.

After reviewing the evidence and repeating his previous arguments, he asked for the rejection of the Bill; his conclusion being :—

My old friend, Mr. Adamson, may perhaps take the opportunity of preparing some joke at my expense, which though coarse is harmless, and not without a certain amount of humour; he may even go so far as to lampoon some of your Lordships, but *that* will be easier to bear than the regret of having sanctioned a scheme which can do but a limited amount of good, and which may result in irreparable mischief to one of the greatest communities of the Empire.

Mr. Littler, on behalf of the Trafford estate, addressed the Committee in a most caustic speech.   He prophesied that either Lancashire would be taxed with ten millions of unremunerative capital, or that ruinous competition would force the canal into the Railway and Canal Conference—a public trust he held to be illusory. "Before many years are over," said he, "you will find Mr. Adamson presiding at the Conference with the railway companies to fix the rates and divide the traffic." He was specially severe on obtaining the fifth part of the fifteen millions of Liverpool traffic which Mr. Marshall Stevens had estimated for the canal, and declared the whole of the traffic of Manchester and 12 miles round only came to 2,400,000 tons, *i.e.*, of Liverpool traffic, and compared Manchester and Liverpool to Kilkenny cats trying to destroy one another, with the former as the aggressor.

Mr. Aspinall followed on behalf of Birkenhead and Bootle, and after pouring a broadside into the promoters and raking their evidence, declared the object of the opposition was to save the port of Liverpool and its harbour and river (the greatest emporium in the world) from destruction.

Mr. Pember then replied on the whole case. What, he asked, was the £16,000,000 of Dock Boards to the £100,000,000 invested in the cotton trade of Lancashire alone? If part of the Liverpool traffic was transferred to Manchester, what harm would be done? He denied that Mr. Hunter's evidence betokened a want of faith by the people of Manchester. Mr. Hunter was alone, and it must be borne in mind that his was also a Liverpool firm—he was an unimportant exception. Could not the energy of the opposition find one man to join Mr. Pope in saying, "This is all moonshine"? Could not Sir William Forwood, with his ingenuity and hate of the scheme, find one man to back him out in all Lancashire? He criticised the Cotton Brokers' monopoly that cost Lancashire £152,600 per annum, and showed Sir William Forwood's error as to cartage. Then one after another he dealt with the evidence given by the witnesses on both sides, making it clear that Mr. Marshall Stevens did not volunteer his views on prospective traffic, but that they were drawn out of him in cross-examination. Then he dealt exhaustively with the question of raising capital, reminding the Committee that the Suez Canal, which cost £8,000,000, was originally considered a wild bubble; yet the cost had gone up to £20,000,000 and still it had succeeded. As a climax, and to show the sincerity of the promoters, he said:—

If you give us this Bill, to show that there is no danger on the subject of capital, you shall put a clause in that we shall not turn a sod until £5,000,000 of money has been subscribed.

Turning to the impossibility of the canal competing with the railways, a point constantly raised by Mr. Littler, and which, said Mr. Pember "that gentleman worries as a terrier does a rat," he pointed out that though the allied railways might have £260,000,000 capital against the £10,000,000 of the canal company, they could not kill the canal. "A waterway once made was made for ever." It did not wear out like the rolling stock of a railway. If a railway carried for nothing, the working expenses would not be reduced a single sixpence; the shareholders would soon call out if their dividend was being frittered away. Speaking of the Liverpool bar and the remissness of the Dock Board in not removing it, he pictured that body as "Secure in a monopoly, secure in the invincible attractions of Liverpool for shipping in consequence of its contiguity to the great consuming and producing districts of this country, the Mersey Board has been apathetic up to this time, and has left the estuary untouched. Startled now by the awakening of Manchester they have become just, so that their apathy, so to speak, and their jealousy both appeal to you; their apathy, if I may be excused the metaphor, yawns, 'Leave the Mersey alone, it will do for us,' and their jealousy almost screams, 'For goodness' sake do not let it be improved so much as to make it good enough for them'." All this in the face of the serious warning conveyed by the fact that, whereas there used to be 17 feet of water over the bar, there are now only 9 feet. When the canal comes into existence, Manchester will be glad to co-operate with Liverpool in anything that may be required for the improvement of the bar, and if the new works should by any accident (really impossible) deteriorate the bar, she will pay any cost that may be incurred.

Mr. Pember spoke during the greater part of three days. It was a magnificent effort to do justice to a most important case, and those who were present will never forget it. He concluded with :—

Surely! surely! my Lords, I am entitled to be somewhat confident in this matter. Upon the whole, surely it is a crisis in which the issue ought not to be seriously in doubt. Great necessities are proved, and great destinies are shown to be imperilled, by one thing and the other, and greater elements than my poor powers of exposition have been able to explain to you are at stake in the commercial interests of England. The contest is one of resolution against timorousness, of energy against apathy, of progress against inevitable retrogression, and your Lordships will not be slow to see it, and I leave with all confidence the determination in your hands. My Lords, that brings to a close, for the present, at all events, the severest task of forensic labour that I have ever undertaken. I am almost ready to hope that I shall never have so heavy a one again. Perhaps you will think that some part of the labour I have gone through is self-imposed, but I was anxious, so far as my poor powers allowed me, to do justice to this vast case. I wanted to be able to say, if I may venture to repeat an old story, what Hyperides, the Athenian advocate, said to the Athenian judges when he drew away the veil of Phryne: "At least I have shown you what you are asked to destroy".

The Committee adjourned, and next day, when the parties were called in, the Chairman gave their decision :—

The Committee are of opinion it is expedient to proceed with the Bill, subject to the insertion of a clause offered by Mr. Pember, prohibiting the commencement of the works until £5,000,000 of money has been subscribed and issued.

CHAPTER XII.

1884.

PARLIAMENTARY PROCEEDINGS IN THE HOUSE OF COM-
MONS—EVIDENCE OF WITNESSES—A RACE FOR TIME—
CASE CURTAILED—LIVERPOOL AND THE DOCK BOARD
GIVE PLEDGES NOT TO OPPOSE A CHANGED ROUTE—
UNFAVOURABLE DECISION.

When we consider the enormous competition to which we are subjected by foreign nations, and the almost costless canal traffic extending in France to every foreign market and centre of industry, he thought we should see the necessity of bestirring ourselves to make the best possible use of similar advantages in this country.—JOHN SLAGG, M.P.

AFTER passing the Lords Committee, the Bill on the 7th July came before a select Committee of the House of Commons, consisting of Mr. Sclater Booth, Chairman, the Marquis of Tavistock, and Messrs. Lewis Fry and James Campbell.

Practically the same senior counsel again represented both parties. The hearing of the promoters' case lasted nearly eight days, and they called thirty-one witnesses. The opponents' case commenced on 16th July and was continued to the 31st July, in all nearly twelve sitting days. On their behalf thirty-five witnesses were called. Mr. Pember alone addressed the Committee in favour of the Bill, whilst there were six leading counsel who spoke on behalf of the various opponents. Eleven petitions were presented against the Bill.

Mr. Pember reminded the Committee this was the eighth speech he had made on behalf of the Bill. He reiterated the Ship Canal case, giving the history of the agitation, and stated the disabilities under which the trade of Manchester laboured in consequence of the toll-bar placed on raw material and finished manufactures by the railway companies and the Liverpool Dock Board. He showed by past legislation that a Ship Canal had been contemplated for at least forty years, and he asserted

(238)

that Manchester had equal rights with Liverpool to the use of the Mersey waterway. He enlivened his speech by referring to some of the evidence given in the previous inquiry "by gentlemen who were good enough to say they were not engineers but yet professed to be critics of engineers". One gentleman, Mr. Stevenson, showed great tact and talent in doing his best for our opponents, but luckily for us the writings of himself and his brother—for he had written books on the subject and so had his brother—were far too strong in our favour, and they corrected very brightly the gloomy prognostications of other witnesses. Mr. Pember humorously alluded to Captain Eads as the gentleman who had come all the way from America to tell them "there was not a particle of sand in the Mersey that was not amenable to natural laws". "When I asked him what works he had carried out himself, he oddly enough said he had carried out training walls at the mouth of the Mississippi to get rid of the bar. 'Did you do it?' said I. 'Indeed I did,' said he. 'I shot the bar and its contents 25 miles out to sea, and we have heard no more of the bar.' I said, "Did you ever hear anything in the shape of prognostication of evil before you began?' 'Oh, yes,' he said; 'a great number of gentlemen told me I should ruin the Mississippi; that at the end of my training wall another bar would start, even if I did not make the original bar worse.' But, he said, 'It is not so'." So Mr. Pember argued that people were fearful lest the Ship Canal works might damage the bar, but he could tell them it was a needless alarm.

He went on to say that just lately he had been receiving instruction from Sir William Forwood, who was no friend of his, but had settled the question of capital as he hoped for ever. After all the adverse evidence as to raising the capital which he had given to the Committee when he spoke in the Liverpool City Council in June last, he said, "the feeling has been that even if Manchester got her Act of Parliament she would not get the money. This is a mistake." There was no better judge than Sir William Forwood, and he, Mr. Pember, was satisfied with his opinion. The great cry of the opponents was that the scheme would be fatal to Liverpool because it would destroy the entrance to the Mersey. "Was it likely," asked Mr. Pember, "that these Lancashire gentlemen would be so foolish as to spend ten millions of money to damage Liverpool, when such damage would mean that they lost every penny of the money they had spent?" He then called on *Mr. Daniel Adamson*, Chairman of the Canal Company, who repeated his evidence given before previous Committees. Sir William Forwood having attacked his evidence as regarded the consumption per spindle of cotton, Mr. Adamson showed his figures were correct, being the result of inquiries from Mr. Ellison, the Liverpool cotton statist, and confirmed by numerous large spinners.

Mr. Marshall Stevens followed with a mass of shipping and commercial evidence, and put in numerous schedules and comparative tables showing the advantage the canal would have over the existing means of carriage. As an instance, on loaf sugar he stated there would be a saving of 12s. 1d. per ton on dues and Liverpool expenses, or more than the railway carriage from Liverpool to Manchester. Again, whilst the present railway rate to London was 40s. and to Plymouth 46s. 8d., the charge if carried all the way by sea would be 15s. To be on the safe side, he had not taken into consideration ship dues charged at all the chief English ports of from 5d. to 2s. per ton. At first, while competition was fierce, it might not be wise to levy ship dues, but he felt sure some day they would be an additional source of revenue.

Mr. Henry Walmsley, cotton spinner, Stockport, objected to dual brokerage, and as an instance of the arbitrary conduct to which spinners were subjected he mentioned the case of Messrs. Ralli Brothers, of Manchester, who opened a very extensive connection in Manchester for the sale of cotton to spinners, and did a large business. The Cotton Association of Liverpool insisted this should be closed, and Ralli's cotton business at Manchester was transferred to Liverpool.

Alderman Bennett said the £10,000 voted by the Corporation of Manchester to the Ship Canal Fund was to some extent in view of advantages that would be afforded to the Corporation in carrying a culvert alongside the river. To his mind Manchester was not adding to the pollution of the river. His opinion was that it was purer when it left Manchester than when it entered the city boundary. He submitted a resolution in favour of the Bill passed at the largest meeting the Chamber of Commerce ever held, there being only two dissentients out of the 400 people present. Also from the Guardian Society for the Protection of Trade, of which he was President.

After hearing *Sir Joseph Lee*, and many other commercial witnesses who were taken very briefly, Mr. Pember, on 9th July, thus addressed the Chairman :—

Sir, there are moments when a leading counsel must act with nerve and take upon himself that amount of responsibility which the moment throws upon him, and he must do it without any hesitation. In this case I have but one anxiety, and that is the question of the effluxion of time. The event of last night (the Government in peril) has not by any means contributed to lessen that anxiety, and I have thoroughly considered the matter with

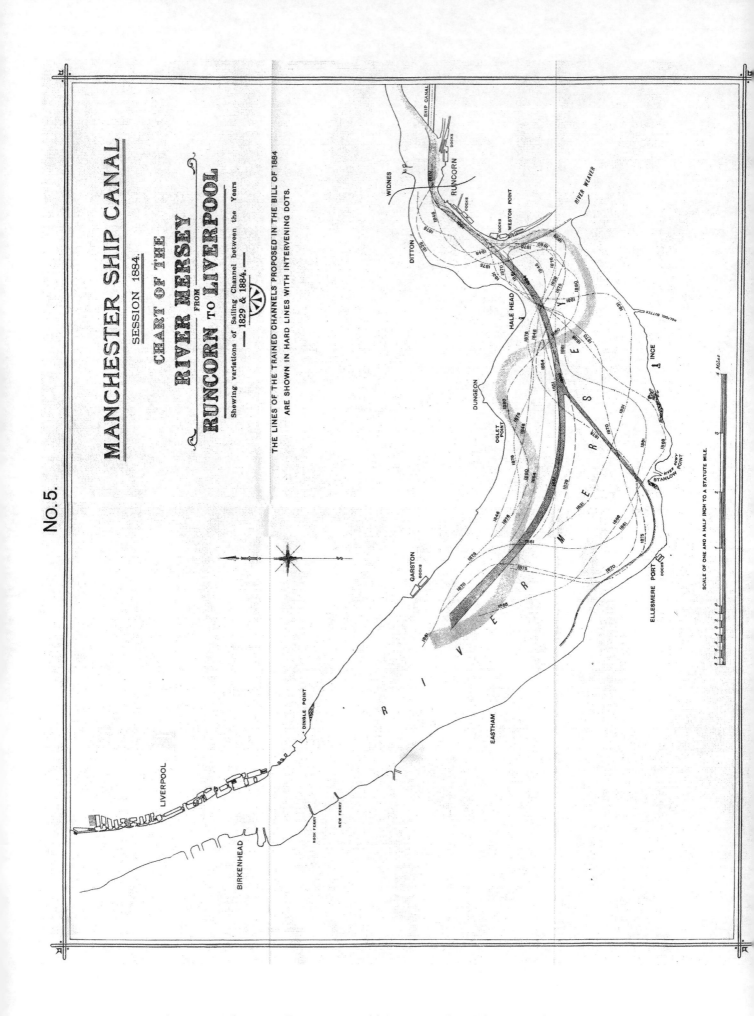

my clients, and given every possible hope of the extension of the session. I have come to
the conclusion that I shall only just have time in the event of your being persuaded upon
another clause of the case that the Bill should be proceeded with to get it through Parliament.
With regard to the commercial evidence we have yet only practically entered upon the
threshold. You have got all the evidence of this year (Lords inquiry) referred to you.
You know what the clause is that was inserted in the preamble in the House of Commons
last year, and you have abundant indirect knowledge of what I might be in a position to
prove. I do not mean to call any more commercial witnesses. I cannot afford the time.
I am determined one thing shall not beat me, and that is time. Therefore, to-morrow I shall
put the engineer in the box. It is perfectly clear to my mind that if I must go through the
whole story, as I did in the House of Lords, my Bill must fail. Therefore, I must run any
risk that I do run by the course I have taken.

Mr. Leader Williams, engineer to the scheme, then repeated his previous
evidence. Exception having been taken to the quantity and quality of the Irwell
water, he showed that even with the low-water flow there was sufficient to pass
twenty-five steamers a day of 2,000 to 3,000 tons each; fifty, from 500 to 2,000
tons each, and 100 barges. Admitting the Irwell was an open sewer, so many
chemicals were poured in that they acted as a kind of mordant. He had been
engineer of the Navigation, and he never knew any men take harm from the smell
even in hot weather. In addition, measures were being taken by all the large
towns, and he hoped shortly to have the river very much improved. The witness
referring to the opponents' contention as to harm arising from the reduction of
tidal area in the estuary, said :—

When the Liverpool people wished to make the docks they never raised this question
of the importance of the tidal area. They have taken in 1,100 or 1,200 acres of land, and
it is only when we want to do something in the Manchester direction that they raise an
objection.

Witness went on to show that erosion in the estuary was going on along the
line of cliffs at the rate of 1 foot per annum, and that a trained channel in the
middle of the estuary would prevent this.[1]

During the proceedings the Chairman more than once intimated that he wished
the engineering evidence shortened.

Pressed in cross-examination, Mr. Leader Williams said he did not see his
way to bring his Ship Canal down to the Mersey below the estuary, because if he
did Liverpool would object to the abstraction of any water that would diminish the

[1] See Plan No. 5.

quantity passing down the estuary, or that would diminish the scour in the upper estuary, and so cause accretion.

A succession of eminent engineers, Messrs. Abernethy, Brunlees, Messent, Wheeler, Lionel Wells, and Giles, M.P., supported Mr. Leader Williams' plans and estimates.

Mr. Fowler, engineer of the Tees, called attention to the grass Sir Frederick Bramwell asserted he had seen growing on land that was being reclaimed, and it was a curious sight to see a crowd of eminent men standing round a box Mr. Fowler had brought up, and discussing if the greenery was seaweed, as he declared it to be, or common grass.

After the engineers came the land valuers and shipowners, all repeating previous evidence.

Captain Pearson, Marine Nautical Assessor of Liverpool, was of opinion that it was as easy to get up to Runcorn in one tide as to dock in Liverpool, and that it was safer to be out of the reach of heavy seas up the river than to be waiting in a rough sea to enter a Liverpool Dock.

The last witness, *Mr. Ernest Deshayes*, of Rouen, declared that Rouen was not an excluded port in bills of lading, that large ships came up the Seine with safety, and that no damage had been done to Havre, a town situated, like Liverpool, at the river entrance.

Amongst the junior counsel for the opponents was Mr. A. M. Sullivan, M.P., the eminent politician and Home Ruler, who had recently joined the English Bar.

The case for the opponents was then taken.

Mr. John Bowden, C.E. for the Trafford estate, said the promoters to appease Salford had agreed to let them have a sewage wharf opposite Trafford Hall, and that it would be a nuisance.

Mr. T. B. Foster called attention to the agreement with Manchester whereby one-fifth of the usual water flow might pass by a culvert to the lower reach of the river. He was of opinion this would lessen the water usable for locks, and increase the nuisance at Trafford Hall.

After *Mr. I. M. Fox*, Medical Officer of Health for Mid Cheshire, had condemned the canal as an unmitigated nuisance,

Mr. Littler, Q.C., addressed the Committee. If in the Lords Committee he had lashed the Ship Canal scheme with whips, in the Commons he now used scorpions. He accused Mr. Pember of making a bid in the Lords, *i.e.*, to plank

£5,000,000, and so to get his Bill by the skin of his teeth. He said with five ways to Liverpool the cry for another meant that a lot of manufacturers wanted to cut one another's throats at the cost of the railway companies. He tried to show there was an attempt to dodge the deposit clause by fictitious shares and deposits thereon. He pictured Sir Humphrey de Trafford turned out of his ancestral home by a company who would become insolvent and not be able to complete the work. Instead of five he was sure the work would take ten years. He complained the usual clause about the reversion of superfluous land was omitted, and that the ninety-nine year lease clause would enable factories and warehouses to be built in Trafford Park "under our very noses". Also that the canal company proposed to take their land for the Corporation to make sewers on it. Because Manchester and Salford hate one another, the former, to suit its vanity, places a public wharf, three-quarters of a mile long, opposite and as a counterpart to the wharves in Salford. "Fancy their coming and perpetrating such a piece of vanity as that! It is simply to please the vanity of Manchester that they say, we will acquire land from Sir Humphrey de Trafford which we do not want, and we will enable others to put that upon his land which will make his house intolerable and uninhabitable." Then Mr. Littler, by figures, tried to show that all private docks were killed by the cost of maintenance, and that the Ship Canal could only get and distribute such goods as came by cart from a limited area or by narrow boats on the canal. Railways would never come and help them. How could this eastern port with £10,000,000 capital compete with the iron-clad of the combined railways with £267,000,000 of capital? The end must be that this poor little canal company when it is bankrupt will be bought up by the railway companies at their own price. The tables of rates put in by Mr. Marshall Stevens and Mr. Adamson were made to suit their own theory, and must have been made by some one who had no practical experience of carrying. The audacity and absurdity of the tables made them hardly worth dealing with. He said the calculator for the Ship Canal was the same Mr. Lawrence who had been the prime mover in the Hull and Barnsley Company. As regarded the latter "there was hardly a house in the East Riding of Yorkshire where the whole of the housemaids and footmen had not sacrificed the whole of the money they had got in the Savings Bank and put it in the Hull and Barnsley". So with the Ship Canal; "they got 50,000 people, with the assurance that there would be a lot of labour required, to demonstrate at the Pomona Gardens in a torch-light procession in favour of this thing. Just in the same way they will induce people all over the country to put their wretched earnings into this miserable concern."

He denied there was any comparison between the Tyne, Tees and Clyde, and the Mersey, and said the latter was a dangerous wild beast, and almost untameable, and ended his speech thus :—

At least see what you are about to destroy, and for whom? You will destroy the Mersey and injure the whole trade of Lancashire all for the sake of a certain number of speculators, flattering their vanity and bringing themselves to ruin.

Mr. A. T. Squarey, solicitor to the Dock Board, after confirming previous evidence, contended that an attempt to improve the bar would be of doubtful result, and would be a burden on the port.

Captain Graham Hills, surveyor to the Dock Board, was under examination for three days. He gave his evidence as before. In cross-examination, he admitted his views as to the formation of sand banks in the Mersey differed entirely from those of Admiral Spratt, the Conservator of the Mersey appointed by Government, and he also differed from that gentleman as to reclamation in the Tees. Attention being called to other reclamations on the Mersey, specially at Ditton Brook in 1873, the witness made a remarkable statement, *viz.*, that the Dock Board objected, and so did Admiral Evans, the then Conservator, but that the Chancellor of the Duchy of Lancaster, being one of the Commissioners and also a landlord interested in the sale of land, was too strong for them, and it could not be prevented.

Mr. George Fosbery Lyster did not think the canal scheme well devised; it would destroy the river and the practical working of the docks. Training walls stopping midway in an estuary ought not to be permitted. "He was assured that the promoters might attain the object they seek, namely, an approach to the upper river docks, by a much better plan than theirs, and by one that would obviate all the difficulty and danger involved in the works they now propose, and not imperil Liverpool." As regarded estimates, his were double those of the promoters —£2,800,000 as against £1,400,000. The moneyed merchants in Liverpool had never asked for the improvement of the bar; they would hesitate to pay increased rates. His opinion was that it would be impracticable to make any permanent effect on the bar by dredging. The only way was by breakwaters, and these would cost millions of money.

Mr. Thomas Stevenson believed in training walls, and said accretion only followed when there was a soft bottom, as in the Mersey.   He believed in the principle, but did not like Mr. Leader Williams' training walls.   He did not agree with what his brother had written about no damage arising from training walls.   He could not for his life tell why the promoters' engineer had come down into this wretched estuary.   There could be no difficulty in carrying a canal to Liverpool without going into the upper estuary at all, by keeping on dry land.

Sir William Thomson, Messrs. Manning and *Leveson Harcourt* all opposed any interference with the estuary; the latter was in favour of bringing the canal along the shore of the estuary, and not in its centre.

Captain Eads again described the means he took to remove the Mississippi bar.   Asked if it was necessary to take training walls through the estuary in order to get from Liverpool to Manchester he replied :—

I do not think it is.   Works could be constructed along the Cheshire shore and the canal brought along there with decided benefit to many interests and at quite as little cost as taking it through the estuary, and it would not be liable to so many contingencies and difficulties.

In his opinion, to remove the bar would cost many millions of pounds.

Mr. Stephen Williamson, M.P., would consider adventurous experiments at the bar as money thrown away.   In Liverpool the Ship Canal was considered a mad scheme that would never be carried into effect, and no serious attention was given to it.   As a shipowner he considered the hazard would be too great to send a ship up if he could avoid it.   The Hall Line of steamers running to Bombay could not come up because of the height of their masts.   Wheat from California, Chili and Australia could never come to Manchester as it was brought in sailing ships.

Sir Frederick Bramwell could see no difficulty in going round the side of the estuary instead of in the middle.

Mr. Thomas H. Jackson, of Birkenhead, believed no time was lost by ships waiting at the bar; supposing they could get over, they must wait in the river till there was water enough to enter the docks.   He believed if the canal were made the navigation would be stopped in the winter.   It would take a ship two or three days to go up to Manchester and the same time to come down.

Mr. T. D. Hornby, Chairman of the Dock Board, had heard a great deal about dredging the bar.   This would have to be done in the open sea.   It would require an Act of Parliament unless the nation did it; the cost would fall upon the commerce of the port, and engineers had talked of millions of money.   He was not in favour of experiments of the kind.   When the Bill passed the Lords a Liverpool deputation went to see the Admiralty with a petition, Admiral Spratt, of the Conservancy Board, being present.   They represented that the clauses prepared at the instance of the Conservancy Commissioners had seriously prejudiced them, and did not give them sufficient protection, and they asked that the company should be called upon to give security to the Board of Trade so as to procure the observance of all stipulations.   The witness in cross-examination admitted that the Conservancy who would, under the Bill, have the power of vetoing the estuary plans had secured Sir John Coode, an eminent engineer, to advise them.

Mr. Bidder, Q.C., then addressed the Committee on behalf of the Mersey Docks and Harbour Board, and applied himself chiefly to reviewing the engineering evidence.

Sir William Forwood spoke seriously of the condition of the bar, and said it was only practicable now by vessels drawing 24 feet of water.   He, too, thought the security of the estuary of primary importance :—

And to show the absolute bonâ-fides of what I have stated, speaking on behalf of the Corporation of Liverpool, I have no objection whatever to this scheme being passed by this Committee as far as Runcorn, and if the promoters will bring in a Bill next year, carrying out their estuary works, as they can carry them out without interference with that estuary, either along the northern or southern shores, we will not, upon principle, oppose that Bill next year.   I can say nothing stronger to show the entire bonâ-fides of Liverpool.

He again challenged the whole of the commercial tables put in by Mr. Marshall Stevens, who, as he said, had assumed to speak as if with the authority of the commercial interests of the Mersey :—

I must say, with my knowledge of Liverpool, and within my own public life in Liverpool, I never heard of Mr. Marshall Stevens until I entered the Committee Rooms of this House.
Mr. Balfour Browne.—It is very sad for Mr. Marshall Stevens.

The witness claimed he was one of the authors of the Railway Commission, and said it gave power to the Commissioners to propose rates over a canal system, but not over canal systems plus the Ship Canal, to Bombay or any other part of the world.   To this Mr. Balfour Browne dissented entirely.

Mr. Aspinall, speaking in the case of Liverpool and Birkenhead, and *Mr. Littler*, for the North Staffordshire Railway, briefly reviewed the evidence.   During Mr. Aspinall's speech, he said that he represented the Dock Board, and was authorised in the strongest possible sense to ratify the offer of Sir William Forwood in the precise words in which he made it.   The effect being that if Manchester would give up training walls in the estuary, and take the canal on the borders on either side, neither Liverpool nor the Dock Company would oppose them.   After this the London and North-Western case was heard.   *Mr. Findlay*, their general manager, recapitulated his evidence before the Lords Committee.

Mr. James Grierson (of the Great Western Railway) believed it might be possible to get full cargoes of timber and grain, but it would be impossible to make Manchester a port for the general trade of the country.

Mr. Scotter (of the Manchester, Sheffield and Lincolnshire Railway) showed the Cheshire Lines Company only got $\frac{7}{8}$ per cent. on an expenditure of £10,000,000, and he considered the estimated profit on the Ship Canal fallacious.   The Amsterdam Canal, 15 miles long, took seventeen years to construct, and cost double the estimates.   It had not paid anything and was now in liquidation.   Hull and Grimsby, with all conveniences, shipped 1,030,000 tons of coal in 1883, and he considered the estimate of 2,000,000 tons at Manchester an impossibility. In cross-examination, witness admitted Amsterdam had been a success commercially.

Mr. Henry Oakley said it was absurd to do a traffic of 5,000,000 tons on 67 acres of dockage and 3 miles of quays, especially when there was no provision for railway connections at the docks.   It meant 2,700 loads of 3 tons each to be brought in and out by horses.

Mr. Francis Stevenson, engineer (London and North-Western Railway), had a firm and deliberate conviction that if the proposed channel was made the Garston Docks would be done for.

Mr. Pope addressed the Committee on behalf of the London and North-Western Railway Company.   He contended that in Manchester and 12 miles round the production and consumption did not exceed 2,400,000 tons yearly, and if the canal filched every ounce from the railways, it would not even then pay; that goods in and out would have to be carted, and that railways would not help, and so bring competition in themselves.   His other main point was that the approach channel to Garston Docks would be silted up.   He asked: "Is the good which this scheme can accomplish for one moment to be measured against the absolute injury and the terrible risk?"   If not, let them amend their Bill next year and profit by the criticisms they have had to endure.

Mr. Pember then made his final reply.   He regretted time was running so very close that he had been compelled to omit a great share of his commercial case,

and pointed out his opponents had not similarly striven to save time.   Forty days
work in the Lords had been condensed into twenty in the Commons.   He ridiculed
the idea that the volume of business of Manchester was only 2,400,000 tons, whilst
Hull, with 283,000 inhabitants, had 4,250,000 tons of traffic.   Much had been said
about the danger of accretion in the estuary, but he pointed out that Mr. Stevenson,
a great authority, and one of the opponents' witnesses, had said, "Accretion or non-
accretion, I never knew one of my works that impaired the bar".   It could not be
disputed that within the memory of living man there had been 17 feet of water upon
the bar, now there were only 9 feet, and at one time only 7 feet.   The cubical
capacity of the estuary, too, on which so much stress was laid, and which was said
to be vital to the river, had been reduced in the last twenty years by 18,000,000
yards.   Captain Graham Hills said the proposed training walls would destroy frets
in the estuary, yet when there had been a phenomenal fret, the estuary had lost
capacity and the bar had gone to the bad.

After all, the Mersey bar was a small affair in area; the opponents admitted
the water to be only 9 feet deep for 100 yards.   In a quarter of a mile on either
side absolutely deep water was reached.   " Is it not a crying shame that with so small
an obstacle to deal with Atlantic Liners are kept waiting close to port sometimes
even for seven hours?"   "Why not do there on a small scale something analogous
to what Captain Eads had done on a large scale on the Mississippi?   Looked at
from the point of view of the enormous industries that would be resuscitated it is a
trumpery and trivial work, and ought to have been done long ago, and the Company I represent would be the very first to co-operate with Liverpool in carrying
it out, and would force her hand if she were sluggish.   Lancashire would rise as
one man to support Liverpool in such a great work."

On previous occasions, Mr. Pember had concluded his speeches with an eloquent peroration, but for once he ended by tamely asking that the Bill might be
passed.   Possibly he felt all through his speech that his wings were clipped by
having to fight against time.

The Committee room was cleared.   Previously it had taken a considerable
time to arrive at a decision, but on this occasion after a short time the parties were
called in, and the Chairman announced:—

The Committee have come to the conclusion that the preamble of the Bill has not been
proved to their satisfaction.

This unexpected decision was an intense disappointment to the canal supporters in the lobby, and was received with much jubilation by the opponents who attributed their success largely to the bait thrown out by Liverpool and the Dock Board to the Committee, that if the Ship Canal would avoid the estuary in a subsequent Bill their opposition should cease.

CHAPTER XIII.

1885.

THE SHIP CANAL BILL RECAST ON MR. LYSTER'S LINES—
LIVERPOOL VIEWS—WITHDRAWAL FROM PLEDGE NOT
TO OPPOSE BILL—BILL PASSED IN THE COMMONS—PUB-
LIC REJOICINGS—PREPARATORY STEPS TO RAISE THE
CAPITAL.

Communications make the trade, not trade the communications.—Sir ARTHUR
COTTON.

DURING the two previous years the fortunes of war had oscillated. In 1883 the Ship Canal Bill was passed by the Commons and rejected by the Lords, whilst in 1884 the Lords passed the Bill and the Commons rejected it. Nearly half a million of money had been spent by the contending parties and no progress had been made. There were signs of exhaustion. The opponents had declared they did not fear any damage to the trade of Liverpool, and would be content if only they could be assured that nothing done to the estuary would damage it, or silt up the bar. Their counsel had put forth a scheme which, as they said, would enable the canal to be made and that would be accepted by them. The promoters had discovered how easy it was to alarm a Parliamentary Committee who naturally wanted to protect Liverpool and the estuary. They had open minds to consider any feasible alternative scheme, and when the dock engineer, Mr. Lyster, proposed the canal should be made alongside the estuary instead of through the middle of it, they gave the suggestion their best consideration.

The result was a new scheme, following to a very large extent Mr. Lyster's lines, and making the canal debouch into the estuary at Eastham.[1] For this plans and specifications were deposited in due course.

[1] See Plan No. 8.

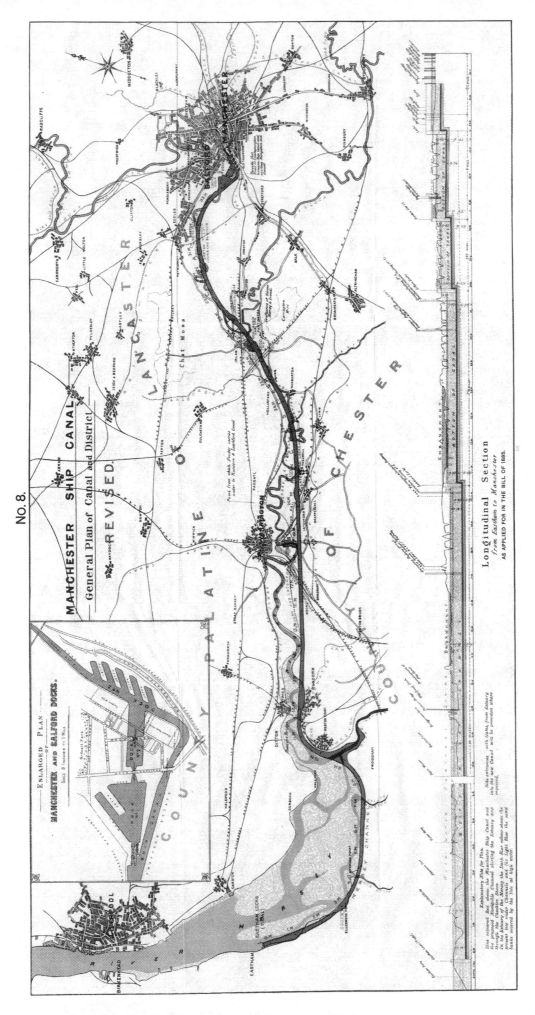

The material originally positioned here is too large for reproduction in this reissue. A PDF can be downloaded from the web address given on page iv of this book, by clicking on 'Resources Available'.

But it soon became evident that Sir William Forwood and his friends regretted the promises they had made under pressure when they feared the 1884 Bill would certainly pass. They showed a disposition to forget all previous assurances and the statement that traders of Liverpool had no fears of damage from the canal, inasmuch as their ships would never go to Manchester.

The year 1885 started with Liverpool, the Dock Board and the Railways being more virulent than ever against the Bill. They were determined to scotch it at all costs, and this notwithstanding their repeated assertions that they did not look upon the Ship Canal as a serious undertaking. The plans and estimates having been deposited, the promoters paid into the Bank of England the usual deposit. This done, the Bill passed through Standing Orders without opposition.

During January "Mancuniensis" (Mr. J. W. Harvey) published some valuable statistics showing how dear transit was in England compared with the Continent. For the same weight and distance raw silk cost £5 in England and 21s. in Belgium; butter, 10s. 1d. against 4s. 10d.; timber, 19s. 2d. against 9s.; pig iron, 15s. against 6s.; and grain 12s. 6d. against 6s. 11d. In France coal was carried at ·28d. and limestone at ·18d. per ton per mile, whilst in the Midland counties the charge on heavy material was as much as 1.40d. per ton per mile.

On the 29th January the Dock Board officially informed the Board of Trade that the Ship Canal was unnecessary, and the works might cause serious injury to the estuary of the Mersey, also might interfere with the approach to the docks and the transit of vessels in the Mersey, and that it would be their duty to oppose the Bill in Parliament.

On the 11th February Sir William Forwood told the Liverpool Council that the pledge not to oppose would have been carried out if the promoters had not sought to make a 800 yards' channel in the estuary from Eastham to Bromborough and to dredge a similar channel 500 yards broad from Runcorn to Eastham. He believed this would cause irreparable damage to the estuary.

This brought an indignant letter from Mr. William Price, of Wm. Price & Co., shipowners, of Liverpool, who objected to the attempt to strangle a new enterprise "by petty jealousies, or any tactics unworthy of a great trading and shipowning community like Liverpool". He complained of the immense waste of money in opposing the Bill, and said it had been spent illegally, inasmuch as a Borough Funds meeting had never been called at Liverpool. He quoted the Act of Parliament to show this was necessary, and undertook, if one were called, to ask for a poll.

In proof that his contention was valid, Birkenhead did have a Borough Funds meeting to sanction opposing the Bill. One of the speakers said he believed a Ship Canal to Manchester would benefit Birkenhead, and another was quite satisfied Liverpool looked upon Birkenhead as a rival, and would not care if the whole of their docks were silted up. Eventually Birkenhead decided to oppose the Bill.

Sir William Forwood's speech to the Liverpool City Council evincing, as it did, a desire to draw back from his undertaking, caused Mr. Adamson to write him the following letter:—

MANCHESTER SHIP CANAL OFFICES,
70A MARKET STREET, MANCHESTER,
February 19th, 1885.

DEAR SIR,

My attention has been called to your speech at a meeting of the Liverpool City Council, as reported in the *Liverpool Courier* of the 12th inst., in which you made the following statements in reference to the Manchester Ship Canal: " But there was still a worse clause, *viz.*, 31, by which the promoters took power to dredge a deep-water channel 800 yards broad from Eastham to Bromborough and to dredge a similar channel 500 yards broad from Runcorn to Eastham, which would not only be as mischievous as the channel proposed last session, but would do infinitely greater damage. . . . He himself believed that it was quite possible to make the canal without infringing upon the estuary at all, and he had still a strong hope that they should be able to get the promoters so to modify the plan as to obviate the necessity on the part of the Corporation of appearing in opposition. But if they persisted in constructing these channels they would inflict irreparable injury to our port and harbour." From these remarks it is evident that you have misunderstood the object of the 31st clause. The promoters have no intention whatever of dredging any continuous channel in the estuary between Runcorn and Eastham, or any channel 800 yards wide between Eastham and Bromborough. The widths mentioned in the clause are merely to define limits within which the company may dredge. The main object of taking this dredging power is to enable the company to maintain efficient accesses to the entrance locks to be provided at Eastham, and for the upper Mersey ports, and to secure river and drainage outfalls. As I am thus able to assure you that the promoters have no intention of constructing the channels mentioned by you, I think if you and your friends would further carefully examine the clauses of the Bill you would come to the conclusion set forth in this letter, that Liverpool has nothing to fear from any act or intention of ours to disturb the fretting action of the Mersey, upon which you founded your opposition to our late application to Parliament. However, should any misapprehension and distrust exist, I am sure if you and your colleagues desire to meet the promoters with the view of coming to a clear understanding, my Committee would be willing to have a conference so as to arrive at an amicable settlement without your incurring further unnecessary expense.

I am, etc.,
(Signed)   DANIEL ADAMSON (Chairman).

Sir W. B. FORWOOD.

This was intended to show Sir William that he misinterpreted the intention of the promoters, and was wrong in his facts. Further, that there was a desire on behalf of the promoters to discuss any differences of opinion at a conference, with the view of saving a Parliamentary fight. Strange to say no reply was ever received, thus showing the opponents had no wish to come to a settlement. They desired to kill the Bill.

There were not wanting, however, warning voices in Liverpool. Sir James Picton asked to what extent the Liverpool Corporation were committed as to the further opposition to the Manchester Ship Canal Bill? The expense hitherto had been so frightful as really almost to make one's hair stand on end, and he wanted to know if this was to go on interminably? It seemed to him that the opposition ought now to be confined to obtaining satisfactory clauses. The *Liverpool Courier* also thought Mr. Adamson's letter deserved favourable consideration, because its tone was moderate and conciliatory, and because it declared that danger to the estuary was to be carefully guarded against in the new project now before Parliament.

If Manchester chooses to convert itself into a seaport—or even tries to do so—we have no reasonable right to thwart the ambition. We do not regard the scheme as being either practicable or profitable, but that is the affair of the people who are foolish enough to put their money in it. It will be altogether a mistake for us to oppose the scheme on the pretence that it may damage the estuary if the real motive is to defeat it by a side wind, not because the estuary is imperilled, but because we fear the competition of a rival port. This is a point which will have to be carefully watched.

On the 24th February, when the Chairman of Ways and Means reported the Ship Canal Bill, Mr. Jacob Bright asked that in order to get it through in time it should be relegated to a joint Committee of both Houses, but Sir A. Otway said such a proposition must originate in the House of Lords, who had the Bill in hand and who could invite the Commons to join if they so desired. No such request being made, the Bill was read a second time on the 3rd March, and passed on to a Committee with Lord Cowper as Chairman.

It was a sensible relief when, after twelve days' hearing of the engineering case, the Bill was allowed to proceed. There can be no question that the Committee were much impressed by the fact that whilst Liverpool was professing to be the guardian of the Mersey, bent on preventing an accumulation on the bar, and with the power of fining a sailor £50 if he threw a bucket of ashes overboard into the river, she herself was pitching into it wholesale the dredgings of the docks and the sewage of the city. And this notwithstanding the protest of Admiral Spratt, the acting Conservator, who urged the refuse should be deposited north of the Rock Lighthouse. On the other hand, if the hoppers had in ten years deposited 3,500,000 cubic yards of dredging, how was it that the bar had got no worse?

The *Liverpool Daily Post*, in its article on the Lords passing the engineering portions of the Bill, wrote :—

This year the canal company claim to have adopted Mr. Lyster's plan. Therefore, unless the Dock Board can show that the new canal differs essentially from Mr. Lyster's suggestions they cannot oppose the scheme on its general merits very creditably.

And it went on to say :—

An eminent man, who had occasion to visit Manchester two or three years ago, was asked on his return to London what he thought of the place? And this was his answer, "Manchester is to let". The fact is Manchester is no longer a centre of distribution as it was ten or twenty years ago. Therefore, Manchester wants to take the place of Liverpool as the only way of retaining its supremacy as a centre of manufacturing industries. Manchester is ambitious. She is ready to make any sacrifice to be the recognised head of Lancashire industries and political influence.

On the 7th May, the Committee of the Lords passed the preamble of the Bill. The news was received with the greatest enthusiasm, but warned by previous disappointments, people seemed determined to reserve their shouting till the Bill was safely through the second House. The coolness of the Manchester papers was such as to attract the attention of the Liverpool Press, and the *Mercury* wrote :—

The Manchester papers are not in an ecstatic mood over the passing of the Ship Canal Bill by the Lords. The leading morning papers, indeed, have never been able to work themselves up to a high pitch of enthusiasm about the scheme. In dealing with it there has been an unmistakable undercurrent of misgiving. Perhaps it is that they are acting on the wise precaution of withholding the song of triumph and the editorial benediction until all the preliminary difficulties have been safely got over.

It went on to say that the *Manchester Guardian* was beginning to take a more hopeful view of the situation, and that the *Examiner and Times*, which had blown hot and cold according to the varying prospects of the scheme, was now relieving itself by a tirade against "the gentlemen of Liverpool" who had broken faith by promising to accept a scheme on the lines laid down by Mr. Lyster, and then refusing to carry out the promise.

This practice of stone-throwing (said the article) will defeat its own ends. The terms of the concession upon which the Bill has been passed are, after all, very much like presenting the promoters with a very huge white elephant.

The *Liverpool Courier*, speaking of Mr. Findlay (of the London and North-Western Railway), said :—

A gentleman tolerably familiar with the ways of the moneyed world is confident the canal company will not be able to raise the enormous sum mentioned.

And further :—

With such prospects before them as we have briefly outlined, is it likely that capitalists will subscribe the five millions necessary to allow the works to proceed?

The *Liverpool Daily Post* said that

The Liverpool authorities should start with the irrevocable determination to kill this ridiculous and mischievous scheme.

The same paper attributed their want of success to the division of the engineering and commercial cases, and went so far as to say that Lords Milltown and Romilly would have thrown out the Bill on the engineering case, and Lords Romilly and Arran on the commercial case, and that then Lord Arran was so anxious to throw out the Bill that he wanted to hark back and join the minority on engineering, but that his colleagues would not allow him to change his mind. This extraordinary statement induced Mr. J. C. Fielden (of Manchester) to write and protest against the unpardonable liberty that had been taken in professing to publish the private views of the Committee, and to say the statement was both absurd and untrue.

It had been said with truth that much apathy existed in Liverpool, and that many thought well of the canal. An attempt was now made to rouse the commercial element to a sense of danger, and persuade them that the future prosperity of Liverpool depended on the defeat of the Bill in the Commons.

Sir William Forwood assured the City Council that the Lords passed the Bill though three members out of five were against it. He said many gentlemen in Liverpool thought there was little need to oppose it, because if carried the canal would never be made He was certain the advantage to Manchester of a Ship Canal would be practically nil, and that if it were made the effect on the bar would be to deteriorate it by the accumulation of silt and detritus, and to imperil the £200,000,000 of commerce that passed through Liverpool. The result of this alarming speech was a resolution by the Liverpool Corporation to vigorously oppose the Bill.   At the instance of Mr. Henry Coke, the Liverpool Chamber of Commerce also passed a similar resolution.

When argument failed, ridicule was brought to bear.   Speaking of the canal, a Liverpool paper called it

The greatest bubble ever thrust upon the credulity of the British public since the South Sea Bubble.

And went on to say :—

"The people who live and work in Manchester have had called before their mind's eye delightful visions of a future when they will be able to gaze on the sea and the ships without taking a trip to Liverpool."   "Liverpool citizens have, as a rule, viewed the design of Mr. Leader Williams with a mixture of wonder and contempt, and have never conceived the possibility of Parliament endorsing a scheme which is certain to fail for either scientific or commercial reasons."   "The question for Parliament is : should the very existence of Liverpool 'the Port of the World' be jeopardised on the very doubtful chance of Manchester people deriving some advantage?"   "The Manchester 'men of light and leading' should do something to put an end to the expensive fizzle on which engineers and lawyers have been fattening for many months."

Such were some of the expressions used in the Liverpool Press with the evident intention of rousing the feelings of the people against the canal, and to a great extent they succeeded.   Though many shipowners believed that the monopolist policy, which made Liverpool a very dear port, was more harmful than ever the Ship Canal could be, yet the masses were alarmed at the war cry that "Liverpool would be ruined" just to please the ambition of Manchester.   The different trading interests which had been in conflict before now were invited to resist the common foe.

The Mayor of Liverpool wrote to the Chamber of Commerce :—

It is not a matter of trade competition.   It is, to my mind, no less a question than whether the estuary of the Mersey shall be so tampered with as to destroy its advantages as a port, and thus ruin not Liverpool only but all interests—and they are many—which depend for their existence upon the preservation of this harbour.   I cannot but think the House of Commons will throw out the Bill, and put a stop for ever to this mischievous project.

The Chairman of the Chamber of Commerce said :—

If the operation succeeded it might be death to Liverpool.   It certainly would be an enormous injury.

Sir William Forwood blamed the people of Liverpool for their apathy.   They had believed the capital to make the canal never would be raised, but he assured them this was a mistake. The men of light and leading and the capitalists in Manchester were not supporting the Bill, but the working classes were enthusiastic in its support. The Chairman was Mr. Adamson, a great promoter of companies, not all of which were successful, and he was supported by Mr. Lawrence of the Hull and Barnsley Railway, a most disastrous concern.

Nearly all the trading companies of Liverpool petitioned against the Bill in the Commons. Alluding to the prevailing excitement, Mr. Pember once said, "they have lost their heads, you can do no good with them. You can't reason with a startled horse."

When the question of again giving evidence against the Bill came before the Birkenhead Corporation, some of the members objected to such repeated wasteful expenditure, and it was urged that it would be better to try and come to terms with the promoters. It was, however, decided to oppose the Bill.

On the 4th June the Bill was read a second time in the House of Commons, and referred to a Select Committee, of which Mr. W. E. Forster (M.P. for Bradford) was appointed Chairman. Full particulars of the Committee's proceedings will be found in another chapter. Suffice it to say that on the 20th July Mr. Forster cheered the hearts of the promoters by announcing that, whilst pronouncing no opinion on the commercial case, there was nothing in the engineering case which would cause the Committee to reject the Bill. The opponents had rather hoped Mr. Forster's idea of pushing the canal farther inland would mean taking it beyond the limits of deviation, and thus cause a collapse of the Bill. When the commercial case came on, Sir William Forwood, who had previously been the chief witness for the opponents, disappeared. His friends said it was not from a disinclination to be called, but because his evidence did not need repetition in face of the strong fresh evidence that had been presented, whilst the promoters attributed it to an unwillingness to face another raking examination by Mr. Pember, who would charge him with a breach of his promise not to oppose the Bill.

The Provisional Committee were very indignant with what they termed the faithlessness of Sir William Forwood and the Dock Board. To expose it they published a list of the pledges given not to oppose the Bill provided the suggested new course was adopted. There can be no question the alternative scheme propounded, which Liverpool said would enable the canal to be made without damaging their interests in the estuary, materially helped in the rejection of the Bill of 1884. When these pledges were given, it was evidently believed that the promoters would have neither courage nor endurance enough to bring forward a third scheme. Space will not allow me to give the evidence in full, but the following witnesses then said they would withdraw their opposition if the promoters would fall in with Mr. Lyster's proposal, *viz.* :—

1. Captain J. B. Eads, America.
2. Sir Frederick Bramwell, London.
3. Mr. Thomas Stevenson, Edinburgh.
4. Mr. G. F. Lyster, Liverpool.
5. Mr. Henry Law, Liverpool.
6. Mr. Leveson Vernon-Harcourt, London.
7. Mr. T. D. Hornby, Liverpool.
8. Sir William B. Forwood, Liverpool.
9. Mr. Stephen Williamson, M.P., Liverpool.

Mr. Bidder, Q.C., and Mr. Aspinall, Q.C., in their speeches both confirmed the pledge, the latter using these words :—

I represent the Dock Board also, and I am authorised in the strongest possible sense upon behalf of the Dock Board, to ratify that offer of Sir William Forwood's in the precise words in which he made it.

Meanwhile, the Liverpool Press were pouring out the vials of their wrath on the devoted canal. They gave a précis of Mr. Collier's evidence on behalf of the Bridgewater Navigation Company to show that a canalised river and a barge navigation would fulfil all the requirements of Manchester, also a letter from Mr. Patterson who said :—

I am quite aware the capitalists in London treat the proposal as sure to collapse from internal weakness.

The *Liverpool Courier*, commenting on the engineering evidence, wrote :—

It is worth remembering that if these expectations of the engineers should be realised, the canal, for which so much is risked, will itself be utterly useless except as a memorial of stupendous folly.

Poor Mr. Lyster, the engineer of the Dock Board, was not spared, because in his innocence he had suggested a canal that would be free from objection :—

What shall we say of Mr. Lyster? Least said soonest mended. The well-intended suggestion of the Dock Board engineer has coloured with perplexity the management of the Liverpool case this year in Parliament.

The 30th July, 1885, was a day memorable in the annals of Manchester, for then Mr. Forster's Committee declared the preamble proved, subject to certain conditions and clauses to be afterwards arranged. These occupied three days, and on the 3rd of August the Bill was formally passed and reported to the House the same night. Seldom has a Bill been pushed through with such celerity. It was read for the third time on the 5th August and received the Royal assent on 6th August, 1885. From the first there had been fears of a dissolution before the Bill could be got through. Thus after three years of incessant fighting, during which the battle favoured each side in turn, a final and decisive victory was achieved by the promoters. Their success was welcomed most heartily, not only in Lancashire, but all over England, for by this time the struggle against vested interests, in order to free the trade of a district from the thraldom of an oppressive monopoly, had attracted general attention, and even those who did not understand the question, admired the pluck and perseverance exhibited by the Manchester Provisional Committee, with Mr. Adamson at their head. The fact, too, that a district should be compelled to spend between two and three hundred thousand pounds in order to conduct its own business in its own way, called forth much comment, and roused some indignation.

Whilst the Manchester papers commented on the success with becoming modesty, some very hard things emanated from the Press of the sister city. The *Manchester Guardian* was sure the public would not be slow to see "the motives which actuate those gentlemen who, until very lately, loudly declared they did not fear the competition of the canal, and who now take every step in their power to prevent its construction". It regretted, too, that when Mr. Adamson offered to meet the opponents, and discuss the altered line of the canal, they had declined to meet him. It hoped the conditions inserted would not be too onerous, urged that the question of capital should be faced boldly and at once, congratulated the opponents on not attempting to defeat or obstruct the Bill on false issues or as regarded time, and quoted Mr. Forster that, "the promoters would not be like themselves if they gave in".

The Liverpool papers did not think it possible for Manchester to raise nearly £7,000,000 of money in two years.

The whole project is visionary in the extreme, and one should not be surprised if it should finally dwindle down into the Bridgewater Navigation extension, which could have been obtained without so much fuss and expense.

A rather rabid prophet wrote to the *Liverpool Mercury* :—

It will be a sickening day on 'Change in Manchester and Liverpool when, after only a short period, the river channel having been induced to flow along the hard-defined boundary of the canal, silting up takes place and a shoaling of the bar follows.

He then pictured a huge ship stranded on it, and the commerce of Liverpool ruined, also further damage done to the upper estuary by the new channel which "it would take millions to restore". Another correspondent in the *Daily Post* advocated paying Manchester off in her own coin by starting cotton mills on the banks of the Great Float and Wallasey Pool.

Manchester proposed to throw £10,000,000 into a big ditch. Let Liverpool put half as much into cotton mills at Wallasey, and long before the big ditch is dug, most of the need for it will have disappeared.

Mr. Adamsons homecoming on the 7th August was made the occasion of a remarkable demonstration. He was met at Stockport by a brass band and a large concourse of people who escorted him home. On the way he passed under triumphal arches which bore the inscriptions "Well deserved success" and "Welcome". A stop was made opposite the Didsbury Hotel, where addresses of congratulation were presented to him by Mr. Kelley (Secretary of the Trades Council) on behalf of the working classes, who were deeply interested in the canal, and by Mr. William Marsland, on behalf of the workmen of Messrs. Adamson & Co., of Hyde. In reply, Mr. Adamson thanked his friends most heartily for their kindness and enthusiasm, and said they had won one of the greatest battles of the kind that had ever been fought. Unfortunately, in the hour of triumph he made some statements which gave his opponents a subject for criticism.

He had no hesitation in asserting that in five years after the canal had been opened £10 shares would be worth £20, and before it had been opened ten years they would be worth £30. Mr. Adamson here was too optimistic, but his prophecy that the canal would save £1,000,000 a year to the trade of the district has, however, been realised.

At Eccles on receipt of the news cannons were fired, the church bells were rung, flags were put out and bands of music paraded the streets. It was also decided to present an illuminated address to Mr. Adamson, prior to which there was to be a trades procession, and an ox roasted in a public place, which was afterwards to be divided among the poor people.

At Warrington a public meeting was held and an address presented. Sir Gilbert Greenall and Mr. Adamson were among the speakers; the latter was very

The material originally positioned here is too large for reproduction in this reissue. A PDF can be downloaded from the web address given on page iv of this book, by clicking on 'Resources Available'.

severe on Sir William Forwood, saying, "he merited all the contempt he got in Committee" for making a promise and not keeping it.

Similar demonstrations were made in Stockport, Oldham, Widnes, and a great many other Lancashire towns where the Ship Canal struggle had all along been watched with the keenest interest. The Corporation of Manchester decided to invite Mr. Adamson to a banquet in honour of the event. Motions of congratulation were also passed by most of the local boards and public bodies round Manchester.

On the 19th August a meeting of the subscribers was held in the Free Trade Hall, Mr. Adamson in the chair. He was received with great enthusiasm, and went on to thank the subscribers and the Manchester Corporation for their unwavering and cordial support, which he trusted would be continued now the capital had to be raised. He recounted the almost insuperable difficulties the promoters had met with, and said it was the first time in the history of a great contention that not a sixpence was paid for getting rid of opponents' claims. He admitted clauses had been wrung from them which they thought unjust, but they were compelled often to settle in view of the limited time at their disposal. He went into the savings to be wrought by bringing a full ship's cargo to Manchester, and combated the contention of his kind friend Mr. Patterson, of Liverpool, that a barge canal would have been sufficient. This meant helping Liverpool ten times as much as Manchester. Unfortunately, he could not resist the opportunity of pouring contempt on his old antagonists, Sir William Forwood, Sir Humphrey de Trafford and Mr. Pope. The first he called the Prince of Prevaricators, the second turned his sewage into the Irwell, and then was the first to seek an injunction against every one else; and as for "The great Northern Pope, Q.C.," as he called him, that gentleman had at any rate done them the justice to say, "they deserved the Bill for their endurance, if otherwise they had no right to have it". He appealed to the working men of Lancashire to help in finding the money, and said that if they smoked and drank a little less, 1s. per week would soon help towards paying for a share. If it did nothing else but encourage thrift and sobriety, the Bill would be a blessing to Lancashire and the whole district. If the subscribers got no other reward they might be sure that they would receive the thanks and prayers of thousands who succeeded them for the good work they had done in their day and generation. Mr. William Agnew, M.P., in proposing a vote of thanks to the Chairman and the Provisional Committee, paid a high compliment to the sturdy endurance and solid weight-bearing power of the Chairman, and also to the close reasoning and marvellous eloquence of Mr. Pember, the leading counsel.   Such service could not be bought for money; his heart and soul were in the business.   He (Mr. Agnew) had never listened with more enraptured satisfaction to any man's utterances in a Committee room than he had done last year to part of the speech which Mr. Pember took thirteen hours to deliver.

Alderman W. H. Bailey as usual could not refrain from a joke.   He observed that a good deal had been said concerning the Liverpool Nebuchadnezzar, Sir William Forwood, but they could not blame that gentleman for refusing to worship at the shrine of their prophet Daniel.   As to the railways reducing their fares and ruining the canal, he assured the meeting this was impossible—they would first ruin themselves.

At another congratulatory meeting he contrasted the wear and tear of railways with the easy passage of ships over waterways, and quoted Byron :—

> Time writes no wrinkles on thine azure brow,
> Such as creation's dawn beheld, thou rollest now.

He then read the following letter which he had received from the venerable Lancashire poet, Edwin Waugh :—

ABERFELDY, N.B.,
August 20th, 1885.

DEAR BAILEY,

I have just been reading with very great pleasure your able review of the extraordinary behaviour of Sir William Forwood in relation to the Manchester Ship Canal. The unscrupulous bitterness of Liverpool at the loss of its great marine toll-bar will need constant watching and careful exposure for a while.   I cannot doubt that the great determination and ability which has achieved the passing of the Canal Bill will also succeed in raising the necessary funds within the stipulated time ; but the opponents of the Bill are evidently determined to stick at nothing to prevent it.   If Mr. Adamson and Mr. Pember never do another stroke of business in their lives, they deserve the gratitude of the whole kingdom for the noble way in which they have fought out that great battle, and their names will be honourably connected with it long after they are gathered to their fathers.

EDWIN WAUGH.

The Liverpool Press commented strongly on Mr. Adamson's attempt to depreciate the value of the Dock Board property, and on the various attacks on Sir William Forwood, also on his optimism in believing in a 20 per cent dividend. The Chairman of the Dock Board (Mr. Hornby) at the next meeting after the Bill

SIR WILLIAM H. BAILEY, DIRECTOR, MANCHESTER SHIP CANAL
COMPANY.

Guttenburg.

To face page 262.

was passed, consoled his colleagues by saying that though they had lost, they had succeeded in introducing monetary and other clauses with which it would be difficult for the promoters to comply, for they were unusually onerous.   In some cases these clauses conferred a freer use of the canal, and in others imposed very serious liabilities on the promoters for carrying out works of various kinds, including keeping open communications with the Mersey and maintaining a fixed depth of water, constructing roads, wharves, locks and working ferries and bridges.   He complained of the evil speaking to which the Dock Board had been subjected when it was said "neither on engineering nor on commercial grounds have the Liverpool Dock Board or Corporation the slightest ground for honourably opposing the Ship Canal Bill in the ensuing session".   He was of opinion if their engineer had not devised an alternative scheme the Bill would have been doomed.   It was hurried and they did not get the clauses for protection they wished, but clauses were given to the Mersey Conservancy Commissioners which it was to be hoped would protect the river.   On the whole, he considered the Liverpool opposition was vindicated by the result, and they had prevented interest being paid out of capital during the construction of the works.   At a subsequent meeting Mr. Littledale, a member of the Board, pointed out that Mr. Lyster's plan, which Sir William Forwood promised not to oppose, had been declared by their own expert, Captain Graham Hills, to be most injurious to the estuary.

Directly the Bill was passed Sir William Forwood, Messrs. Patterson, Coke and others commenced a correspondence in the Liverpool and Manchester papers, with the evident intention of disparaging the canal and preventing the capital being subscribed.   These letters and the replies to them are very interesting, but space will not allow of their insertion.

When Mr. Forster reported the Ship Canal Bill, Sir A. Otway moved that the report be received, and the Standing Orders having been suspended, the Bill was read a third time, and thus, after three years' hard fighting, the right was obtained of making a Ship Canal to Manchester.

The Provisional Committee then began seriously to form plans for raising the money.   As power to issue £1 shares could not be obtained, they were fixed at £10 each.   It was proposed to make the payments easy; the calls were not to exceed £1, and not more than £2 10s. was to be called up in any twelve months. It was felt that the refusal to allow payment of interest during construction would be a serious block, and it was determined early on to apply for that power in the next session of Parliament, especially as it had been granted elsewhere. Various suggestions were made for raising the capital. The *City News* urged that Manchester and the surrounding towns should contribute liberally, and if there were legal difficulties in the way of their doing so, that an application should be made to Parliament to give the necessary powers. Suggestions came from several working men that means should be taken to allow them to contribute small sums, and it was quite evident that the working classes were thoroughly in earnest, and willing to help according to their means. To meet their case it was arranged to issue 1s. coupons in books of ten each, the Ward Committees to collect the money till there was a sufficiency to purchase a share.[1]

The Civic Banquet in celebration of the passing of the Ship Canal Bill took place on 6th September, Sir John Harwood, the Mayor, in the chair. He was supported by his colleagues in the City Council, the Mayors of surrounding towns, the Consular body, the local M.P.'s and many leading citizens. Mr. Adamson and the Ship Canal Directors and staff, with their leading counsel, Mr. Pember, Q.C., were the guests of the evening. The Chairman heartily congratulated the Ship Canal promoters (and especially Mr. Adamson) on their success; they deserved well at the hands of Manchester, and he hoped a good providence would spare them to complete their labours. "If there be any man of wealth in our midst who will not come forward at this time to the help of those who have this project in hand, and will not do something beyond his own selfish interest and self-gratification, something in the interests of humanity" (to use the words of Sir Walter Scott), "he shall go down to the dust from which he sprung, unwept, unhonoured and unsung". Mr. Adamson, in responding, paid a high compliment to his colleagues, the engineer, the solicitors, and the four counsel (of whom Mr. Pember was the leader), and asked for the monetary support of the mercantile men of Manchester—of the support of the working classes he was already well assured.

Mr. Pember made a very happy speech. Manchester had now been married to the sea, or rather he should say betrothed, for her true marriage morning would be that on which the rising flood tide on the Mersey rippled through Mr. Leader Williams' flood gates at Eastham. He had been so long a witness to the heroic refusal of Lancashire and Cheshire to acknowledge or tolerate defeat, that he could not imagine they would now basely repudiate victory. All great enterprises must be started to some degree on chance, and must involve some risk, but he looked hopefully on the monetary prospects of the canal. In the future of England there

[1] See Specimen.

No Coupon is valid without the Signature of the Chairman of the Provisional Committee, DANIEL ADAMSON, Esq.

Daniel Adamson

MANCHESTER SHIP CANAL.

PROGRESSIVE Nº OF COUPON 008815

MANCHESTER SHIP CANAL.

SUBSCRIPTION TO £100,000 PRELIMINARY FUND.

ONE SHILLING COUPON.
NON-TRANSFERABLE.

WHEN NOT LESS THAN TEN OF THESE COUPONS HAVE BEEN ACCUMULATED, THEY WILL BE ACCEPTED BY ANY OF THE BANKERS TO THE SHIP CANAL PROJECT, AND WILL BE EXCHANGED FOR AN OFFICIAL RECEIPT, ENTITLING THE HOLDER TO A PRIORITY IN THE FIRST ALLOTMENT OF SHARES.

James Collins & Cº Manchester

Specimen of the One Shilling Coupon.
Obverse and Reverse.

Manchester Halfpenny, 1793.
"Success to Navigation".
Obverse and Reverse.

To face page 264.

would be room enough and to spare for both Manchester and Liverpool, and he trusted both of them would advance along the paths of commerce, at once the greatest rivals and the greatest friends.

Mr. Jacob Bright thought it childish to suppose the capital required for the scheme would not be forthcoming, and the Mayor in replying to his health said he considered the 2d. in the £1 added to the rates for the Bill had been well-spent money.

On Saturday, 3rd October, the people of Manchester gave vent to their jubilant feelings in a great trades procession, with subsequent public meetings. From the first the various trade guilds of Manchester had been some of the heartiest supporters of the canal; they saw in it more work, cheaper food and greater general prosperity, and they were anxious to further the cause and show their gratitude to the men who had so successfully fought the battle in London.

So it was arranged to have a general holiday, and that the Trade Societies of the town, with their flags, banners and insignia of office, should meet in Albert Square and walk to Belle Vue Gardens. They met at noon to the number of about 30,000, and were headed by Mr. Adamson, who had with him in his carriage Mr. Pember, Q.C., and Mr. Leader Williams. Following were Mr. Jacob Bright, M.P., Mr. Houldsworth, M.P., the Provisional Committee, and many of the leading subscribers. In the procession were also the Mayors and Corporations of Manchester and Salford, and bringing up the rear were the various Temperance Societies of the city. For want of room it was arranged the latter should hold their meetings in Alexandra Park. The procession of the societies was original and characteristic. The boilermakers and shipbuilders carried a model of a large screw steamer. The engineers, 3,000 in number, hoisted various emblems. The bakers bore aloft an enormous loaf, with the name of Daniel Adamson on it. The tinplate workers had made a suit of armour for their standard-bearer which was much admired. Each of the glass-workers wielded a glass sword, and many had glass helmets which sparkled and were very effective. The Orders of Foresters and Oddfellows, arrayed in their insignia, with green and red sashes, made a most imposing sight. They carried a large model of a powerful steam-tug bringing a ship freighted with cotton up the canal. The bookbinders held aloft an enormous volume entitled *The Revival of Lancashire Industries*, by Daniel Adamson. The umbrella makers appropriately carried umbrellas, which were of various colours, and were a striking feature in the show. Unfortunately, the weather was somewhat boisterous:

with this exception the procession was a conspicuous success; it was four miles long, and the van had reached Belle Vue before the rear had left the Square. The crowd in the streets was most enthusiastic, especially in the reception they gave to Mr. Adamson and Mr. Pember. Speeches were made by Mr. Adamson, Mr. Pember, the Mayor of Manchester and several Trades Union leaders, but the noise and weather combined prevented them being heard by every one.

Afterwards, Mr. Adamson and party hurried off to Alexandra Park, where the temperance contingent had assembled, with the Mayor of Salford as Chairman. Here another round of speeches was delivered by Mr. Adamson, Dr. Pankhurst and others.

The demonstrations in favour of the Ship Canal were continued on the succeeding Monday night in the Free Trade Hall, a citizens' meeting being held to hear addresses from Mr. Adamson, Mr. Pember, Mr. H. M. Stanley and other eminent citizens. The hall was densely packed. Mr. Adamson desired the audience to give a hearty welcome to Mr. Pember, who had fought the Ship Canal battle with great zeal and ability; he then went into the question of the large savings that would be effected to the traders of Manchester and the surrounding towns. Speaking of his times of depression and encouragement, he mentioned that when he returned from London sick at heart at losing the 1883 Bill, the very next morning's post brought him a promise of £500 from Mrs. Jacob Bright and of £1,000 from Mr. John Rylands, and this encouraged him to go on. Mr. Pember followed, and astonished his audience by his eloquence and wonderful memory. He went through all the figures of his commercial case, and the particulars of the principal witnesses' evidence almost without a note. He indulged in some pleasantry about Sir William Forwood who, in 1884, put the traffic of Liverpool at 25,000,000 tons when he had no reason to say otherwise, and then when he wanted to show that Manchester could get no traffic, he tried to back out of his previous statements. Mr. Pember would not wonder if for once in his life Sir William was right in his 25,000,000 tons, adding, " It is a perfectly legitimate thing to be taught even by an enemy".

He recounted how when first he was retained as counsel his friends used to say of the canal, " Of course you do not like to say so, but it is all moonshine, is it not? Parliament will never pass it—it is a mischievous idea." Others said, "Why the deuce can't they let things alone" (laughter). This was said by excellent people who knew nothing of the nature or merits of the project.

Mr. H. M. Stanley said when he came to Manchester the previous Saturday in the same carriage with a Frenchman, there was so much stir in the town, crowds, flags, etc., that his companion got anxious, thinking there was a revolution, but on asking a young man at the station, "What is the matter?" the reply was, "Oh, Manchester has gone mad" (laughter). "In what way?" "Oh, don't you know —have you never heard of the Ship Canal?" He really had never thought of it till that time, though he recollected, when on the Congo, reading of some Ship Canal inquiry from some interior town called Manchester. If he could only persuade Mr. Adamson to assist them in building the Congo railways, *that* and the cheapening of carriage by the Ship Canal would assist in realising his beautiful vision of the millions of yards of cotton cloth and other materials that would go to the Congo, and they would then have a better chance of beating their German competitors.

It could not be expected that the jubilation in Manchester could pass by without comment in the Liverpool papers. The *Echo* poked fun at the ox roasted at Eccles, and said:—

If the promoters are going to raise their millions through the instrumentality of popular shows, they should capture the great Barnum, and profit by his experience in working up sensations and appealing to the imagination of the mob. There should be a substantial practical outcome of all this exuberance : processions, banners, bands, feasting, and a profuse interchange of compliments are all very well in their way, but much more will be necessary before the sea finds its way to Manchester.

The *Courier* wrote :—

But why Manchester should go into ecstasies over the Canal Bill in chill October may not perhaps be intelligible to the unsophisticated public, yet the reason is very plain. They have got the Bill, now they want the cash. That Manchester will ever see masts of the great steamships of the Atlantic and the East Indies mingling with its chimney stacks is a picturesque scene never to be realised. This sober fact should be borne in mind by those who put their money in the gutter.

Mr. Henry Coke told the Liverpool Chamber of Commerce that the demonstrations in Manchester reminded him of a Bombay company in 1864, got up to cut a new channel to Back Bay with £2,000,000 capital. Shares worth £500 went up to £2,500, then the bank failed : it was a great fiasco. The channel never was made, all the illuminating and tom-tomming was a fraud, and the people lost all their money. Evidently he expected the same fate would befal the Ship Canal.

Two controversies at this time occupied much space in the Manchester papers. One between Sir Joseph Heron and Mr. John Patterson, of Liverpool, and the other on "What is a Port?" Mr. Patterson stated that it was under the lead of Sir Joseph Heron that Liverpool was compelled to buy the Birkenhead Docks, and so increase the dock debt. This Sir Joseph indignantly repudiated. There was not the slightest ground for such a statement. The £1,143,000 paid by the Liverpool Corporation for docks in 1855 to get rid of the opponents of town dues, was the act solely of the Liverpool Corporation, and the after expenditure was largely under Acts obtained as early as 1853, whilst the first Act, for which Manchester was responsible, was dated 1857. He admitted it was a mistake that Act did not stipulate the dues should cease when the money paid by the Corporation had been recouped out of the dues collected, but it was always expected this would be done. Mr. Patterson in reply argued that though the Liverpool Corporation had the then moribund Birkenhead Docks to deal with in 1855, it was not till 1857, and at the instance of Manchester and other parties, that the Dock and Corporation properties were merged. He taunted Sir Joseph with never having attempted to correct the mistake he admitted was made as to the dues.

Sir Joseph Heron rejoined that in 1855-56, and before Manchester actively interfered, the Dock Board and Corporation were negotiating as to the Birkenhead Docks, and that the former knew they were incomplete and would need a large expenditure.

Mr. Patterson then explained that if Manchester did not interfere prior to 1857 it was not their fault, for they tried to get a *locus standi* and were refused, and he again charged Sir Joseph with want of prevision about the dues which had remained to be a blister on the trade of Liverpool. He ended by saying, "To this statement I firmly adhere".

The "What is a Port?" controversy was started by a letter from Mr. Howard Livesey, of Lancaster, to the London *Times*. He maintained there must be a good waterway and an existing market to make a port:—

To suppose the merchants of Liverpool would, in case the canal were made, vacate and establish their base of operations in Manchester is too absurd, and yet nothing short of such an issue would give Manchester the smallest chance of becoming a port.—There is no room or occasion for more marts for foreign produce, at least in this part of the Kingdom. In the nature of things centres for the sale of foreign produce cannot be too numerous. If Manchester is ever connected with the sea, and if anything is ever imported there, the least likely of any foreign produce to come is cotton, and yet the importation of cotton is made the basis of the hope of the canal enthusiasts.—We have one port in this part of England which is quite adequate and which cannot be rivalled or opposed. We have one exchange for the home cotton trade which is established in Manchester.   This is exactly what I say in regard to the importing trade, which cannot be taken from Liverpool where it is firmly and permanently established.

The above are extracts from a long letter, and were not encouraging to promoters just about to raise £10,000,000 of money.

Mr. Jacob Bright, M.P., in an excellent letter, showed the transparent fallacies in Mr. Livesey's letter, and said that the same argument could have been used even by Lancaster and Chester, which towns Liverpool in bygone times had supplanted.

Mr. W. H. Raeburn, shipowner, of Glasgow, also wrote showing Glasgow's being farther up the Clyde had not prevented her progressing much faster than Greenock, and that trade would go where there was population to manufacture goods and consume food.   Also that the sea carriage was as cheap for the longer distance to Glasgow.

Many other correspondents joined in the prolonged newspaper war.   The London *Times*, in an article summarising the correspondence, said :—

Direct sea communication with a town like Manchester, with the consequent saving of the break in the conveyance of the material of its industry, is like an invention of improved machinery for the sole benefit of cotton spinners.   They will be enabled to work more cheaply.   Liverpool does not endeavour to mitigate the strain of the yoke.   Its tolls on the goods which have to pass its docks are heavy and complicated.   Its cotton rings irritate to frenzy the Manchester spinners, who find their industry subjected to the schemes of knots of speculators whom they regard as no better than gamblers at their expense.   In the Ship Canal project they see the means of breaking the Liverpool chain from off their necks. It has awakened in Manchester an enthusiasm not inferior to that for the Anti-corn Law League.

On the 15th October the Salford Town Council agreed to contribute £250,000 towards the Ship Canal capital, and deposited a Bill for that purpose.

The Press and the Public were divided as to the best means of raising the £8,000,000 required.   On one hand, it was held that the working classes ought not to be asked to risk their earnings, and that Corporations had no right to help what was termed a private enterprise that might end in disappointment and loss, and that unwilling ratepayers ought not to be compelled to contribute.   On the other hand, it was contended that the wage-earning classes were more interested in the Ship Canal than any other, and that even if no dividend were earned, cheap living, good wages, and a good supply of work would be advantages that would yield an indirect dividend. As a reason that Corporations should take shares, it was urged that the canal would be a great highway upon which the future prosperity of Lancashire would depend, and that they should have such an interest in the canal as would give them influence and voting power, so as to prevent it becoming merely a money-making concern or passing into the hands of monopolist railway companies. It was further urged that if the various Corporations of Lancashire had an interest, this would facilitate the canal gliding into a public trust.

Mr. Adamson had an idea there would be no difficulty in raising the capital, and on the 8th October, 1885, a private and confidential preliminary prospectus was sent out. No portion was to be underwritten, and no brokers were retained to assist with the capital.[1] The prospectus was issued to capitalists and others, and asked for £8,000,000 in £10 shares. Application was to be made to the National and Provincial Bank of England, and other Manchester and local banks. The estimated cost put before Parliament for the works was £6,311,137. A contract, however, had been entered into with Messrs. Lucas & Aird to execute the work for £5,750,000, and the contractors had engaged to pay 4 per cent. interest on capital during construction.

Each subscriber to the Parliamentary Fund had a prior right to have shares allotted up to twenty times the amount of his subscription, and what he had already paid up was to be taken as part payment of the shares. When the Bill was passed most of the subscribers availed themselves of their right, and the £750,000 allotted shares afterwards referred to, were largely taken up by original subscribers to the Parliamentary Fund.

When the Commons had passed the 1885 Ship Canal Bill, there was a general feeling that the Board should be strengthened, and in the 1885 prospectus the following gentlemen were stated to be willing to join the Board at the first ordinary meeting of shareholders, *viz.* :—

> William Henry Houldsworth, Esq., M.P., Knutsford.
> Sir Joseph Cocksey Lee, J.P., Manchester.
> Alderman W. H. Bailey, Salford.
> John Rogerson, Esq., Durham.
> (The latter an old friend of Mr. Adamson's.)

The result of the first prospectus was very disappointing; the response was most feeble, only £750,000 of ordinary shares were applied for, and the issue was at once withdrawn; it became very evident that to obtain the capital it would be

[1] See Appendix No. IV.

necessary to get legal power to pay interest during construction, so on the 13th November a Bill was deposited for that purpose.

The necessity for a Ship Canal to Manchester has often had to be justified. It received confirmation, however, from an unexpected quarter. Sir A. B. Forwood, addressing the Liverpool Dock ratepayers on the 11th December, told them, "He did not believe they would have heard one word about the Manchester Ship Canal scheme if the Dock Board had not kept the rates so high".

During November a most determined attack was made by a professedly nautical man who wrote under the name of "Navigator," and who declared the water in the canal would be inadequate for a large trade, that it would be frozen for weeks together, and that when the ice broke up the injury to screws and paddle-wheels from floating ice would make the canal unusable. Mr. Leader Williams and Mr. Jacob, the surveyor of Salford, gave convincing answers founded on evidence given before the Parliamentary Committee, but still "Navigator" went on trying to alarm the public by the contention that because Rostherne Mere froze in winter the canal must do so likewise. The result has shown what illegitimate arguments were used to damage the canal.

The year closed on the Directors as they were busily engaged in preparations to obtain the necessary capital to make the canal.

CHAPTER XIV.

1885.

SHIP CANAL BILL IN THE HOUSE OF LORDS—EVIDENCE BEFORE COMMITTEE — SPEECHES OF COUNSEL — BILL PASSED WITH ONEROUS CONDITIONS.

I am one of those who have never had any jealousy whatever on the question of competition between water and rail. I believe there is a trade for the water and a trade for the rail. In addition to that, I think there is a traffic sometimes carried by railway which might be carried more profitably by water.—Sir E. W. WATKIN.

ON 12th March, 1885, the Ship Canal Bill was for the fifth time brought before a Parliamentary Committee. It consisted of the following members of the House of Lords: Earl Cowper (Chairman), Earl of Milltown, Earl of Arran (Baron Sudley), Lord Harris and Lord Romilly. The promoters practically made no change in their counsel, but inasmuch as the alteration of front as regarded the estuary involved fresh interests, several new counsel appeared for the opponents, notably Mr. Reader Harris for the Salt Chamber of Commerce, and Mr. Meysey Thompson for the borough of Widnes, and Wigg Bros. of Runcorn.

The Committee being quite ignorant of the case and its varying aspects, it was necessary to begin *de novo* with the engineering and commercial evidence. This will not be recapitulated, the intention being to direct attention to evidence regarding the changed course of the canal, and any fresh matter connected with the case.

The hearing commenced on the 12th March, and the decision was given on 6th May. The promoters had seventy-six engineering, commercial and other witnesses, the chief engineer being under examination for four days. The opponents had forty-two witnesses. There were thirty-one petitions against the Bill, of which twelve were supported by counsel.

Mr. Pember, Q.C., after alluding to the history and vicissitudes of the Ship Canal Bill, said that last session the opponents, in their desire to get rid of the proposed channel down the middle of the estuary, promised that if it were withdrawn, and next year the promoters would come for a Bill to skirt the side of the estuary, they would not oppose it.   He quoted the evidence of Sir F. Bramwell, and Messrs. Eads, Lyster, Thomas Stevenson, Law, Vernon-Harcourt, Hornby, and last, but not least, of Sir William Forwood, who had pledged Liverpool not to oppose.   He expressed himself much astonished that, after practically adopting the plan of Mr. Lyster, the dock board engineer, he should still meet with opposition, even more intense than before.

After describing the new line of canal from Eastham to Runcorn, Mr. Pember explained the system of sluices at Weston, whereby the waters of the Weaver, after crossing the canal, passed into the estuary, and showed that the total tidal abstraction by the new scheme would not be more than a tenth of that caused by making the walls of the Liverpool Docks; it only amounted to about 3,000,000 cubic yards, out of a total tidal capacity of 960,000,000 yards.   The estimates for the works exceeded those in the last Bill by £388,000.   Speaking of the necessity of cheap freightage, he said England's commercial supremacy often rested on a balance as fine as a razor's edge, and that the transit of goods between Manchester and Liverpool represented $2\frac{1}{2}$ per cent. on the value of the manufactured article, and $7\frac{1}{2}$ per cent. on the labour it took to produce it.   Further, that whilst on the Continent the carriage of manufactures averaged one penny per ton per mile, between Liverpool and Manchester the average was over threepence.

At this point the Chairman suggested that in view of the commercial advantages being acknowledged twice by the House of Commons and last year by the Lords, there was no need to go into that branch of the case.   To this the counsel for the Railways and the Trafford Estate strongly demurred.   It was pointed out that the Lords spent twenty-one days the previous year entirely on the commercial portion, but after the protest the Committee decided to hear the whole case.

Mr. Pember went on to prove there was a reasonable prospect of the canal paying, "but," said he, "we should be prepared, and I will undertake to say the whole of Lancashire will be prepared, to see this canal made, and never return a single sixpence upon its working expenses for the sake of the emancipation they would get by it".   He then dealt with the petitions.   Of the one from the Dock Board he said it complained of damage to the estuary, and seemed to be a copy of that of last year, notwithstanding the promoters had virtually taken the canal out of the estuary for a length of 13 miles.   It stigmatised, too, the provisions in the Bill regarding the return of the land water above and below Runcorn as illusory.   In reply, he pleaded he was giving three or four times more capacity for passing away water than had previously existed at the Frodsham openings.   As to the complaint of encroachment, Liverpool in making her docks had taxed the estuary to the tune of about ten times what the promoters proposed to take, and themselves said they had done no harm. Of the Weaver petition, he said the opponents complained of loss of access by the interposition of the canal in front of the Weaver and the works at Weston Point. All he could say was that the promoters were giving far better openings than before; the entrance to the Weaver itself would be safer and deeper, and if there was any difficulty the canal could be used by paying for it.

The Bridgewater Navigation Company for two years' running had come to a formal agreement with the promoters to sell them their undertakings.   These gentlemen had now wakened up to the fact that for £350,000 they could make their navigation 9 feet deep, and said they did not feel bound by any arrangements they had previously made to sell their property.   Mr. Pember denied such a waterway as they proposed to make would be satisfactory.   The petition was simply that of a company objecting in order to make terms.

Here the Chairman again intervened and asked the opponents :—

If they really denied that this canal, if it can be made, will be a great commercial advantage to Manchester ?

Mr. Pope.—My Lord, we assert, and think we can prove conclusively, that it is as great a delusion as the South Sea Bubble.   We are prepared with a great deal of evidence.

Earl of Milltown.—Because you are asking us to reverse a decision of a Committee of this House last year given after a prolonged investigation, enormous expense to the promoters and a great taking up of public time.

Mr. Pope.—Which we are quite entitled to do.   All I can say is, that one House did undoubtedly, as your Lordship says, last year, after inquiry, pass the Bill by a majority of one.

Earl of Milltown.—I beg your pardon.   The Committee last year were agreed as to the commercial advantages of the proposal.

Mr. Pope.—Members of that Committee did not make any secret of what their opinions were, and we know, therefore, as my learned friend very rightly said, that three members of that Committee were of one opinion and two of the other.

Earl of Milltown.—That was on the engineering question.   Pardon me, I also have the means of ascertaining their views in the same manner as you have, and I do not believe there was the smallest difference in opinion on the part of the Committee as to the commercial question.

Birtles, Warrington.

FRODSHAM MARSH.

To face page 274.

Mr. Pope.—I undertake to say that the Duke of Richmond at all events was entirely with me in my argument on the commercial question.

Earl of Milltown.—At any rate the evidence is before us.

After a prolonged discussion it was agreed to take the engineering case first, and the Chairman asked Mr. Pember to shorten the case as much as possible.

The Earl of Milltown then asked Mr. Pember a puzzling question :—

Are you going to call Messrs. Eads, Lyster, Law, Vernon-Harcourt and Sir William Forwood as your witnesses? They stated last year that if you adopted your present line they would be satisfied. If they are still of that opinion, do not you think it very material to your case to call them first?

Mr. Pember in reply said he feared such a course would be impossible. Mr. Eads was safe at the bottom of the Mississippi—at least he was not in England.

Great as would be my delight to find myself on the same side for once in my life with Mr. Lyster, he is hardly the gentleman I take him for if I could call him as a witness without the use of something approaching physical force, and I have not the slightest doubt he will still come and say something on the other side, when I shall have the pleasure of dealing with him in cross-examination. But Mr. Vernon-Harcourt, who was what I may term another independent engineer, I shall have great pleasure in calling on behalf of the promoters.

Touching the petition of Liverpool "that the tidal scour would be diminished which might lead to the silting of the bar, and that by training a new channel to the Eastham Locks the course might permanently be diverted to the Cheshire side," all the promoters wanted was to keep their channel open, just as Garston and Runcorn were in the habit of now doing, by dredging. They were willing to adopt clauses limiting the dredging to be done. To the *locus standi* of the Salt Chamber of Commerce, the North Staffordshire Railway and the Rochdale Canal Company, Mr. Pember offered opposition with the result that the *locus* of the former was disallowed, as was also that of the London and North-Western Railway in the North Staffordshire Railway petition, but the *locus* of the Bridgewater and Rochdale Canals were allowed. Mr. Pember's opening speech occupied two days.

Mr. Leader Williams was then called, and gave the early history of the Bridgewater Navigation undertakings. He mentioned that in 1827 the Liverpool Corporation, as owners of the Liverpool Docks, failed in their litigation when they tried to prevent the Mersey and Irwell Navigation taking water from Latchford to Runcorn for local purposes. Also that when, in 1846, the Mersey Conservator and the Admiralty fixed the height of a Mersey bridge, they inserted a clause to make all crossings by swing bridges, and this was the case a few years ago when the Trafford Bridge was erected at Old Trafford.   Before the Ship Canal was thought of Parliament insisted that, if the Liverpool water pipes crossed the Mersey it must be at a depth of 20 feet below datum line, so that the river could be deepened for navigation.   He corrected the opponents' evidence, and said eight hours would be ample time for lockage and passage up the canal, and explained that the plans of this year changed the location of the docks, and brought them three-quarters of a mile nearer Manchester.   In consequence, the very costly race-course would not be required, and the areas of Trafford Park to be taken would be reduced to 115 acres. At Partington the disused Cheshire lines on both sides of the river would be made available to convey Lancashire coal on one side and Yorkshire coal on the other to special tips to be erected.

Mr. Williams stated that this year's plans showed an entirely changed treatment as regarded the estuary.   When in Committee last year Mr. Lyster, the engineer of the Dock Board, said that if the promoters would so alter their scheme as to skirt one side of the estuary with the canal and do away with training walls in its centre, the Dock Board would no longer oppose the scheme.   Admiral Spratt, the Acting Conservator, speaking of an alternative plan designed by Mr. Lyster, and which that gentleman believed would be more feasible and less costly, said he had examined it by the desire of the Dock Board, and he was enabled to say that he saw no reason for the Conservancy Commissioners opposing it on the ground of interference with the navigation of the estuary.   This had caused the promoters to see if they could fall in with the suggestions of the opponents, and they had adopted Mr. Lyster's views, practically in their entirety, and now proposed to bring the canal on the Cheshire side to Eastham, whence they would, by dredging, widen and deepen the present channel, and make a good entrance into the deep waters of the Sloyne. The ground was favourable, and the whole of the work could be carried on for twenty-four hours a day, if necessary, instead of working in a treacherous estuary and only in calm weather.   They could, too, use the most modern French and German machinery and work by electric light.   In the past, whilst the Mersey estuary was under the care of the Dock Board, he had never known them spend a penny on improvements or in stopping the erosion of the cliffs.   One reason why Eastham had been chosen as the terminus of the canal was that it was outside the zone wherein the Liverpool Dock Board could collect dues.   In 1875 the Upper Mersey ports collected £105,000 and extinguished all town dues on the upper part

Birtles, Warrington.

LATCHFORD CANTILEVER BRIDGE.

To face page 276.

of the river.   Really his (Mr. Williams') plan was almost identical with Mr. Lyster's. If there was any difference, he trenched less in the estuary, and whilst Mr. Lyster proposed to take the whole width of one of the London and North-Western Runcorn arches, he only proposed to take half.   On a high tide he would only abstract from the estuary 2,000,000 cubic yards against 12,000,000 proposed by Mr. Lyster.   Along the banks of the estuary, and especially on the Cheshire side, the sea was making constant encroachments, carrying away great quantities of soil, and often large trees and shrubs with it, and these helped to obstruct the estuary.   By placing stone embankments in front of the crumbling cliffs, the promoters would largely benefit the navigation of the river.

The witness contended that water being indestructible the carriage thereby must be much cheaper than carriage by rail or road; once get a waterway with paved slopes and the expenses would remain in an almost normal condition, whilst the traffic receipts steadily increased.   He instanced the Suez Canal, the traffic of which in 1870 was 436,000 tons, whilst in 1883 it was 5,775,861 tons, with working expenses very little increased.   The maintenance of a railway was over 50 per cent. against under 10 per cent. on the Suez Canal.   There was no intention to make a continuous channel in the estuary and so divert water from the Garston side.

Cross-examined by *Mr. Bidder*, Mr. Leader Williams said the utmost abstraction entailed by his plan was $\frac{1}{4}$ per cent. of the contents of the whole tidal estuary, and could not affect the bar.   The witness then underwent a severe and critical examination on plans and estimates in turn by Messrs. Bidder, Aspinall, Saunders, Littler, Dugdale, Meysey Thompson, Ledgard and Pope on behalf of their respective clients, but without being shaken in any respect.   He was in the box nearly four days, during which time he successfully confronted some of the most able counsel at the bar.   At times he carried the war into the enemy's camp, as when he stated that Sir Humphrey de Trafford was a direct contributor to the sewage nuisance about which he complained, also when he showed the Committee that opening out a channel from Eastham to the Sloyne deeps was similar to what the London and North-Western Company had done when they connected their Garston Docks with a deep channel in the estuary some distance away.   Again, that the limited abstraction proposed could not affect the bar, because whilst the Dock Board were themselves abstracting tidal area in order to construct docks, at that very period the sectional area of the bar increased.

Mr. James Abernethy said he had not changed his mind about the soundness and practicability of the original scheme; but if it were necessary to adopt other lines he was of opinion the next best course was to form a line of channel connecting the canal at Runcorn with the sea at Eastham. He believed the Weaver and its navigation would be enormously benefited by the change, also the bar.

Mr. Leveson Vernon-Harcourt was of opinion the promoters' bill of this session carried out the alternative plan he suggested when last session he was giving evidence for the opposition. The interference with the estuary was insignificant compared with the proposition of last year. Putting a stop to cliff erosions would be a great benefit.

Mr. Fowler (of the Tees) said that the channel to the Sloyne deeps could not possibly deflect the flow from the Lancashire side, nor was it at all probable that the water, passing through the lock and tidal openings, could form a channel on the Cheshire side. In this opinion Mr. Messent, of the Tyne, and Mr. Deas, of the Clyde, concurred.

Mr. Giles, M.P. (Southampton), put in plans of the Mersey made in 1822 by his father, Mr. Francis Giles, showing that the tidal capacity of the estuary had altered very little, notwithstanding about 13,000 acres of land once flooded at high tides had been reclaimed, and he was quite sure the very small abstraction now comtemplated would not be felt. Inasmuch as the tidal openings would discharge at right angles no channel could be formed on the Cheshire side.

Mr. Duckham (of Millwall Docks), to prove the capacity of the locks for business, showed that in 1884 he had passed 58,498 ships and barges through one lock with a tonnage of 5,795,266 tons; he estimated with three locks, side by side, Manchester could pass through a set of locks 15,000,000 tons her annum.

At this point there was a discussion between the Committee and counsel as to what was to be included in engineering evidence. Mr. Pember was quite willing to follow the suggestion of the Chairman, and have the engineering case taken and decided upon first. Some of the opponents now wanted to change their minds; this induced Mr. Pember to say, "My learned friends are like horses at a ford, pawing before they begin to drink and then complaining the water is muddy". After clearing the room the Committee maintained their previous decision, that everything connected with the estuary and bar must be kept separate. Various pilots were then examined, who all said it was impossible that fresh water from the openings could make a fresh channel.

Mr. Leader Williams recalled, showed that though he embanked 417 acres, or

Barningham.

EXCAVATIONS AT THE ENTRANCE TO EASTHAM LOCK, LOW WATER

To face page 278.

5,904,950 cubic yards, he gave additional tidal capacity to the extent of 3,530,439 cubic yards, and thus all he abstracted was 2,374,513 cubic yards.

On behalf of the opponents the first witness called was *Captain Graham Hills;* he was of the opinion that the result of the work now proposed would be similar to that of last year. It would bring a fixed channel on the Cheshire shore. It would stop the erosion of the banks by the shifting channels in the estuary, but only to the extent to which that channel was carried. The loss of that erosion would cause the silting up of the bulb of that great receptacle, the estuary, and that loss would re-duce the depth upon the bar and would bring banks into the lower part of the Mersey, extending them much lower down than they were at present and would eventually affect the foreshore in front of the Liverpool and Birkenhead Docks. It would take away from Garston whatever were its present chances of a natural deep channel. He estimated a reduction of 1,000 acres in the tidal area. Channels were always shifting and affecting ports; his conviction was that the Mersey was so thoroughly impartial, that if it served one man's interest to-day it would serve another man's to-morrow, and if Ellesmere Port got its turn for a few years, Garston on the other side would get its turn in due time. He was quite sure Mr. Williams' channel from the lock entrance to the Sloyne would form a new channel on the Cheshire side. It was ancient history, but he admitted Liverpool, since the first dock was made in 1710, had been constantly encroaching on the estuary. The dock wall, north, was extended in 1863, but whether it had been for good or bad, what they had done had been sanctioned by Parliament with the belief that the estuary would not be damaged. It was true that when Birkenhead wanted to enclose tidal area Liver-pool opposed them. Of 12 miles of clay cliff liable to be eroded the proposed canal would protect only 1 mile 2 furlongs, and he estimated it would take 1,100 years to make a saving equivalent to the tidal area abstracted. In cross-examination, witness admitted he was in error, having calculated certain embankments as abstractions, and they must be eliminated.

On Mr. Pember asking the witness if the openings from the Ship Canal into the estuary were not what Mr. Lyster also proposed, he at once disowned the plans of the Dock Board engineer saying, "But Mr. Lyster's plan is not my plan," also "As regards Mr. Lyster's plan, I should recommend that it should be opposed by any body interested in maintaining the estuary of the Mersey as much as yours". The plan he should favour would be that Manchester should abandon the Mersey and take her canal to the Dee.

He was aware that last year when it seemed extremely probable that very serious damage was about to befal the estuary of the Mersey (passing the Bill), a suggestion was made that less mischief would occur if a scheme could be arranged along the shores of the Mersey, but Mr. Lyster did not put forward a complete scheme for the Manchester Ship Canal.   He certainly was in the House when Mr. Lyster suggested an alternative plan, *but he took care not to know anything more about it until it was necessary* to compare Mr. Lyster's plan with the one deposited. Witness said the total of Liverpool's Dock estate was 1,600 acres, and of this 1,300 acres had been taken out of the tideway at one time or other, but he did not think that the encroachment had done any serious damage to the estuary.   The loss of water inside the bar had a bad effect on it.   Also he admitted that channels had been dredged to Garston and to some Liverpool Docks in the same way that it was now proposed to dredge from Eastham to the Sloyne.   Last year it was the intention of the promoters to stereotype a fixed channel in the estuary; this year he was sure the results would be the same, though it was not intended.

Mr. George Fosbery Lyster, though he had designed an alternative plan, believed the canal to be the most Utopian idea he had ever heard of in his life; it was not a desirable, and could not be a remunerative undertaking.   By his plan he meant on the average to abstract nothing from the estuary; he put back as much as he took out.   Neither had he intended to use any water for lockage, except what was pumped; then he had fixed the level of his locks 12 feet higher than Mr. Williams, and this would interfere less with the estuary.   Also he had intended to have a tidal basin and six locks against three side by side on the promoters' plan. Whilst Mr. Williams was intending to make the excavation on the foreshore at a level coincident with the sills, he proposed to excavate the channel 4 feet deeper outside.   He believed the effect of the promoters' plans would be to stereotype a new channel on the Cheshire side, and damage the Lancashire side.   He had never heard of a sea-going ship being lifted 24 feet, and if that could not be done their deep sill would be perfectly useless.   Also he did not believe Mr. Williams could prevent his sills being blocked.   He could not conceive him making either towing paths or roads on his embankments.   Cross-examined, witness did not deny that as he followed the contour of the coast and the promoters went inland and took a straight course they abstracted less of the estuary than he did.   His idea was to put a small lock on the embankment opposite the entrance to the Bridgewater Canal, the Weaver Canal, and Ellesmere Port, to enable an interchange of traffic between the

ROCK CUTTING, LOOKING TOWARDS ELLESMERE PORT.

To face page 280.

Birtles, Warrington.

Lancashire and Cheshire shores of the Mersey, but he thought it very unfair to put an embankment in the front of those undertakings, and not to give them the freedom of the canal. He believed the interest on the cost of putting down and maintaining very expensive locks would, if the money were not spent, provide a subsidy for allowing small boats to use the canal free of charge. Such small traffic had now the free use of the Liverpool Docks.

Mr. Lyster in the Session of 1884 placed on the walls of the Committee room a plan prepared by himself for construction of a canal along the Cheshire shore, by adopting which the promoters would (he said) be able to avoid injury to the estuary and not imperil Liverpool. The promoters, before the plan was removed from the wall, took a careful copy, and it was fortunate they did so, as evidently the opponents regretted the exhibit, and did not afterwards allow the identical plan to reappear in the room. They had, however, previously submitted it to Admiral Spratt, the Acting Conservator, for his approval, and the report of that official saying he saw no reason for opposing the plan had been issued. The Dock Board subsequently were anxious to give the go-by to the plan exhibited by Mr. Lyster, but the promoters pinned them down to it as being the one on the adoption of which Sir William Forwood and others had made their promises not to oppose a future Ship Canal Bill.

Mr. Adamson, writing to the Liverpool papers on the 19th February, 1885, on the subject of the action taken by Mr. Samuel Smith, M.P. for Liverpool, said :—

Mr. Smith has evidently overlooked the engineering evidence and the undertakings given by counsel on behalf of the Dock Board and Corporation in Parliament last Session with reference to Mr. Lyster's alternative plan, a copy of which evidence I shall be happy to furnish on application to any of the shipowners and merchants of Liverpool who may be sufficiently interested in the question.

I am also prepared, on behalf of my Committee, to present to the Liverpool Royal Exchange for public exhibition in their rooms a copy of the deposited plans and sections of the Manchester Ship Canal of the present Session, provided the Mersey Dock Board will agree to hang up with them a copy of the plan prepared by their engineer, Mr. Lyster, for a Ship Canal along the Cheshire shore, which was referred to in the evidence given by the Dock Board's engineering witnesses in the House of Commons last Session, and to which scheme, it was stated, no objection would be raised by Liverpool.

I will leave it to the shipowners and merchants of Liverpool to decide whether the promoters of the Manchester Ship Canal have not made a *bonâ fide* effort to meet the objections of Liverpool. I am confident that any unprejudiced person on inspecting the two sets of plans and sections will be of opinion that those deposited by the promoters of the Ship Canal interfere even in a less degree with the estuary than those prepared by the engineer of the Mersey Docks and Harbour Board, and which by their instructions were submitted to the Acting Conservator of the river Mersey, and approved by him in his annual report lately published.

Mr. John Wolfe-Barry thought Eastham the proper terminus for a canal like Mr. Lyster's, but that Mr. Williams' canal with a lower sill ought to have gone to Bromborough, where deep water could more easily be reached.   When reminded in cross-examination that the promoters must get the sanction of the Mersey Commissioners for their works, he said he did not think Liverpool interests should be placed in their hands.   He believed the proposed works must destroy Garston as a dock.

The Committee then inquired of the promoters if the depth of their lock at Eastham was essential to them?   In reply Mr. Pember offered to raise the lock sill 4 feet if the Committee thought it expedient.

Sir Joseph Bazalgette gave his experience on the Thames.   There it was found that when dredging was carried out on the north side the south side of the river silted; in fact there was an equilibrium as Barking side became dredged the Crossness side shoaled, and the same result had followed dredging in the Erith reach.   He was of opinion if dredging was resorted to for the $2\frac{1}{2}$ miles between Eastham Locks and the deep water of the Sloyne, the Garston side would be starved. In cross-examination, it was elicited that he had given evidence in 1880 that dredging in one particular part did not affect the general régime of the river and was comparatively of small consequence.   Witness admitted dredging below the level of low water did not increase the quantity of tidal water or interfere with its flow.   He did not agree with Mr. Lyster in dredging 4 feet below lock sill.

Messrs. Henry Law, G. F. Deacon, Francis Stevenson and *A. M. Rendel* confirmed the engineering evidence previously given.

Sir Frederick Bramwell, for the Bridgewater Company, objected to limiting the tidal area at Runcorn Bridge.   In answer to a noble Lord he said if the promoters' lock sill was raised 4 feet it would make a considerable difference, but the dredging would still be a serious thing.

The Chairman then announced that the Committee would hear counsel speak on the question of the bar and Garston Docks.

At the adjournment on 14th April Mr. Pember complained that though Mr. Lyster had offered a conference where some engineering figures and facts could be adjusted, yet when the promoters tried to arrange a time of meeting Mr. Lyster replied that he was engaged, and that there was no need of the conference which he himself had first suggested.

Mr. Bidder was astonished that notwithstanding Mr. Lyster had pointed where and how to make the canal, Mr. Williams should have exercised such perverse ingenuity and made his scheme as objectionable as possible. The promoters were drawing on their imagination if they conceived he was going to throw Mr. Lyster overboard, even though Mr. Barry, Sir Joseph Bazalgette and others had criticised his scheme and thought that gentleman too venturesome. It was an individual opinion of Captain Graham Hills when he said he was as much opposed to the plans of Mr. Lyster as to those of the promoters. If Mr. Leader Williams wanted to shelter himself under the wing of Mr. Lyster, he should have accepted his conditions, which were essential, but he had not done so. Mr. Lyster had been amiably benevolent in drawing a plan of how a canal should be made, and putting notes of instruction on it, and had taken a higher view of his duty to his opponents than he, as an advocate, would have done. He complained that the promoters had never realised the magnitude of the work, or the danger there would be in dredging from Eastham to the Sloyne. Further, it was a certainty that the outfalls of fresh water would aid the dredging in making a new channel on the Cheshire side. Mr. Leader Williams had laid it down as an essential condition that the land water should not be interfered with, and had actually suggested putting down pumps to provide sea water for lockage. Now he was proposing to utilise the fresh water. Again, he was going to put his dock sills 12 feet deeper than Mr. Lyster. It was all very well to say that the Mersey Commissioners and Admiral Spratt had a veto, and could stop any work calculated to damage the estuary, but that body had showed themselves weak-kneed, and poor Admiral Spratt had been so worried that he had abrogated his functions. Besides, he had no funds to fight with even if he were disposed to do so, and there was a clause in the Bill to have an eminent engineer to appeal to if the Commissioners were against the promoters.

Whilst there were conflicting engineering opinions, was it right, was it safe, to pass a scheme that might imperil, nay ruin, the port of Liverpool with its enormous interests all over the world? He renewed his pledge of last year, that if the promoters would—as they could—bring forward a plan that would not jeopardise the estuary, his clients would not oppose it. If they had raised the level of the canal and gone down to deep water at Bromborough, he would have been happy and contented ; the reason this had not been done was the cost and a desire to avoid the Liverpool town dues.

Mr. Aspinall, for Liverpool, said :—

Talk about the enormous importance and greatness of Manchester! They anticipated that if the scheme were carried, an equally great city (Liverpool) would not only be disastrously affected, but virtually destroyed. Damage to the bar and estuary meant little to Manchester, but it was utter destruction to Liverpool.

Mr. Pope, Q.C., for the London and North-Western Railway Company, said it was a hydrostatic law that no river or estuary can permanently maintain a larger sectional area than the forces of nature have enabled it to maintain ; therefore, if a deep-water channel were dredged from Eastham to the Sloyne, followed by a new channel on the Cheshire side, the sectional area would be altered, and injury to Garston and its channel would be certain.

The whole scheme was reckless in its character, reckless of any outside interests, carried forward, I admit, with a courage, I may almost admit with an enthusiasm, which probably is not often equalled in the promotion of public undertakings, with a determination to bear down every obstacle and every cause which may interfere with the crotchet they have chosen to adopt as a sort of national work.

Mr. Saunders, for the Bridgewater Navigation Company, having also addressed the Committee, *Mr. Pember*, in reply, was amused at Mr. Bidder's self-complacency in speaking of the education his side had provided for the promoters' eminent engineers, in teaching them how to make the canal, and then rebuking them for not following instructions. Speaking of damage to the estuary and bar, Mr. Pember asked, would not Manchester, when she had spent £10,000,000, have the strongest interest in preventing damage to either? The great engineer Rennie said for the expenditure of half a million (and Admiral Spratt confirmed him) the pier perdu of the bar could be swept away. "Give us our Bill, I say, and there shall be no bar to maunder over." Mr. Bidder asked, "Why did you not bring in a Bill acceptable to Liverpool?" That would have been impossible. Whatever we do there will be a chorus of complaints. "Go to deep water at Bromborough," says Mr. Bidder, and shakes the phrase before us as a terrier does a rat. But this is deviating from Mr. Lyster's plan which, he says, we should adhere to. As to the Mersey Commissioners' clauses, if Mr. Bidder and Mr. Pope cannot agree, what are the promoters to do? Mr. Pember alluded to his having been twitted about his "*bids*," and said surely Mr. "*Bidder*" imagined he ought to change patronymics

with him. Though he did not think it necessary, he had offered to meet the opponents by raising his dock sill 4 feet, and the evidence showed the new channel
would be most beneficial in preventing erosion of the land on the verge of the
estuary. If the estuary and the bar reacted upon one another, as Messrs. Hills
and Barry affirmed, how was it that at the very time Liverpool was abstracting
large areas from the estuary to make the docks the bar improved? This very
fact upset the theory. Mr. Vernon-Harcourt, a man of the greatest authority, and
last year opposed to the canal, said of the Eastham dredging, "Whatever it
might do to Garston, the elongation of the Sloyne could not affect the bar the least
bit". And further, "I would not have come if Garston Dock had been affected".
Captain Graham Hills admitted the Pool Hall rocks extended into the estuary, and
these would render a channel on the Cheshire side impossible. It would seem that
Liverpool, Garston and the Bridgewater Navigation Company, may all enclose or
deal with estuarial lands, "but the moment Manchester lays a finger on the mere
rim of this estuary it is fatal". Now for Mr. Pope's dictum, "A good Sloyne
makes a poor Garston". Recently his clients, the London and North-Western
Railway, put in a plan by which they were going to dredge a fixed channel on the
Garston side—just what they object to us doing on the other side. Again, Mr.
Rendel's theory that there was a kind of see-saw between Garston and Eastham
was proved to be a fallacy, because he admitted both were affected when a central
channel appeared. Mr. Pember ended :—

> If I cannot convince, I do not care to persuade. My learned friend asked you not to
> pass the Bill if there is a risk of harm ; in other words, if eight or ten adverse witnesses come
> forward, they are to put an end to any enterprise, however valuable. If this is to be the
> case, good-bye once and for all to the material progress of Great Britain.

The Committee retired, and on re-entering, the Chairman said, "We allow the
Bill to proceed, and I have a suggestion to make that in the commercial case the
opponents take the initiative, and state their objections". The Earl of Milltown
further said "that the commercial advantages of the case had been already established before a Committee of this House". Objections being raised it was decided
to take the Bill in its usual course.

Daniel Adamson was, as usual, the first witness for the promoters. He repeated
his previous evidence. Speaking of depression in the iron trade, he said that iron
in 1872-73 fetched £6 per ton ; it could now be bought at 33s. to 34s. per ton. To

show that Manchester had long been seeking a cheap waterway, he produced a Manchester halfpenny dated 1795 with the inscription on one side "Success to Navigation".[1] He stated that to convey the cargo of a 3,000-ton ship from Liverpool to Manchester would require 1,500 carts or 1,000 railway waggons, and that there would be the same number of loadings and unloadings. He showed, too, that the canal would effect a saving in carriage of 51·85 per cent. on wheat, 56·75 on potatoes, and 58·53 on timber, and he estimated a traffic of 3,000,000 tons per annum, realising £750,000 in freightage. Competition with water carriage by railways was impossible when by the former goods could be carried at one-tenth of a penny per ton per mile. Since the Ship Canal agitation commenced, the railway reductions in the Lancashire district equalled £80,000 per annum. Eventually he believed there would be a saving to the cotton trade of £450,000 per year, and to Lancashire, Yorkshire and Cheshire of one million to one million and a quarter pounds annually.

Mr. Marshall Stevens showed the insufficient and costly accommodation at Liverpool, and that 95 per cent. of the vessels using that port had a tonnage of less than 2,000 tons. Also that Garston was now doing a traffic of 2,000,000 tons a year, which was increasing. Further that the Weaver Navigation did about the same weight as Garston. If a vessel arrived at the Liverpool Bar about high water and was too late for the tide to get into dock, she would have to lie outside ten hours, sufficient time for her to get to Manchester by canal.

At this point *Mr. Pope* intervened, and asked (on behalf of the opposing counsel) if the impression which had been conveyed to them that their Lordships would be bound by what transpired last year, was or was not correct, because it would be useless to go into a detailed inquiry that would be to some extent perfunctory. The Chairman assured counsel that whatever expressions of opinion had come from individual members, the present Committee would not be bound by the decision of previous years: they would not in any way cut short the opponents' evidence.

In cross-examination, Mr. Marshall Stevens said his statement about limited accommodation in Liverpool was but a reflex of the openly expressed opinion of many members of the Dock Board. Till the Bridgewater Navigation Company bought the Bridgewater Canal, it was quite free from railway influence. Once the rates between Liverpool and Manchester were 2s. 6d., now they were 9s. His statement was virtually correct that the same charge was made for conveyance when

[1] See Specimens, page 264.

goods were passed over the ship's side into barges as when they were carted to a railway station.

The Chairman again appealed to the parties to shorten the evidence. The Earl of Milltown said, "If the engineering question had been decided in the Commons in favour of the promoters last year, that Bill would now have been the law of the land". The opposing counsel, however, would not consent to any abridgment. The case must be fought out to the bitter end.

Mr. W. H. Cornforth estimated the brokerage of both buyer and seller in the Liverpool cotton market at £400,000 per year, and believed that in most cases one brokerage would be saved if the trade was done in Manchester; with the Ship Canal in existence he considered this quite feasible. To protect the middleman, the Liverpool Association rules prohibited a merchant selling direct to a spinner. This rule obliged the firm of which Sir William Forwood was a member to give up their office in Manchester, but they still had an agent who did a large business without the intervention of Liverpool brokers. In Manchester there was no such thing known as trades unionism among brokers. Yarn and cloth agents had no restrictions, they enjoyed free trade.

Mr. Samuel Ogden said the cotton trade was so bad in the Rossendale Valley that one-fourth of the mills were absolutely stopped, or working short time. India spun yarns competed closely with English spun coarse yarns for export, and $1\frac{1}{2}$ to 2 per cent. would turn the scale. Speaking as an overseer he might say that whilst years ago the assessments in Manchester averaged an increase of £6,000 per annum, they had lately receded to an average of £900 yearly, so that they were going to the bad at the rate of £7,000 per annum. He felt sure the canal would attract capital—the Suez Canal did so; the Manchester Chamber of Commerce was laughed at for recommending the Suez Canal, but they proved to be right.

Mr. Samuel Andrew said the Liverpool Cotton Association had boycotted the Oldham Cotton Buying Company, and the Liverpool brokers were not allowed to sell to them.

Mr. C. E. Ross gave the cost of freight of ten bales of yarn from Bombay to Liverpool as £2 15s. (6,500 miles), whilst from Liverpool to Manchester (31 miles) the cost for the same goods would be £2 8s.

Mr. Charles Holt, corn merchant, Manchester, brought cargoes from all ports in North and South America, and paid in 1883 £4,000 for Liverpool Dock and town dues, importing 400,000 sacks of flour and 50,000 tons of wheat. Roughly speaking, the expenses in Liverpool (independently of warehousing) were 5s. per ton.   In addition he paid 8s. 2d. per ton carriage to Manchester.   At these rates the Ship Canal would have saved him £21,000 in the year.   Possibly he might have to reduce his prices 1s. per sack, but this would be a gain to the public.   In 1883 he had ships waiting for ten days to discharge at Liverpool, and when he wanted warehousing it was scarce and dear.   The Alexandra Dock was 7 miles from the Corn Exchange, and it took him as long to get to his ship as it did to come to Manchester.   The Corn Exchange in Manchester was commodious and well arranged; it did a large distributing business.

When *Mr. Pope* came to cross-examine this witness there was considerable friction.   Pressed as to why millers did not use Fleetwood, he replied, "Common-sense would tell you that they would have bought there if it had been to their advantage," which brought the rejoinder, "Kindly answer my questions, and neither compliment me nor chaff me".

The witness was asked if the millers intended to revolutionise the trade and take all the breadstuffs to Manchester?

In reply he said :—

It is time the trade out of Manchester was revolutionised; they have been taxing our produce for the last fifty years, and it is time we untaxed them.   First look at the trade heavily taxed by Liverpool.   I tell you this, that if an angel from heaven had traced out the Ship Canal, and no human being could say it was wrong, you would represent a clique in Liverpool who would endeavour by hook or by crook to throw out the Ship Canal.

Mr. Pope.—I have not yet recognised the finger of heaven in Mr. Leader Williams!

A.—I hope you have.

Q.—Heaven has derived a good deal of assistance from the other place in the last two years at any rate?

A.—I know more about Liverpool than you do.   I am older than you are.   I have been fifty years in trade in Liverpool, and I must know more than you.

Q.—Not much?

A.—I have known you for many years, and I know you are very clever, but practical experience counts for something.   I know this, it is high time that the Ship Canal did come to Manchester, not only for corn but for all provisions as I say.   Just look to-morrow in the Corn Exchange and see the grocers' market—it would surprise you—and even you would come round to us then.

Later on Mr. Pope and witness got very angry.

Q.—Would you rather make impertinent speeches or answer questions?

A.—You have been as impertinent as anybody, and you ought to know better; I am straightforward enough, and I have evidence for all I have said.

Mr. John Burgess, Alderman of Warrington, bore witness that for ordinary carriage the river between Warrington and Liverpool was not made much use of. The tides prevented any regularity, and it was only when they were high that ships could get up. Somewhere about seven tides each fortnight allowed vessels to do so, and then they were simply barges of 70 to 80 tons. Between Runcorn and Warrington there was very little traffic by river.

Mr. Joshua W. Radcliffe, Mayor of Oldham, said 1,400,000 tons of traffic came in and out of his borough, of which 350,000 tons were cotton.

Mr. John C. Fielden reiterated the valuable and interesting evidence given before previous Committees, and maintained the Ship Canal would effect a great saving in heavy traffic. He disputed Sir William Forwood being an authority on the cotton trade; his attempt to prove Lancashire cotton spinning was prosperous because the shares in certain cotton mills had gone up, was a ludicrous failure. If shares nominally worth 5s. in 1879 were quoted at £3 in 1884, he called it prosperity, whereas calls of £4 10s. may have been made in the interim, and the shares be at a discount. Anyway Mr. Ellison, a Liverpool authority, put forward £19,000,000 as the loss in the cotton trade between 1877 and 1880. Sir William Forwood stated last year in public that he had no doubt the capital would be raised. The same gentleman said to the Select Committee on Railways in 1881 that he had seen an offer (if 1,000 tons each way were guaranteed) to cart from Manchester to the ship's side at Liverpool at 8s. per ton for 35 miles; *now* the opponents profess to disbelieve the promoters can cart 19 miles for 6s. 8d.

Mr. Bosdin T. Leech bore witness to the disgraceful state of the Irwell, and to the efforts that were being made by Manchester and other towns to purify their sewage. He believed "the manufacturers and merchants would get a return for their money if they did not get any dividends; the saving in the carriage on the Ship Canal would be so considerable that it would be a handsome dividend if the canal never paid a penny".

Mr. Gustav Behrens was of the opinion that a 12-mile radius was no criterion of the trade of Manchester; and to show the disadvantage under which Lancashire laboured he said the same vessel charged 15s. from Glasgow to Calcutta that charged 20s. from Liverpool to Calcutta, though a shorter voyage.

Mr. James W. Southern said the carriage of timber from Hull to Manchester was 12s. 6d. for 89 miles, whilst from Liverpool to Manchester (little over one-third the distance) it was 9s. Dear carriage pressed heavily on a cheap article like timber, often it was 25 per cent. of its value.   He estimated 300,000 tons came into Manchester and district, and on this he expected there would be a saving of £75,000 per annum.   Australian goods were sent to London to be packed, because casing timber could be bought more cheaply there.

Mr. J. T. W. Mitchell, Chairman of the Co-operative Wholesale Society, with its head establishment in Manchester, said that his society had a share capital of nine million pounds, and had a banking business with a turnover of sixteen million a year.   The society paid £75,000 per annum for carriage, and he expected the canal would effect a saving of at least £5,000 per year.   This induced a noble Lord to ask in amazement, "What are the objects of your society?"   Without a moment's hesitation, in stentorian tones and almost in a breath, the witness replied: "The objects of the society are to buy the produce, everything that is required by the members, at the cheapest possible cost by the use of their own capital.   We have a soap works at Durham, and we manufacture dry soap at Crumpsall, and there we have biscuit works, and there we manufacture different kinds of spices.   Ours is the Co-operative Wholesale Society of Manchester.   If your Lordships will not mind, I will hand you this book—it will give you a particular account of who we are and what we are, and I leave it with you."   Suiting the action to the word, the witness dived into a black bag he had close by and produced a large book.   Without more ado he stalked across the room and astonished the Chairman by placing the volume in his hands.   Addressing the other noble Lords, he told them he would be glad to send each of them a copy.   He kept repeating, "Ours is the Co-operative Society," and told them if they would read the book they would know a great deal about the society.   The Committee enjoyed the episode, and evidently thought Mr. Mitchell a typical Lancashire man who had little fear of dignitaries.

Mr. Balfour Browne humorously said to the Chairman, "It is a very handsome volume, but I don't want it to be considered as bribery and corruption".

Mr. William Berisford, wholesale grocer, Manchester, estimated that £4,000,000 was the value of the sugar consumed yearly in his district.   From Glasgow the mileage rate per rail was seven-eighths of a penny per ton, while from Liverpool to Manchester it was 3⅞d.

The Chairman here intimated that no evidence as to the impurity of the water would induce the Committee to throw out the Bill.   The opponents' case then began.

Mr. Dugdale, Q.C., on behalf of the Shropshire Union Company, said the new route would entirely cut off Ellesmere Port from the sea, and prevent his clients

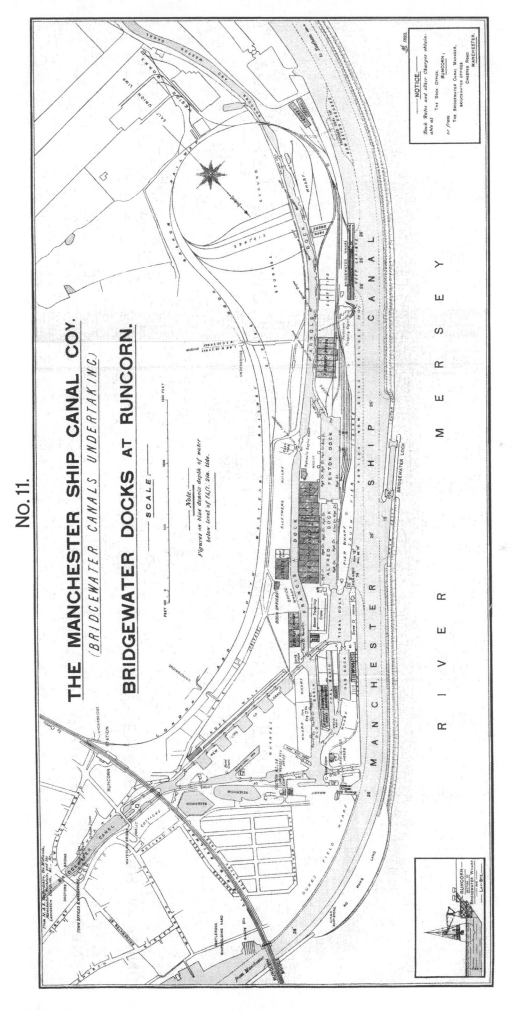

The material originally positioned here is too large for reproduction in this reissue. A PDF can be downloaded from the web address given on page iv of this book, by clicking on 'Resources Available'.

from utilising the land they had bought near the Pool Hall rocks, with the view of taking their navigation down to deep water. Several witnesses were called to prove this. One of them said a ship of 420 tons had come into Ellesmere Port. On cross-examination, it was elicited that in six months only sixteen vessels of over 100 tons had entered that port.

After *Mr. Saunders* had spoken for the Bridgewater Navigation Company, he called *Mr. William H. Collier*, who gave the history of that company, and said that in 1884 they carried 2,815,018 tons over their system. The minimum depth of the Mersey and Irwell Navigation was 4 feet 6 inches, and the state of repair was good. In the two previous years they had contemplated an arrangement with the promoters, and had not actively opposed the Canal Bill, but this year they had a scheme of their own for deepening the river to 10 feet and carrying with large barges hauled by steam power. These would each carry 300 to 400 tons. The estimated cost was £300,000 to £350,000. Cross-examined, witness defended the power of charging a bar-toll of 1s. 2d. per ton to the Rochdale Canal as a means of preventing suicidal opposition. The carriage of cases to Liverpool was now 9s. 2d. per ton, notwithstanding the maximum toll on the river was 3s. 4d., and on the canal 2s. 6d. The maximum freight from Runcorn to Manchester was 6s. per ton, the extra 3s. 2d. was for taking goods from Runcorn to Liverpool.

Mr. Bartholomew, of the Aire and Calder Navigation, did not believe it was an economical means of conveyance to take a large ship up an inland navigation. Barge traffic would be cheaper.

Mr. George Findlay, general manager of the London and North-Western Railway, admitted he had been pressed by the traders for more accommodation at Garston, and he would have liked to go to deep water nearer Liverpool, only that then he would be subject to Liverpool dues, and Garston was just outside the limit. In face of the 1s. 2d. rate given by the London and North-Western Railway Company to collieries, he did not believe the Ship Canal would ever carry more coal than was now carried on the Bridgewater Canal. If any line of steamers ever came to Manchester, they must come into the Liverpool conference and agree to rates, or they would be run off the road by the rich companies. He thought Mr. Marshall Stevens' estimate of an average of 5s. per ton an extravagant one; that gentleman formulated his information like a schoolboy with a rule of three sum. At the end of last year Sir William Forwood's Committee of Liverpool traders approached his company with their grievances, especially asking for equal mileage rates, and he had reason to believe these would be met, and they would be satisfied. The cost of conveyance in England and France was not comparable, because in the latter country the means of conveyance were owned by the State.

In cross-examination, witness admitted that he had built his criticisms on the trade of the "Town of Liverpool," whilst Mr. Pember had spoken of the "Port of Liverpool," which made a difference of close on 5,000,000 tons. Also that it would be fallacious to estimate the business of a town by the trade done within a radius of 12 miles. Having given the working expenses of the Ship Canal Docks at 77 per cent., he was confronted with the cost at Glasgow being 23 per cent. and at Bristol 27 per cent. He still believed that (with the exception of London) Liverpool was the dearest port in the Kingdom.

Mr. James Harrison, timber merchant, of Liverpool, felt sure that any ships coming up the canal would require an extra 2s. 6d. per ton, and as the 10,000 tons of timber used in Oldham would all have to be carted, even to supply that town would not be practicable.

Mr. Henry Coke, of Liverpool, was the Chairman of a Liverpool Committee when an effort was made to reduce the dock charges in Liverpool, and he had tried to get these reduced, also the railway charges. He was quite sure Manchester had never helped them in their efforts, and that the Ship Canal would not tend to the reduction of charges at Liverpool.

In cross-examination, witness agreed that since 1879 there had been a reduction in Liverpool dock rates and dues of 15 per cent., and that the reduction of 10 per cent. in railway rates might have been since the introduction of the Ship Canal scheme. Also he admitted that he went to London with a deputation, who suggested that the way to emancipate Liverpool from the railway monopoly was to have a new independent line to Manchester.

Mr. C. W. Cayzer thought the delays in bringing a 3,000 to 4,000 ton steamer to Manchester would be such that it would be far better for a shipowner to pay the railway carriage than bring his ship up the canal. Whole return cargoes could not be got from Liverpool, and there was less chance of getting them in Manchester.

In cross-examination, witness admitted that after deducting carriage from Manchester, Liverpool and Suez Canal charges and dues, and Bombay rates there only remained 3s. 4d. for carrying a ton measurement to Bombay in twenty-five days, or a little over 1d. per day, and therefore the idea of charging 2s. 6d. per ton for the single or double journey up the canal was singularly disproportionate.

At this point Mr. Pember informed the Committee that the promoters had agreed to buy the Bridgewater Navigation property for £1,710,000.

Admiral H. D. Grant, of H.M.S. *Serapis*, said ships exceeding five knots an hour could not steer in the canal, and not at any time so well as in the Suez Canal, because the sectional area was less. A vessel rarely went through the Suez Canal without grounding *eight or ten times*, but little damage was done because the bottom and sides were soft sandstone. The banks of the Manchester Canal being of masonry would make it very risky for navigation, because the heavy winds ships would encounter must drive them on to the sides. Under the most favourable circumstances it would take twelve hours for a ship to come up, and if the tide did not suit for entry, a vessel *might* lose two days. Two ships meeting and passing one another would be attended with great risks. He disagreed with a witness on his own side who said it would be impossible to steam under six knots on the canal.

Captain French had seen screw steamers on the Clyde throw a wave 20 feet up the banks, and no vessel would be safe at its moorings in the canal. He would say the insurance would be four times as much as the usual Mersey rate, and it would take *nineteen hours* from Eastham to Manchester. Cross-examined, he admitted he had never been through the Suez or Amsterdam Canals.

Mr. Samuel L. Mason, late traffic superintendent of the Great Northern Railway, was sure short distance traffic did not pay railways, because of costly terminal expenses; they swallowed nearly the whole rate. Mr. Marshall Stevens had calculated that a ton register was equal to 40 cwt. His investigations in various trades showed that in Glasgow a ton register averaged 15 cwt. and in Liverpool 19 cwt. This was the cause of the promoters' mistakes as to the traffic that could be got. His estimate of the tonnage of the port of Liverpool was 12,000,000 tons, and he differed both from Sir William Forwood and Mr. Alfred Holt when in their tramway scheme they stated that the inland traffic only was 15,000,000 tons.

Mr. Alfred Holt was of opinion that railways were superseding water traffic where there was competition, and he gave reasons for preferring a barge to a ship canal. There was no analogy between the Suez and Ship Canals. If the former had only been meant to take goods to Suez it would not have been made. A ship would be delayed at least three days each voyage in coming to Manchester. He estimated 9d. per ton per day on the registered tonnage for demurrage, and ships must charge a higher freight. He thought this a foolish, useless, and wasteful scheme likely to do injury. In cross-examination, witness owned he had written "As for the present canal project, were it not for standing in the way of the plateway, I am in favour of it".

Sir William B. Forwood, cross-examined by Mr. Pember, said that in the autumn of 1884, thinking the Ship Canal was dead and buried, and not likely to be revived, he wrote a letter to the Chairman of the London and North-Western Railway Company and asked them to take into consideration the railway rates between Liverpool and Manchester, with the result that a *modus vivendi* had been happily hit upon.   "On the supposition you can get a happy *modus moriendi* for the promoters of the Manchester Ship Canal," replied Mr. Pember.

Witness said in the interview following the letter, no mention was made of the Ship Canal, and he denied the statement about the interview in the newspaper which said, "The Manchester Ship Canal formed an important element in the discussion". Twitted by counsel with having entirely changed his opinions since 1881 in order to try and defeat the Ship Canal Bill, witness replied that though he had then said that it was immaterial in nine cases out of ten if goods took twice as long by road as by rail, there had been a great change since then.   He could not agree with Mr. Findlay when he said, "I do not think I know of any port that is so favourably placed with regard to railway rates as Liverpool".   It was true when he said, "Practically Liverpool is the milch cow of the railway companies; they get a large profit out of their Liverpool traffic which they spend in developing other ports".   He believed Liverpool was overcharged £400,000, and though they worked with the railways against the Ship Canal, there was no combination.

He still maintained that the practice of sending goods from Blackburn to Liverpool was a growing one.

Mr. Henry Oakley, general manager of the Great Northern Railway, believed the canal would be a failure; there was neither ground space nor warehousing, and as there were no railway connections, it was a practical impossibility to remove 3,000,000 tons by carts.   He did not believe there were 200,000 tons carted out of the Victoria Docks in a year.

Mr. Aspinall, Q.C., addressed the Committee for the Liverpool Corporation. He characterised Mr. Adamson as—

An honest, able, energetic man of business, a man who, when he once forms an opinion, holds to and adheres to it with the strongest tenacity, but who himself would be most difficult to convince that any opinion of his was not founded upon the strongest grounds.

He was the sort of man who could persuade people, and no doubt the energetic measures he had taken, promising success and a good dividend, had roused the country.   But in Liverpool no one believed in the scheme, not because of competition, but because they believed the river Mersey and its navigation were imperilled.   He asked, "Who is going to establish a line of vessels for any purpose to Manchester?" For small purposes, and for the use of the immediate neighbourhood, such as bringing potatoes from Glasgow, he admitted it might be used.   "Can your Lordships conceive that for years and years after the completion of the canal, even if it were an absolute success, there will be such a trade formed at Manchester as would induce shipowners to establish a line to go to Manchester and back from anywhere upon the face of the earth?"   To make shipping profitable a whole cargo must be ready to go back to the port whence the ship comes, or to some other port.   The idea of the promoters was wild and visionary, and calculated to be delusive in the strongest possible way.   Unless the Committee were satisfied the canal would pay in a reasonable time, and not hang over and stop other improvements, or fall into the hands of the present monopolist-carrying companies, he asked for the rejection of the Bill.

Mr. Pope, Q.C., said there was hardly a period in his own career when some one or other of the promoters had not exhibited to himself some personal friendship or kindness, therefore he would like to treat them with great consideration, and he must respect the feelings of Manchester.   But he must object, by raising the railways, to the construction of a wall between Lancashire and Cheshire at least 70 feet high, also to dangerous swing bridges affecting not only any future but the existing lines which carried an enormous traffic from Scotland and the northern counties to the South of England.   It was claimed this was an alternative to the swing bridges reserved by ancient Acts of Parliament, but he believed no Board of Trade would ever listen to the enforcement of such antiquated obligations.   With the greatest respect to Sir William Forwood, he doubted his estimated 25,000,000 tons as the traffic turnover of Liverpool; but even admitting the promoters' estimate of 15,000,000 tons, the evidence he had adduced proved they could only get 1,140,000 tons instead of the 3,000,000 they built upon.   He ridiculed the idea of the £5,000,000 which the promoters were willing to find before they started, or the protective clauses they offered; these gave no security in case of a constructive or monetary breakdown. He believed if the money was locked up, as it should be, till the completion of the Bill, the promoters would never go on.

Mr. Littler, Q.C., said the Cheshire lines only carried 669,000 tons of traffic.

How could the promoters hope to get 3,000,000 tons in face of fierce competition? The figures were absurd. The promoters' case was a beautiful picture seen at a distance, nearer it was ordinary, closer still the paint was laid on too thick, and on the spot it was a simple daub. It was very tempting, and its glamour may have attracted their Lordships as it had many Lancashire people, who ought to have commercial knowledge. He ventured to say few people were more ignorant of the carrying business than traders themselves.

Mr. Adamson knows absolutely nothing about railway carrying; Mr. Marshall Stevens, with great submission, knows less. Mr. Marshall Stevens' figures are simply a collection of arithmetical puzzles, and, as Mr. Findlay said, not very good at that. That is the class of evidence on which your Lordships are asked to sanction an expenditure of £10,000,000. The moment the canal is opened, all that the London and North-Western and other railways have to do is to reduce their rates, and they can drive these people out of the field. No Bill was ever so encumbered with protective clauses for everybody. They have bought off opposition right and left so that they might triumphantly say, "We have settled with the various landowners and interests". The probable result will be that having forced the hands of the railway companies, not one inch of these works will ever be carried out at all.

Mr. Littler trusted the Committee would feel justified in throwing out the Bill.

You are dealing with interests counted by hundreds of millions, with the interests of shareholders all over the country, many of them exceedingly poor, persons who are to be sacrificed simply because some of these rich people in Lancashire think they can make a little more money out of their cotton than they do now. This scheme, concocted during the unfortunate illness of Mr. Adamson, is nothing more than the frenzied fancy of a sick man's dream.

After clearing the room for consultation the Chairman said :—

We will allow the Bill to proceed subject to the insertion of the clause with regard to the £5,000,000 which has been agreed to. We are also disposed to impound the 4 per cent. interest deposit mentioned in clause 39, amounting to £276,539, either by omitting clause 41, or in any other such way as may satisfactorily effect that object. With regard to this last point, we are ready to hear any objections which the promoters may have to urge.

Mr. Pember objected to impounding the 4 per cent. deposit, and after a long discussion the Committee agreed not to insist upon the clause.

CHAPTER XV.

1885.

BILL BEFORE MR. FORSTER'S COMMITTEE IN THE COMMONS —EVIDENCE—MR. PEMBER'S POWERFUL APPEAL—OTHER SPEECHES BY COUNSEL—FINAL DECISION—CLAUSES.

With regard to the canal itself, I repeat that I think a greatly improved waterway between Manchester and Liverpool is bound to be made, and that it would be a great blessing to Manchester and no damage to Liverpool.—Sir EDWARD WATKIN.

THE Ship Canal Bill having passed the Lords on 15th June, 1885, was introduced to a House of Commons Committee consisting of the Right Honourable W. E. Forster (in the Chair), Lord Eustace Cecil, Mr. Dalrymple and Mr. William Fowler. Practically the same leading counsel again appeared for the promoters and opponents. Many of the petitioners in the Lords, having been satisfied by protective clauses, dropped out, as did the Bridgewater Navigation Company whose concern the Ship Canal had agreed to buy. Commercial fear, however, prompted some dozen Liverpool Trading Associations to oppose the Bill, and they were represented by Mr. Potter, Q.C.

Again it was virgin soil, and the new Committee had to begin at the A B C of the case and have everything explained to them. The very able Chairman said it was their duty to hear fully both counsel and new evidence, but in view of the great costs that were being incurred and the danger of not getting through before the session closed, he asked counsel on both sides to do their best to shorten the case, and he promised his Committee should read the past evidence and give it the same attention as if it had been produced in the room.

The case lasted from the 15th June to 3rd August when the decision was given. There were thirty-four witnesses for and forty-five against the Bill. Of the petitioners seventeen of them appeared by counsel.

Mr. Pember in opening his case pointed out how his Bill had been played

shuttlecock with in the two previous years by the Commons and Lords.   He gave
the history of the Bill from the commencement, and showed how the business of Manchester was suffering from dear railways and a dear seaport, so much so that trades
were leaving the city, which was fast going to decay.   He complained that having
given up the channel through the estuary and virtually adopted the scheme of Mr.
Lyster, the Liverpool engineer, they were still bitterly opposed notwithstanding
repeated pledges given on the part of Liverpool in the previous session that
opposition would cease.   He showed by diagrams on the walls that the canal would
be deeper and wider than the Suez and Amsterdam Canals, and said that vessels of
over 5,000 tons were now navigating the former.   He stated that to meet public
requirements his company had given up their right to swing bridges and contented
themselves with a 75 feet headway.   Also that the railway gradients to be given
were fairly easy.   After explaining the Weaver sluices he showed that the abstraction of water from the estuary was small and non-important, indeed it was only a
tenth part of what Liverpool abstracted in order to make her dock walls.   He was
of opinion that in the end 50 per cent. of the costs of transit would be saved by the canal,
and he pointed out how, on the Continent, goods were carried at an average rate for
1d. per ton per mile against 3d. at home.   He also showed the unfairness of the rates
charged on Manchester goods, thus giving a great advantage to foreign competitors.
The result was that the selling prices of staple articles of food were 5 per cent. dearer
in Manchester than in some other ports.   He argued that if Sir William Forwood's
estimate of 25,000,000 tons (being the weight passing through Liverpool) was correct,
Manchester, as the great consuming centre, might justly hope to get 3,000,000 tons
of traffic.   Again that as the working expenses of the Suez Canal varied from 5 to 9
per cent., it was a reasonable assumption to place 15 per cent. as the working expenses
of the Ship Canal.   He treated the sheaf of petitions, which the indefatigable hostility
of Sir William Forwood had produced, as simply moves to wreck the Bill.

Mr. Daniel Adamson put in the following tables :—[1]

1.   Population of Manchester and District.
2.   Value of imports and exports of chief ports of United Kingdom.
2a.   „   „   „   „   „   „ London and Liverpool.
3.   „   „   „ of foreign and colonial produce from Liverpool.
4.   „   „ principal exports of British produce from Liverpool.
5.   Shipping statistics, tonnage of vessels conveying cargo in and out of principal
        ports in 1883.

[1] For further particulars see Appendix No. III.

6. Comparison of railway rates—grain.
7.   „   „   „   „   of square timber, long and short distances.
8.   „   „   „   „   raw cotton.
9.   „   „   „   „   pig iron to Manchester.
10.   „   „   „   „   „   „   „   „   conveyed from different ports
11.   „   „   „   „   undamageable iron.
12.   „   „   „   „   machinery in cases or frames.
13.   „   „   „   „   machinery and boilers.
14. Dock charges on imports at certain principal ports.
15.   „   „   on exports „   „   „   „
16. Growth of gross revenue at the Liverpool Docks from 1864 to 1884.
17. Cotton statistics.   Number of factories, spindles, looms, workers.
18.   „   „   Distribution of spindles.
19.   „   „   „   „ looms.
20.   „   „   Deliveries of cotton at home and abroad.
21.   „   „   *Résumé* of cotton consumption and production.
22.   „   „   Value of exports abroad in 1883.
23.   „   „   Exports of cotton, yarn and cloth from British India, 1872 to 1884.
23a.   „   „   Exports of Indian cotton to the Continent.
23b.   „   „   Depression in cotton trade, decline in profit.
24.   „   „   Cotton brokers' cash statements.
25.   „   „   Charges for cotton to Newton Moor Mills.
25a.   „   „   Summary of savings to Newton Moor Company.
26.   „   „   Estimate of annual savings within carting distance of Manchester.
27.   „   „   Summary of entire savings in cotton trade.
28. General trades.   Comparison of charges *via* Liverpool and Ship Canal (inwards).
29.   „   „   „   „   „   „   „   „   „   „ (outwards).
29a. Shipping statistics.   Increase tonnage, Glasgow and other towns.
30. Coal statistics.   Production of chief fields in 1882.
31.   „   „   Exports abroad in 1883.
32. Shipping statistics.   Shipbuilding in 1884.
33. Timber statistics.   Imports of foreign and colonial timber in 1883.
34. Shipping statistics.   Number and tonnage of English ships (under 3,000 tons).
35.   „   „   Applied to Liverpool in 1884.
36.   „   „   Examples of large vessels able to use the canal.
37.   „   „   Tonnage, also arrivals and departures, Dutch Canal, 1877
                             to 1883.
37a.   „   „   Arrival of vessels at Antwerp, 1873 to 1884.
38.   „   „   Increase of tonnage entering Liverpool 1864 to 1884.
39. Bridgewater Navigation Company.   Statement of tonnage, capital and nett revenue.
40. Working expenses.   Comparison with the Suez and Amsterdam Canals.

Also table showing the growth of tonnage of vessels passing through the Suez Canal and receipts derived therefrom from 1870 to 1884.

Witness gave evidence in support of the above tables, and said Manchester had subscribed £18,000, Salford £6,000 and Warrington £2,000 out of their rates towards the Parliamentary expenses of the current year.   He also pointed out that already the dock dues had been reduced, and 1s. per ton had been taken off the carriage of cotton.   Speaking of the coal business to be built up, he said Dublin alone took nearly 1,000,000 tons of English coal.   In cross-examination, Mr. Pope said they did not intend to contest the sufficiency of the estimates.

Mr. S. T. Bradbury, of Messrs. Gartside & Company, Limited, who had been a witness before on each previous Committee, but had been taken briefly, now underwent a long and searching examination both by the opponents and the Committee, the latter evidently being anxious thoroughly to master the cotton business, especially as to the distribution of cotton and cotton goods by carts, and the system of dual brokerage in Liverpool.   The Chairman, at the conclusion, said though it had been long, it had been useful, and he did not think it had been a waste of time.

Mr. Reuben Spencer, of Manchester, and *Mr. F. H. Bowman*, cotton spinner, of Halifax, followed.   The latter said he ran 130,000 spindles, turning out 3,000,000 lbs. of yarn yearly, and that he had the largest mill in Yorkshire.   He showed how the canal would benefit the cotton trade, and denied the assertion of Mr. Findlay "that there were no cotton mills in Halifax at all".

Mr. Michael Spitz, of Manchester, estimated the canal would save on the shipment of cotton waste 4s. 7d. per ton as compared with Liverpool, and 8s. 9d. as compared with Hull.

Mr. Charles Holt, corn merchant, of Manchester, repeated his evidence.   *Mr. Pope:* "I shall not cross-examine Mr. Holt".   This was no doubt the result of the sharp passage between them in the Committee of the Upper House.

Mr. John C. Fielden, of Manchester, astonished the Committee by estimating the canal would save £3,000,000 sterling to the community, and he certainly gained their ear by the arguments he used to make good his statements.   As regarded food he said :—

We should develop the growth at home instead of getting it from abroad ; they can bring food from New York as cheaply as from the South of England—they can bring it cheaper.   They can bring worked timber, doors, etc., from Chicago to London, 700 miles by rail and 3,000 miles by sea, cheaper than they can carry the same goods from London to

WILLIAM J. CROSSLEY, M.P., DIRECTOR, MANCHESTER SHIP CANAL
COMPANY.

Franz Baum.   *To face page* 300.

Manchester, and they are taxing our goods deliberately 15s per ton in order to carry foreign trade.

Mr. Marshall Stevens bore witness to vessels at Liverpool having sometimes to wait ten hours in the river before they could dock, and at other times having to complete loading in the river; in both cases because of tidal difficulties. Loading and discharging in the river was free in London, but in Liverpool ships had to pay dock dues. It had been said that a traffic of 3,000,000 tons could not be done on 64 acres of dock land, but Garston did 1,800,000 tons on 14 acres. Again, Mr. Oakley had spoken of the impossibility of distributing by cart and water from Manchester; yet at the Alexandra Dock, Liverpool, in 1884, on 44 acres, 3,250,000 tons of goods were distributed by similar means. Liverpool arrangements were such that merchants there need not pay quay attendance, but this had to be paid by inland merchants. Garston was crowded out, and ships sometimes had to wait two or three weeks for a berth. It was 1s. per ton cheaper to bring goods there and pay the railway carriage and expenses back to Liverpool than to unload in Liverpool direct. If cotton could come to Garston, there would be a saving of 3s. 6d. per ton. In 1822 cotton to Manchester cost 15s. per ton. After the advent of railways and in time of competition it went down to 2s. 6d. per ton, and now it was 9s. 2d. by canal or rail. It had been said that the railways by reducing their rates would easily ruin the canal, but he pointed out that as the present cost of cotton freight, dues and expenses was 14s. 6d. per ton, out of which railway freight was 8s. and the total cost by Ship Canal 7s.: the railway companies must practically carry for nothing in order to compete with the canal. The terminal and Liverpool charges had even more to do with dear freightage than the mere haulage. To run the canal off the road the railways would have to give the haulage, and the shareholders would certainly object to no dividends.

The *Chairman* here asked to hear the engineering evidence at once. He did not wish to shorten the case unduly, but he thought both sides must wish to bring this costly inquiry to an end: the promoters had yet 119 witnesses to call, and possibly the opponents as many. If they did not quicken, the chances of a decision were *nil*. He asked for a few leading commercial witnesses to be called that day, and that on the next they should go on with the engineering case. To this Mr. Pember at once agreed.

Mr. Stevens, the witness, said he had taken no revenue for animals, but he believed there would be large shipments of cattle. It was quite fallacious to draw a zone of 12 miles round Manchester and limit the canal traffic to it. Take the traffic 12 miles round Liverpool and it will not include 5 per cent. of the traffic done in that city. Though the opponents declared the railway companies would not convey goods brought by canal, he was perfectly sure directly the canal and docks were made there would be an actual race between the companies for the traffic. It was said the bulk of the timber came by sailing ships unsuitable for the canal; he believed in ten or fifteen years it would all come by steamers.

When *Mr. Littler* cross-examined witness he took a personal tone, and said the London and North-Western Railway did not know of Mr. Stevens' agencies, and that as he had made scandalous charges, he ought to give the name of any of their servants who had influenced traffic against him. This the witness at once did, and said the manager had. Upon this the Chairman said Mr. Littler had brought the answer upon himself by his sneers. In conclusion Mr. Littler said after witness's attack on Mr. Findlay's figures, he would not attempt anything. To which the retort came, " I think you had better not ".

Mr. Edward Leader Williams gave a lengthy and exhaustive description of the canal works, and explained how they differed from the scheme of the two previous years. He was encouraged to do this by the Chairman, who said he commenced the inquiry literally knowing nothing of the merits of the case. Witness said one of the principal objections was that there would be a small abstraction of tidal water. The same argument was used when Birkenhead wanted to build docks similar to Liverpool. This necessitated stopping the tidal flow up the Wallasey Creek. " The fight was intense—as bad as this." Liverpool said the Wallasey Creek water was of great value to the bar, and there must be no abstraction from the estuary. Birkenhead replied: " You have been always doing it yourselves. Why not we? " It has been admitted that neither this abstraction, which was allowed, nor the 20,000,000 cubic yards taken from the estuary by Liverpool to make her docks, had done any damage to the bar. This latter abstraction is five times as much as we propose to take. " It is only when Manchester comes into the upper estuary and takes a minor quantity that mischief is done."

The *Chairman* again interposed and asked if the opponents would follow the promoters with the estuary case; he did not consider they would be damaged thereby. After a night's deliberation counsel for the opponents refused to take a course, which, as they said, meant taking their case piecemeal; they would, however, consent that if the consideration could not be finished, the Bill might be carried over as a remanet to be heard by the same Committee next session. This suggestion being found impossible, the case proceeded.

The *Chairman* here made the announcement that one of the Committee, Mr. Dalrymple, had accepted office under the Crown, and either Parliament must appoint a substitute, or, with the consent of both parties, the remaining three could finish the case. The latter course was at once adopted.

The *Chairman* also asked Mr. Pember if the promoters could adopt Mr. Lyster's plan instead of their own? In reply the witness showed how far the two plans could be assimilated, and further that if abstraction was the bugbear, Mr. Lyster's plan on a 14 feet tide abstracted 5,000,000 cubic yards of water, whilst the promoters only abstracted 4,250,000 cubic yards. But all negotiation was cut off when Mr. Pope, on behalf of Liverpool and Garston, refused assent to Mr. Lyster's plan. Concessions seemed useless. When the engineer offered to meet an objection as to curves, Mr. Bidder said, "I do not know what you would not do to get the Bill," to which Mr. Williams replied, "I do not know anything Liverpool would stop short of to prevent us getting the Manchester Ship Canal". Witness said it was evident more dock accommodation was wanted somewhere; Liverpool could not supply it. The question was whether Garston or Manchester had to do so.

Mr. James Abernethy gave his full support to the plans and estimates, and to his mind it was sheer nonsense to say because a sunken ship blocked the Suez Canal, only wide enough for one vessel, that of necessity the same mishap would block the Ship Canal, made wide enough for two vessels to pass.

Mr. G. M. Cunningham, of Edinburgh, did not believe the small abstraction would damage the estuary or bar; the proposed embankment would be most beneficial.

Mr. James Deas, of Glasgow, *Mr. Messent*, of the Tyne, *Mr. Fowler*, of the Tees, and *Mr. G. H. Hill*, of Manchester, highly approved of the scheme, and *Mr. H. J. Martin*, of the Severn, who had declined to support last year's scheme, gave his hearty assent to a canal skirting the Cheshire side of the estuary.

Mr. David Cunningham, of Dundee, was sure the proposed works would have a beneficial and not a prejudicial effect on the Mersey, and *Mr. Lionel B. Wells* had a similar opinion.

Mr. Giles, C.E., of Southampton, was sure no one could detect the abstraction. It meant 1 inch in 25,000 acres of estuary, and no one could be sure to an inch in his soundings. He approved of the engineer's deep sill at Eastham, and he pointed out that at one time Mr. Lyster, in order to get under Runcorn Bridge, placed his sill 6 feet 4 inches deeper than that of Mr. Williams, of which he now complained.

Mr. W. H. Wheeler, of Boston, supported the scheme.   The Pool Hall rocks and other headlands must prevent a channel on the Cheshire side, and any abstraction was infinitesimal compared with Liverpool abstractions, which would do much more harm :—

The neck gets wider as you near the bulb of a bottle.   If I want to fill the bottle it would be easier if the part near the cork was wider; but the Liverpool people have contracted it just where you want to fill it, therefore they stop the water getting in; and therefore the abstraction is one which is likely to prevent water coming into the upper estuary.

Mr. T. S. Hudson, shipowner, West Hartlepool, expressed himself satisfied as to the canal being safe for navigation.   Cross-examined as to why only 650,000 tons of coal were shipped at Liverpool against 10,000,000 tons at Cardiff, witness said it was largely owing to defective apparatus.   At Birkenhead the North Wales coal was loaded by barrows, and it took double the time to fill a ship.

Mr. L. F. Vernon-Harcourt believed that just as the Liverpool Docks trained the channel, so would the regulated sides of the Ship Canal; and he believed making the Liverpool Docks, if there was any sensible abstraction, would be more damaging to the bar than an abstraction in the broad part of the upper estuary.   He could not believe the fresh waters of the Mersey and Weaver would be diverted from their usual course and form a new Channel on the Cheshire side.   Had he thought so he should not have supported the Bill.

The opponents then put forward their case.

Captain Graham Hills, marine surveyor of the Dock Board, described the Liverpool bar as being one-fourth of a mile across at the top and about half a mile at the base, and predicted as the tide rose higher at the Weaver than at Liverpool there would be a stream coming through the openings and down the canal that would prevent the opening of the Eastham gates, just the contrary of what was expected.   As regarded abstraction damaging the bar, witness said that in 1873 the landowners reclaimed 240 acres from the estuary, and he admitted after that date the bar improved.   The Chairman asked if this was so, why was he so afraid of the 11 acres now to be absorbed?   Witness went on to say that in consequence of the crowd of ships at anchor it would be dangerous to navigate the Sloyne, and in flood time ships would have to go stern first into the Eastham lock.   He had been in the North Holland and Amsterdam Canals, and they were both failures, and he thought

To face page 304.

SHARP CURVE NEAR RUNCORN BRIDGE.

Killon.

the very look of the place sufficient to cause Dutchmen to run away.   He believed
past abstractions had seriously damaged the bar, that last year's scheme would have
destroyed half the inner estuary in time, and this year's plan would defer the injury
for a longer period.   He had not changed his view of Mr. Lyster's plan, to which
he objected in the Lords Committee.

Mr. George F. Lyster said whilst Liverpool and Birkenhead had twenty-four
graving docks for 533¾ acres of dockage, Manchester had made no provision at all.
Cartage was the only way by which in Liverpool they could remove goods from the
quays to the stations and warehouses; and the railway managers found this the
most convenient way of removing them.   Railway lines had been laid along some
of the docks, but after three or four years' experience as they were never used they
were pulled up, with the concurrence of the railway people.   He presumed the railways charged for the carting and included it in their charges.   Both Liverpool and
Birkenhead had been fitted out with the best coal loading appliances at a cost of
£280,000.   The promoters had made no such provision.   He considered they had
been guilty of very serious plagiarism in copying his plans, but unfortunately they
had mutilated his ideas, and what they proposed was both objectionable and
dangerous.   The promoters could not keep their lock entrances (18 feet deep) open;
the sand would overwhelm their platforms and gates, and they would neither open
nor shut after a gale of wind.   A passage through the Sloyne would be equivalent
to destroying it as an anchorage.   In cross-examination, witness was reminded of
his statement in 1867 at the Surveyors' Institution, that the abstraction of the whole
of Wallasey Creek, when making Birkenhead Docks, had not damaged the bar
though it was thought at the time it would do so.   This he fully admitted.   He did
not believe a canal on the Cheshire side would injure Garston.   Time had convinced
him that his other objections in the Lords Committee still held good.

Mr. G. F. Deacon at once began to puzzle the Committee by highly technical
terms, such as "hydraulic mean depth" and "wetted perimeter,"—against which
the Chairman mildly protested.   Witness ended by supporting Captain Graham
Hills' views.

Sir William Thomson compared the action of sand on the sea-shore driven
by currents and wind to the dust of the air which is driven about and seeks shelter in
all kinds of crevices and recesses, to be moved possibly by the next wind blowing in
a different direction.   The tide takes the place of wind in the Irish Channel.   He
believed the promoters' scheme would have a tendency to make a channel on the

Cheshire side, and would be disastrous to Liverpool. The tide would follow the line of least resistance. If the question of damage to Garston did not exist, he could see no objection to deepening the Eastham channel. He feared Garston would in time be silted up.

Even if Mr. Pember consented to reduce his dredging from 20 feet to 16 feet, the result would still be disastrous; Mr. Lyster's 12 feet he thought might be perilous. Cross-examined, witness said he had never done any estuary work. In reply to the Chairman he maintained his theory of sectional area, *i.e.*, if the same quantity of water came into a river, a channel made deeper on one side must lessen it on the other. If, however, the water coming in deposited silt, it might slightly alter the equilibrium of the channels. If the entrance bar, against which the estuary may contend, was raised say 5 feet by extra deposit, he was of opinion the same amount of water would still come in, because the tidal capacity inside the bar still remained the same. This induced the Chairman to ask, " Do you admit, Mr. Vernon-Harcourt, what Sir William Thomson has just said?" "No, I do not," was the reply. Witness objected entirely to letting tidal water into a canal 24 miles long, and he did not believe flap boards to the 14 feet tidal openings would mitigate the current. Carrying the proposed canal further inland would not prevent interference with the mouth of the Weaver, and this he considered the most serious part of the whole thing.

Mr. Thomas Stevenson, of Edinburgh, thought it a most dangerous thing to interfere with the mouth of the Weaver. He objected to the proposed abstraction, but could not account for the fact that the bar had improved after previous abstractions; it might, after all, be more of a theoretical than a serious evil. He approved of the canal being taken alongside of the estuary, and the fact that there were only eight extra high tides in the year made him think better of the scheme.

Sir William Thomson having written two letters to the Chairman to alter his evidence, Mr. Balfour Browne objected, and it was arranged the witness should be recalled, also Mr. Vernon-Harcourt, and it was found they were entirely at variance as to the effect of dredging from Eastham to the Sloyne, whether it would or would not cause a channel to form on the Cheshire side, and thus damage the Garston side.

Admiral H. D. Grant recapitulated previous evidence, and believed the velocity of the current would prevent the navigation of the canal, and there would always be danger of collision in the Sloyne. It would be a perfect whirlpool at the proposed lock gates; they could not be kept closed, a flood tide would force them open. Dock gates ought to be at right angles to the stream to be safe. Going down against tide a ship would never steam through a lock, the resistance and pressure of water being so great. Eastham was the most exposed spot that could possibly be fixed upon for an entrance, and he did not support Mr. Lyster's plan. In cross-examination, the Admiral admitted he had never navigated a ship through a lock in his life, though he had seen lock gates working, and also said that at Hull there were dock entrances similar to those of the Ship Canal.

Mr. Henry Law had advocated that the Ship Canal should come along the Cheshire shore, and last year he had had a glimpse of Mr. Lyster's plan but he did not approve of it.

Sir Frederick Bramwell repeated the substance of former evidence. On cross-examination by Mr. Cripps, that gentleman probed a weak point in the knight's armour. Sir Frederick had given evidence in the Lords for the Bridgewater Navigation Company, and now for the Mersey Dock and Harbour Board, and to further the interests of either clients had given diametrically different evidence. The witness knew he was in a corner, and for some time cleverly parried the barrister's questions. At last Mr. Cripps put the pertinent question, " Do the laws of nature, as regards the Mersey, differ when giving evidence for the Bridgewater Navigation or for the Mersey Dock and Harbour Board?" This brought the angry rejoinder, " I think that a very improper question; but if you, as counsel are inclined to put it, I say, no". For once this very clever witness was fairly caught.

Mr. F. C. Stileman, of Birkenhead, believed dredging to the Sloyne deep would coax a channel on the Cheshire side as it was in 1822 and 1882. He objected to this and to the impounding of the Weaver and Gowy. A large ship crossing the bar could not possibly get to Manchester by the same tide. In cross-examination, witness confessed he had not allowed for the fifty minutes' difference in the tide between the bar and Eastham.

Mr. E. R. Peel, nautical assessor, believed it would not be safe to moor a vessel alongside the canal banks; she must be moored in the centre of the canal and thus block it. In good weather it would take $11\frac{3}{4}$ hours to steam to Manchester.

Mr. A. M. Rendel thought the Ship Canal would destroy Garston. Admiral Spratt, the Conservator, in approving Mr. Lyster's plan, had shown a most complete contempt for the interests of Garston. He had the Sloyne on the brain. The *Chairman* said Garston had dredged (just as it was now proposed to do) on the other side. How was it Garston had been allowed by Ellesmere Port to do the very thing she now objected to. All this time the Sloyne had been suffering from Garston. Witness replied, "Still the Sloyne is master". Mr. Lyster was highly enamoured with his own plan, but he did not agree with him.

Mr. J. Wolfe-Barry was in favour of the canal being placed on the north side of the estuary, and Mr. Pope did not express *his* views when he said the opponents approved the south side. Tilburn was the only place that had sills 26 feet deep.

A number of sea captains and pilots then gave evidence; they supported one another in saying crossing the Sloyne was dangerous, and that it would take at least twenty-four to forty-eight hours for a large ship to come up the canal.

Mr. John Hughes, of Liverpool, shipowner, said it would be cheaper and better to discharge in Liverpool and tow barges than send a ship to Manchester, as it would take two extra days to get there. Shipping rates were so much lower to Liverpool than to other ports that the difference would pay the railway carriage, and he instanced a case where they could deliver *via* Liverpool cheaper than to Garston direct.

Mr. A. T. Squarey, solicitor for the Mersey Dock Board, defended the retention of town dues on the ground that the same money would have to be raised by some other assessable tax, and this meant £25,000 more to pay in rates and taxes. Liverpool had managed to do the largest traffic in the world, and now to try and remove the bar would involve serious risk, and the cost would be enormous. The Dock Board never thought it desirable to meddle with the bar. The urgency of the case did not justify such a tremendous experiment.

Q.— The Chairman.—In fact whatever may be the fluctuations in politics generally, there is no doubt about Liverpool being conservative as regards the bar.

A.—And the bar is conservative also, for it always lets our traffic pass in and out.

Q.—Nobody does anything to revolutionise it—it seems as if it would remain in that sense conservative.

A.—Yes.

Q.—Mr. Fowler.—You seem very uneasy about it?

A.—The Liverpool people, about the bar! Not at all.

Mr. Francis Muir, cotton broker, said it would be physically impossible to do the carting from the docks contemplated by Mr. Adamson. There were 508,000 tons of cotton alone, and he estimated shippers would require 2s. 6d. to 5s. per ton to come up the canal. The reason why the Cotton Buying Company of Oldham could not join the Cotton Exchange was that they wanted to do the business of fifty large limited liability companies and pay only one subscription. He believed the idea of a cotton market in Manchester was visionary.

At this point the *Chairman* informed counsel that if the Committee had to decide the question on engineering evidence they would allow the Bill to proceed, but after consideration of Mr. Lyster's plan, the promoters' limits of deviation, and Mr. Pember's statement about abstraction and dredging, they wished attention to be given to the necessary clauses on these matters. They desired Mr. Pember to give a plan showing what he could do to meet the opponents within his limits of deviation, and to define exactly what he could do (under a promise made them on the 26th June) to limit abstraction and dredging. This Mr. Pember undertook to do, supplying the other side with clauses, plans, sections, etc.

Mr. Henry Coke said every previous effort to secure freight competition had tended to increase the charges. Manchester had been a party to make Liverpool spend six to eight millions on the Birkenhead Docks, and this cost had ever since been a drag, and instead of the new Midland line cheapening carriage, the contrary had been the case. He had no faith in a costly Ship Canal helping them. Trade would have to pay the interest on twenty-six instead of on sixteen millions.

Mr. John Patterson, of Liverpool, speaking for the corn trade, thought the Bridgewater Canal immensely valuable and capable of great improvement. He objected to its sale to the Ship Canal, because it was a great public highway, and in the new hands the object would be to divert trade from Liverpool. He strongly advocated a barge canal, and was sure a barge could carry at one-eighth to one-tenth of what it would cost to bring an ocean steamer to Manchester.

Mr. John Hargreaves, provision merchant, said in 1882 the value of the provisions imported to Liverpool was £12,000,000, and this did not include cattle, dead meat, or eggs. The latter, strange to say, were termed green fruit, according to the custom of the trade. The value of the American hog crop was £80,000,000, or twice the value of the cotton crop. Freight to Liverpool was 5s. to 7s. 6d. per ton less than to any other port.

Mr. Harvey C. Woodward was sure that as corn could be landed in Liverpool 2s. 7d. per ton less than in Manchester, the distributing trade must remain at the cheaper port. Besides, Manchester had no warehouse accommodation. "You might as well compare a donkey-cart with a four-in-hand as Manchester with Liverpool for accommodation for grain."

Mr. E. Brownbill admitted that some of the general brokers in Liverpool said, " Let the canal be made; we want a competing line for traffic". Also that he was ignorant of the business of Messrs. Vickers of Manchester, and Crosfield of Warrington, as regards phosphates and manures.

Mr. Pember, on 23rd July, produced the promised plans and clauses as offered by the promoters. The Bill showed the Eastham Dock sill 28 feet below the old dock sill. He proposed now to make it 23 feet, dredging 15 feet instead of 20 feet at Eastham. Mr. Lyster's depth was 12 feet. As regards the line of the canal it was proposed to carry it farther inland to the line of deviation. By these alterations the tidal abstraction would as a rule be less than that proposed by Mr. Lyster.

Mr. George Findlay still maintained his 12 mile radius theory, and that the only means of distribution from the docks would be by cart and limited to that distance. Railway connections they would not have, and he did not think the canals would be available, except for coal and local traffic. Mr. Marshall Stevens and the promoters were utterly at sea about their traffic estimates. Tramp steamers might come to Manchester, but regular liners would not, because the latter were as a rule in trade conferences. It was not fair to stigmatise Liverpool and Manchester railway rates as very dear, because, say, on iron, they included unlimited storage, and on cotton two months' free warehousing. Docks were not remunerative investments—those at Garston did not pay 2 per cent. Cross-examined, witness admitted that over and above the 12 mile radius, Manchester would get traffic from Scotch and east-country ports, and from Barrow, etc. ; also that he had not included Runcorn and Warrington traffic. He was amazed at Mr. Leigh, M.P., for Stockport, speaking of the import of Irish cattle. Witness admitted "there was a considerable trade, but not an ounce will ever go by the Ship Canal".

The evidence of other railway managers was by agreement taken as it had been given in the House of Lords Committee.

Mr. Bidder, Q.C., then addressed the Committee on behalf of the Dock Board. He could not disguise from himself that his task was a difficult one, because of the frank confession that the Committee did not take his views on the engineering question; but he was glad they had an open mind as regarded the estuary. The promoters were constantly changing their ground to suit circumstances :—

I know my learned friend is very indignant whenever we accuse him of bidding for a Bill, but he has developed talents as a Parliamentary huckster which are quite unparalleled. He is very anxious, and not unnaturally anxious, to get a Bill of some sort.

They wanted to go back with, at any rate, a measure of success, and possibly they deserved it for their pluck and perseverance. It was true they had reduced their abstractions from 6,000,000 cubic yards to 5,000,000 cubic yards by pushing the canal inland; also the dredging by 5 feet, and so the evil would be reduced, but only for a time. This simply modified the danger. He proceeded to combat Mr. Leader Williams' contention that the rocks at Pool Hall rendered a channel on the Cheshire side impossible, and asserted the Weaver and Gowy, entering at right angles, would impinge on the outflow of the Mersey, and joining it fix a channel to skirt the shore to Eastham and then join the new cutting from the Sloyne. He denied that in entering Liverpool Docks great delays took place, and said that in consequence of cross currents at the tidal openings the navigation of the canal would be dangerous; worse than that of the Suez Canal, where 12 per cent. of the ships stranded. He claimed to have discovered that at the first ebb of the tide the water would run into and down the Ship Canal, and so rob the estuary, and he severely commented on the absurdity of the flap boards proposed by Mr. Williams to remedy this. He could not understand an engineer venturing to commit himself to a scheme so novel, so wild, and so full of risk :—

The way to make a good work being put before Mr. Leader Williams, he goes out of his way and makes a bad one, and gains nothing by it, because all that nonsense about the vessels being taken in at all times is perfect rubbish.

The Committee were asked to incur a terrible risk and responsibility, for if the canal were allowed it would be practically useless, inasmuch as the currents in it would be such as no sensible shipowner would dare to encounter. The promoters had been warned, and had the opportunity of making a good scheme. By the present Bill there was a chance that Liverpool might be destroyed. He therefore asked for its rejection.

Mr. Pope, Q.C., would address the Committee on the commercial aspects of the case. As a man born and bred in Manchester, and for many years engaged in commerce there, he had a personal interest in the city, and could scarcely approach the question with the indifference of an advocate. However, it was his duty to deal simply with the evidence that had been given. For thirty-two days he had (like the late Lord Abinger—who gained his great success by trying to consider himself a juryman instead of an advocate) tried to imagine himself the fourth Committee man in this case.

The Chairman (slyly).—I have heard he obtained his success because he tried never to talk for more than half an hour, on the principle that every extra word drove out something he had said before.

Mr. Pope.—Lord Abinger had another great secret.   He would never cross-examine a witness when he found he was well up in his subject.

He himself would try and take the Chairman's hint and be brief.   His own opinion was that the scheme meant a wasteful expenditure of enormous capital, and that it would end in disappointment, injury and risk.   Comments had been made that this was the first time the Liverpool Chamber of Commerce and the Liverpool traders had come forward at an inquiry, but they had always been to the forefront in all efforts to cheapen freights, and now they were bearing fruit, inasmuch as they had a promise of a reduction in the rates, which no Ship Canal agitation could have accomplished.   Liverpool wanted to prevent the Bridgewater Canal getting into the hands of an adverse interest, and for £300,000 the Mersey could have been dredged and deepened.   If the Ship Canal got both navigations they would charge maximum rates in order to transfer traffic to their own canal.   He argued that the new undertaking, like railways, should be prevented from buying the Bridgewater Canal. Speaking for the railways, his fear was that the Ship Canal Company, when they found cash hard to raise, would buy the Bridgewater undertaking, improve the river and never make the canal.   The Committee ought to be satisfied it would save money and how much it would save.   Further, that the saving would continue and not be like the case of a battle of rates when freights go down to an abnormal figure and then go up higher than before.   Success in Manchester depended on being able to get return cargoes, and railways preferred carrying to Liverpool because they got more out of it.   Shippers now pay 2s. 6d. per ton out of their freight to deliver at Hull rather than go up the river to Goole; so it would be on the canal.   Mr. Marshall Stevens' idea of a cotton market in Manchester was one of the absurdities which only that gentleman would be capable of conceiving.   He declined to be handed over to the Mersey Commissioners and the decision of their Conservator, Admiral Spratt, to whose salary the Ship Canal contributed.   He knew his predecessor well, and a more agreeable "old lady" he never sat next to. Parliament had allowed the Dock Board to borrow close on £20,000,000 on bonds. Were they going now to damage those securities and destroy the trade and port of Liverpool?   Those who proposed to tamper with a great river ought to show clearly they were not endangering vested and valuable interests.

Mr. Aspinall, Q.C., was sure the tradesmen of Liverpool did not fear commercial competition. Call a meeting to-morrow to oppose on these grounds and not a hundred people would attend. They did fear the delay and suspense which would paralyse all other efforts to cheapen and benefit the port. He heartily endorsed the remarks of Mr. Bidder, and Mr. Pope.

Mr. Littler, Q.C., repeated his arguments in previous speeches. Capital would not be forthcoming, so the canal could not be made; but even if it were made, it could not secure the traffic to pay; and the cost of working and for maintenance would eat up the revenue. He brought forward numerous instances to prove his case. Mr. Marshall Stevens' evidence as to traffic was absolutely ludicrous from end to end. At his instigation the promoters had adopted a guess-work scale of half the previous charges. This would be found too low to pay, and then, as in many other cases, there would be an appeal to Parliament for permission to raise the scale. The Hull fiasco would be repeated; people who could not afford would put their money in the canal, but he would recommend no sane man to put a halfpenny in. He could not deny there were enthusiasts who believed in it, men with the faith of Mr. Adamson, who regarded disbelief in this scheme as almost as great a crime as disbelief in the Christian religion. To controvert the figures of Mr. Marshall Stevens, who conceived himself the apostle of the business, and fit to teach Mr. Findlay and Mr. Oakley, he showed that old ports like Hull, Grimsby and Bristol did not equal the business expected at Manchester. He vouched no South Yorkshire coal would come to the canal, because heretofore not an ounce had come by the Manchester, Sheffield and Lincolnshire Railway, and as for tea, Mr. Stevens' revenue of £1,600 for it was absurd. Mr. Littler went on to argue on behalf of the Trafford estate, that the stagnant water in the docks would render Trafford Hall unbearable, and be a danger to public health. He concluded by showing how easy it would be to dodge the authorities as to raising the £5,000,000 deposit and the £1,710,000 for the Bridgewater Canal, and he submitted that as the scheme was both doubtful and dangerous it ought not to proceed further.

Mr. Pember, Q.C., said the subject divided itself into (1) Abstraction, (2) danger of fixing a permanent channel in the Mersey, (3) danger to Garston, and asked if it was necessary to take up time by dealing with the question of actual abstraction? To this the Chairman replied, " I do not think, considering the value of time, that this need be dealt upon ". That being so, he would first deal with the Weaver question, it being the point on which Mr. Bidder had concentrated his attack. It was singular the many knights who now supported that gentleman should all at once have blossomed out on a subject on which at least one of them had been examined before, and professed he could not understand. Their contention was that as the tide rose higher, up the river, than at its mouth, it would at its ebb rush in by the openings and go down the canal instead of flowing back down the estuary; in other words, flow into and not out of the canal. It was absurd to say the Weaver water would not flow into the estuary through the side openings provided for them. No one doubted this would be the case when the receding tide had fallen to 14 feet or below it. All they could question was the top layers of the ebbing tide for the first hour or so. The bar could not be directly affected by Weaver water. The opponents' witness, Mr. Stevenson, had said that it is on an ebb tide and during its latter portion that the influence is most beneficial, and ought to be conserved. The Weaver weir was objected to as being vertical; this was valuable as giving force to the current. " Like as with a hunting man (which I know you are) a vertical fall means a horizontal position afterwards." True, in 1885 Mr. Lyster talked of passing the Weaver under the canal, but in 1884 he himself proposed regulating weirs. Mr. Pember then compared the evidence of Sir William Thomson and Mr. Deacon with that of Mr. Henry Law, and showed how their levels were at complete variance as regards the Weaver estuary. Further, he pointed out the admission by Mr. Deacon, that the closing of Mr. Leader Williams' lock gates to turn the tide would alter all his calculations.

As regarded the Eastham entrance from the Sloyne, they were simply going to put a tail to the tadpole; in other words, to add a narrow channel to connect the lock entrance with the deep water of the Sloyne, and this would be 2 miles below the Pool Hall rocks. If by any chance the channel could be carried to them, it must end there. It could not become a channel entirely on the Cheshire side. If such danger was to be apprehended, how came it that when the entrance to Garston was dredged no channel was formed on the Lancashire side? Mr. Pope's dictum "Good Sloyne—Poor Garston," and vice versâ, was not maintainable or true, for his own witnesses had clearly proved that they had been good and bad together.

Replying to Mr. Pope's speech, Mr. Pember expressed surprise that at the eleventh hour the Liverpool commercial element, who had hitherto professed to despise rather than fear the canal, should now come forward to oppose the Bill. He did not wonder that Sir William Forwood should give vent to his heat and spleen, using every weapon in his armoury against them, but he ought to remember

Birtles, Warrington.

WEAVER SLUICES.

To face page 314.

that he had said, "I have no objection whatever to this scheme being passed by this Committee as far as Runcorn". He thereby assented to the purchase of the Bridgewater Canal, which he now opposed. It would have suited Liverpool to have barges on the river with a maximum of 3s. 4d. against 2s. 6d. per ton on the Bridgewater Canal because they would have retained all their dock charges, even though it remained as now in the railway conference. Then it was said the Ship Canal would buy the Bridgewater Canal to destroy or transfer its traffic. Was it likely they would wish to do away with 1½ to 2 million tons of traffic (mostly local) that could not be transferred? How could the offer to deposit £5,000,000 before starting the canal be a sham? Were his clients a group of penniless, obscure, semi-fraudulent speculators? No! They were some of the most important and wealthy representatives of every trade in Lancashire, backed up by the chief Corporations. They had spent £150,000 in the struggle. Could this be for a toy? No! There had been nothing like it since the agitation for the Corn Laws. They meant to win the victory.

The Chairman.—A general may lose all his men in the victory.

Mr. Pember.—True, but he does not lose all his will to succeed, nor if his victory be costly does he value it the less. It is said by our opponents Manchester will never be more than a little local port, with a trade radius of 12 miles. The statement is absurd. Can you suppose Sir William Forwood, or the Liverpool cotton brokers, can stop trade going to Manchester when there is a good and cheap port there? Bristol, Glasgow and the Tyne ports have all prospered with small populations round them compared with Manchester. Is it not in the nature of things that Manchester, seated as she is, will concentrate trade? How came Liverpool to be the vast emporium she is? She has no manufactures, with a sparse population on one side and only the sea on the other. Manchester has groaned long enough under the burden of the old man of the sea, and she means to get rid of it. Even if she wastes her £8,000,000 it is her own affair, and nobody else's. Any way she must benefit the vast population round her. Mr. Marshall Stevens' 9,000,000 tons of traffic has been attacked on all sides, but remember this was to be the result of growth and gradual development. The promoters have only built for 3,000,000 tons, and they consider this should make the canal remunerative.

Now (said Mr. Pember) I have done, sir. If my purview of the case will not make you pass the Bill, provided you are satisfied about the engineering, I am afraid nothing will. If your nerves are to be shaken by the gravity of Mr. Bidder, or the impassiveness of Mr. Pope, I am afraid I cannot hope to overcome their influence. Heaven forgive me for seeming to be contemptuous of other men's methods, but I do despise all forms of personal appeal. Runcorn, Widnes, Warrington, the proprietors of all the different systems of inland canals, the corporations of Salford, of Manchester itself, the population of Lancashire, are all interested in the estuary of the Mersey and its bar. Their interests combined outweigh in magnitude many times those of the Liverpool cotton brokers, *et hoc genus omne*.   Why do you not find all those vast interests I have represented before you, as bound up in the Mersey, oppose this scheme?   Why do most of them actively support it?   If the danger was real—if the fear were shared by all, or any one of them, do you suppose the protest against this measure, with three long years for it to swim in, would not have rung out so loud that it would have drowned the voice of misguided enthusiasm?   But were the thousand-to-one chances, spoken of by Mr. Aspinall, to come about, would not all these vast and wealthy interests at once combine to rectify it?   Engineers can pierce canals from ocean to ocean at Panama and Suez; and the Americans shoot into the ocean like a bar of dirt the bar of the Mississippi, to which the bar of the Mersey is as a shovelful of dirt.   And if this Bill becomes law, and the canal is made, injury or no injury, shoaling or no shoaling, depend upon it the wealthy populations that will then have a hold upon the river will not sit down long and apathetically acquiesce in a condition of it, and its bar, which at present is a disgrace, though it is a disgrace to Liverpool alone.

The room was then cleared.   When counsel were called in again the Chairman said :—

The conclusion we have come to, I am very glad to say, is unanimous.   We consider the preamble proved, upon the following conditions : That the limits of deviation be so made use of that the canal will come upon dry land after entering the lock at Eastham ; that what Mr. Pember has two or three times stated would be possible should be carried out, namely, that the dredging should only be 12 feet; it was proposed to be 15 feet, but we consider it will be enough if it is 12 feet.   With regard to the capital clause (38), Mr. Pember made a suggestion, which, I believe, was only carrying out what had strongly occurred to each one of us—that this clause about the £5,000,000 of capital being raised ought to be entirely in-dependent of, and in addition to the purchase of the Bridgewater Canal.   We think that the time in which this capital is to be raised should be the time fixed for the Bridgewater Canal, instead of three years, two years.   We cannot consent to clause 42, that is to the alteration of the deposit.   There is only one other thing we wish to state, that is, that we think the clause for the protection of Ellesmere Port ought to be very fairly considered.

The succeeding two days the Committee met to discuss clauses.   Amongst other matters they decided the deposit for the works was to be impounded for three years.

BARTON AQUEDUCT DURING CONSTRUCTION.

Killon.

To face page 316.

CHAPTER XVI.

1886.

ATTEMPTS TO RAISE THE CAPITAL—BILL TO PAY INTEREST
OUT OF CAPITAL—ASSISTANCE OFFERED BY NEIGH-
BOURING TOWNS — ROTHSCHILD'S ATTEMPT TO RAISE
THE CAPITAL—FAILURE—GREAT DISAPPOINTMENT—
CONSULTATIVE COMMITTEE APPOINTED.

There is no cheaper traffic in the country than the traffic between Manchester and
Liverpool; it is a constant flow both ways.—Sir WILLIAM B. FORWOOD.

AFTER the passing of the Ship Canal Bill in 1885 came an interregnum,
broken only by attempts to ascertain what amount of support would be
accorded towards raising the necessary capital. Whilst everybody must
give unstinted praise to Mr. Adamson for the dogged persistency and admirable
fighting power displayed in securing the Bill, it cannot be denied that when he
stepped into the region of finance he was apt to get out of his depth. Over and
over again he alarmed his best friends by his prophecies of canal dividends, and by
his belief that the aid of capitalists was not essential, and that the masses of the people
would rush in and provide what was required. The beginning of the year 1886
found the Ship Canal directors fully occupied with devising plans to raise the capital,
and to assist them a Bill was promoted in Parliament to pay interest out of capital
during the construction of the works.

Its justification was that no great public undertaking had ever been carried
out without—either directly or indirectly—providing for payment of interest out of
capital during the construction of the works. The London and North-Western Railway, the Lancashire and Yorkshire, the Great Northern and nearly all the other railways had more or less enjoyed exemption from the onerous Parliamentary restriction
placed on the promoters. So had the Mersey Dock Board itself and the city of
Liverpool as regards its costly waterworks. Also the Metropolitan Board of Works.

(317)

The Government railways in India had interest during construction added to the capital cost, and so had many private companies.

Salford this session was asking Parliament for permission to take up shares in the canal to the value of £250,000.   Twice the Council, by a majority of about 5 to 1, recommended this course, but when in January, 1886, it came before the Borough Funds Meeting, Mr. James Heelis strenuously opposed granting civic help, and was told "he was an agent of the Liverpool Corporation".   In the end the Salford Bill was approved by an overwhelming majority, on which a poll was demanded.   This was taken some weeks later, with the result that 16,653 votes were given in its favour and 2,443 against it.

When the 1885 Prospectus failed to secure the necessary capital, it became evident that the assistance of London financiers must be sought, and as Messrs. Lucas & Aird, the intended contractors for the canal, had been used to work in conjunction with the eminent financiers Messrs. Rothschild & Sons, the directors of the Ship Canal decided to approach that firm with the view of obtaining advice and assistance. They were most courteously received, but were told that the help of capitalists could not be secured unless interest was paid during construction.   An immediate application to Parliament was advised, and hopes were held out that if permission to pay interest out of capital was obtained, Messrs. Rothschild would be able to pilot the proposed issue of £8,000,000 Ship Canal shares.   Hence the before-mentioned application to Parliament.

Opponents of the Ship Canal Bill were not idle.   They ridiculed the canal, and prophesied all manner of evil in order to frighten investors.   The *Liverpool Daily Post* said they were told millions of money were waiting in the pockets of Lancashire men ; these millions would flow in, and in a year or two great Atlantic steamers would be moored at Throstle Nest.   Now the Chairman admitted that only about £750,000 has been subscribed.   He added, it is true, that he had arranged with Messrs. Rothschild to find the remainder at 4 per cent. interest and 1 per cent. commission, and that it was also proposed to raise a million from shilling contributions of working men.   But before Mr. Adamson could do this he must have the consent of Parliament, and that consent, we need not say, was not likely to be given.   "We believe Messrs. Rothschild will think twice before fulfilling their bargain."   It was not to be supposed that they themselves wished to invest upwards of £4,000,000 in this gigantic ditch.

On the 1st of February, 1886, the First Ordinary Meeting of the Ship Canal

SIR JOSEPH LEIGH, DIRECTOR OF THE SHIP CANAL COMPANY;
CHAIRMAN OF THE BRIDGEWATER COMMITTEE, 1904 *SEQ.*

Lafayette, Ltd.   *To face page* 318.

Company was held. The Chairman, Mr. Adamson, after giving financial statistics, said the balance-sheet would have been less satisfactory had not most of the engineers and other professional men taken half their usual fees, and he reminded the meeting of Mr. Pember's words: "It is a question of endurance; if you stick to this and fight on, you will certainly get your Bill, you have established your position; you have promoted the strongest commercial case that has ever been presented to Parliament, and there is no doubt of your ultimate success". Speaking of an interview the directors had had with Lord Rothschild, he quoted a speech of that nobleman :—

Don't *you* believe that we are going to do anything but what is right and economical for you. We are satisfied that the negotiation of the finances of this great national enterprise will do our house great honour, and that we, with the strength of our name and associations, will be able to find you all the money that you require.

At this meeting Messrs. Daniel Adamson, Henry Boddington, Junr., Jacob Bright, William Fletcher, Richard Husband, Charles P. Henderson, Junr., Richard James, Joseph Leigh, James Edward Platt, Samuel Ratcliffe Platt and John Rylands, were continued in office as directors, and Sir Joseph C. Lee, John Rogerson (Durham) and W. H. Bailey were appointed new directors. Messrs. George Hicks and Bosdin T. Leech were elected shareholders' auditors. The remuneration of the directors was fixed at £2,000 per annum for the whole Board.

During the Ship Canal fight the company's engineer and solicitors courageously cast in their lot with the promoters and received no adequate remuneration. When the Bill was obtained their five years' persistent and laborious work was recognised, and a handsome honorarium was presented to them by the Board of Directors.

So anxious were trades unionists and working men of the district to support the canal that on their own initiative they issued early in February the prospectus of the "Co-operative Shares Distribution Company," the object being to enable persons to acquire shares in the Canal Company by weekly instalments of 1s. each. The leading trades unionists were directors and the Wholesale Co-operative Society acted as bankers. This was carrying out the pet idea of Mr. Adamson, but though a considerable sum was raised by working men it was soon evident that capitalists must find the bulk of the money.

On the 23rd February an influential deputation, introduced by the Lancashire members, waited on Mr. Mundella at the Board of Trade, to seek Governmental assistance for their payment of interest during construction Bill. After many speeches and explanations, Mr. Mundella gave the deputation to understand he was much impressed, and that their request would have his favourable consideration, though at the same time he reminded the deputation that both Mr. Adamson and Mr. Pember had repeatedly said the promoters could get the capital without such a clause.

As a kind of counterblast to the Manchester deputation, Mr. Mundella was waited upon by an influential body of Liverpool gentlemen, chiefly members of the Dock Board, Chamber of Commerce, etc. ; they were introduced by Lord Claud Hamilton, who urged the canal would do irreparable damage to the river Mersey : that the promoters were not likely to get their capital and might leave an unfinished work ; that the principle was a bad one, to pay interest out of capital, and that the plea of charity in finding work for the unemployed ought not to be considered ; he therefore asked Mr. Mundella to oppose the Bill on second reading. Sir William Forwood urged that when Liverpool wished to reduce railway rates Manchester held back, and said they had no grievance. Mr. Mundella replied Government would not oppose Liverpool having a *locus standi* on the Committee, but that they must take an impartial attitude on the second reading of the Bill.

At the beginning of March news came that Sir Henry Meysey Thompson had undertaken to move the rejection of the "Canal Interest out of Capital Bill," it was supposed on behalf of the railway companies.

Mr. Houldsworth moved the second reading of the Canal Bill on the 9th of March, recapitulating its history, giving the reason why the payment of interest clause had previously been withdrawn, quoting precedents, and stating it was an absolute necessity the clause should be passed in order to obtain the capital.

Lord Claud Hamilton then moved that the Bill be read again that day six months. He maintained the Ship Canal directors had broken faith with the Committee who passed the Bill, and sneered at the response that had been made to the first appeal for capital, telling a tale of the respectable Manchester artisan, who, being asked why he did not take shares, replied with a wink, "I shout for't canal but they sees none o' my brass". His honourable friends, Mr. Houldsworth and Sir Robert Peel who were anxious to speak, were like the artisan, ready to shout, but the canal saw little or nothing of their brass. As to the plea of impoverished Lancashire he pointed to Mr. Houldsworth and Sir Henry Roscoe (both well-favoured men), and asked, "Did they look as if they had passed through a time of privation or starvation?" But when Lord Claud went on to attack Sir T. Farrar for his Board of Trade report on the Bill, calling him the "Bismarck of Whitehall," Mr. Mundella sharply intervened and the speaker had to withdraw his remark.   Jeering at the £750,000 raised, he said, "Poor impoverished Manchester!"   He could assure them that they in Liverpool looked upon them with the greatest commiseration.   He contended that the payment of interest out of capital was bad in principle, and he asked, if the House decided to pass the second reading, that the opponents should be granted a *locus standi* on Committee.   Sir Henry Meysey Thompson seconded the rejection of the Bill.

Mr. Mundella defended Sir Thomas Farrar, and wished to state why the Government thought the Bill ought to receive the assent of the House.   He showed that whilst railways were now precluded by Standing Orders from paying interest out of capital, there was no such prohibition as regards canals, and even the House had relaxed in several cases as regards railways; indeed, many believed the standing order did more harm than good.   As regarded a *locus standi*, it rested with a Committee; the House could not interfere.   A canal, unlike a railway, could not be opened in lengths, and people were hardly likely to invest money that might not bring any return for seven years.   He could not agree to the Bill being sent to a Select Committee where the case might be reopened, and cause another large expenditure of money.

Mr. Courtney was in favour of a Select Committee if only the question of finance were dealt with.

Mr. Sexton intimated he should challenge the votes of all railway directors who took part in the division.

Ultimately the second reading was passed without a division, on which Lord Claud Hamilton moved for a Select Committee.   Many speakers urged that the Bill should take its ordinary course, and the proposition of the member for Liverpool was defeated.

For the motion	61
Against	375
Majority against	314

This conspicuous triumph was minimised by the Liverpool Press; they said Mr. Adamson had baited one of his cunning hooks with his usual astuteness, and caught the Irish members by making them believe the canal would regenerate the Irish trade.   They consoled their readers: "It is felt in all probability the canal will never be completed, and if it is, no great harm will be done.   Many people, on the contrary, are disposed to think that, should the big ditch be dug, increased trade will flow into Liverpool, and reduced rates will promote the prosperity of the district whatever they may do to the canal shareholders."

On the 25th March the Bill came before the Committee on Unopposed Bills, and being passed was ordered for a third reading.

At a meeting of the Liverpool Engineering Society about this time, Mr. Alfred Holt, of Liverpool, was very free in his condemnation of the canal. He had disbelieved all along in the Ship Canal project. He thought that the South Sea Scheme of 1711 was a sort of Bank of England scheme compared with the Manchester one for stability and security, and in his opinion canals would be eclipsed by improved railways.

On the Salford Bill (enabling that Corporation to subscribe £250,000 in aid of the canal) coming before the Local Government Board, an unfavourable report was made. It held the Committee of the House ought to decide if the docks, etc., would be a direct advantage to the borough, and then that they ought to be very fully informed as to the financial success of the undertaking; they must not support a speculative project. The London and North-Western Railway Company also petitioned against the Bill on the ground they would be taxed to support an opposition scheme. The report and opposition caused the rejection of the Bill when it came before a Select Committee of the Commons, presided over by Sir Edward Birkbeck.

The Interest Bill having passed the third reading in the Commons, the next move was to the Standing Orders Committee of the House of Lords. Here was a direct Standing Order against the payment of interest out of capital, passed at the instance of Lord Redesdale, which had already blocked one railway Bill, and the only hope was to get Parliament to vary the Standing Order and this, it was well known, Lord Redesdale would oppose.

Strange to say, just at this juncture death carried away the noble Lord, who for so many years had been a terror to all those who appeared before the Standing Orders Committee with any flaw in their Bill. The Duke of Buckingham took his place.

The Lords then appointed a Select Committee to consider the wisdom of varying the Standing Orders of their Lordships' House as to paying interest out of capital, and this Committee took evidence on the subject. A witness, Lord Rothschild, instanced good railways in India which never could have been made if the privilege had not been allowed, and quoted what he had told the Ship Canal Company.

Killon.

MR. PLATT'S YACHT *NORSEMAN*.

To face page 322.

"When they asked my advice, I told them that the only thing for them to do was to go to Parliament and get power to pay interest out of capital during construction." In his opinion the simple payment of interest would not induce the public to subscribe to an undertaking which was criticised in the Press for its unsoundness. As a result of the inquiry the Select Committee reported to the Lords that, subject to certain restrictions, and under peculiar circumstances, the payment of interest out of capital during construction might be allowed, one condition being that two-thirds at least of the share capital authorised by the Bill must be issued and accepted, and a certificate to that effect he obtained from the Board of Trade before commencing the works.

The Bill was read a second time without opposition and remitted to a Committee consisting of the Earl of Milltown (Chairman), the Earls of Ducie, Strathmore and Howth with Viscount Hood.   The London and North-Western Railway Company, the Dock Board and the Liverpool Corporation were the only petitioners against the Bill. Mr. Pember, for the Ship Canal, objected to the *locus standi* of the petitioners, but it was allowed by the Committee.   He then proceeded with the Bill on its merits, and stated that though the promoters' estimate of cost was £6,300,000, the greatest contractors in England, Messrs. Lucas & Aird, had offered to complete the works for £5,750,000.   This was confirmed by Mr. John Aird.   Sir Joseph Lee, Deputy Chairman of the Canal Company, went exhaustively into the canal figures, and was severely cross-examined by Mr. Bidder, who pointed out that of the £2,000,000 expected from Manchester only £750,000 so far had been promised, and that heretofore all railway companies who wished to pay interest out of capital had only been allowed to pay 3 per cent.   He also criticised the £60,000 commission to be paid to Messrs. Rothschild.   The witness explained that the interest to be paid was limited to £752,000, and that no more interest could be paid even if the canal took seven years in construction.   The Committee, after two days' hearing, passed the preamble and decided 4 per cent. might be paid.

Though feeling ran very strongly in favour of the Ship Canal, there were a few bitter opponents and passive resisters.   A Mr. W. H. Adams went so far as to refuse to pay the rate of 2d. in the £1 that had been raised in support of the Ship Canal. He paid all his other rates, and allowed himself to be summoned for 2s. 4d. by the overseers of Cheetham.   Though the defendant protested the rate was illegal, Mr. Headlam, the stipendiary, decided against him, on the ground that he should have appealed at the proper time, and made an order for payment.

About this time Mr. C. P. Scott, of the *Manchester Guardian*, was contesting

North-East Manchester, and was severely tackled for the hostility displayed by that paper towards the Ship Canal. He pleaded that criticism had done good, and secured a sounder scheme, and he claimed credit as an originator. He said that at an early stage in canal history one of the persons most interested in the scheme came to him and asked, "Who shall I get to work this Ship Canal?" He thought for a moment, and then he had a flash of genius—such things come to the biggest fools sometimes—and he said, "Go to Daniel Adamson". Now Mr. Daniel Adamson was the father of the Ship Canal, and he claimed in this matter to be the father of Mr. Daniel Adamson.

The Interest Bill having become law, the next step was to secure the capital. All the arrangements were left in the hands of Messrs. Rothschild. It was thought their great name would secure the sum required without much exertion in Manchester and the neighbourhood. The idea was, " Leave it to them, they best know how to go about getting the money". This was a fatal mistake, the more so as they were not given a free hand, and the policy of the directors was cheeseparing in its character.

Instead of paying Rothschilds for underwriting the whole issue, the directors thought they themselves could raise £2,000,000, and arranged to pay 1 per cent. to Rothschilds as commission for raising the remaining £6,000,000. Could this have been carried out it would have been a cheap issue of stock. But the 1 per cent. to Rothschilds did not give them scope to enlist the Stock Exchange brokers on their side, they could not afford to pay them a good commission and make each broker an agent for securing capital. When the Ship Canal stock came on the market, the whole of the London and Provincial share markets were hostile to it, and besides, there came the rumour that the £2,000,000 expected in Manchester had only been subscribed for to a very limited extent. It would seem that while Messrs. Rothschild have a world-wide fame for raising British and foreign loans, they were scarcely used to finding capital for Lancashire companies, and it was felt that closing the list in four days gave too little time for investors to make up their minds.

Anyway, it was evident from the first moment the Ship Canal stock was put on the market that there was a dead set made against it by powerful hostile interests. The shares, which were at first quoted at a premium, quickly went down to par, and the news circled round the London Stock Exchange, " the thing's a dead failure ".

Jobbers and brokers interested in railways did their best to block the canal issue, and others rather prevented their clients subscribing to the capital. On the third day, seeing they could not stem the opposition, Messrs. Rothschild & Sons withdrew the stock and made the following official statement :—

The directors of the Manchester Ship Canal Company, having been advised by Messrs. N. M. Rothschild & Sons of the number of shares applied for, have decided to withdraw the issue for the present, the amount being less than is required by the Acts of Parliament to enable the company to proceed with the construction of the canal. The sums deposited on application will therefore be returned without delay.

It was an intense disappointment to Lancashire; only a few days before, Parliament had refused to let Salford contribute £250,000 and so had prevented other towns assisting. Now the canal was blocked for want of support. Too late the directors saw their mistake in not having the stock underwritten and in not having been more liberal with commissions and thus enlisting the help of the stock market. They determined to take breathing time before making a second appeal for capital. Of course all kinds of reasons were given for the fiasco. It was said there was much local irritation, and the Ship Canal officials had offended many subscribers by breaking faith, and compelling them to apply for shares in the ordinary way instead of granting shares up to twenty times the amount of the subscription. The financial papers gave three reasons for the failure : The directorate was not strong enough; the terms offered were not good enough; the wheels of the Stock Exchange had not been sufficiently oiled—no large commissions had been offered, and there were not sufficient pickings to be made.

At the July meeting of the Manchester, Sheffield and Lincolnshire Railway the Chairman, Sir Edward Watkin, who had hitherto been very magnanimous in his view of the canal, made a strong attack on the Ship Canal Board. He had not helped to oppose the undertaking, and he was in favour of improved water communication, but he believed the design was bad. The Ship Canal ought to have no locks, and the bar should be removed prior to an improved water communication. He was in favour of cutting through the Birkenhead peninsula and making a new way to deep water. The Ship Canal was a great delusion, and was about the worst engineered business he ever remembered. The scheme had been a failure, and would be a failure, and he ventured to predict the promoters never would get the capital. If they did, their works would close up the port of Liverpool. The whole scheme had been a gigantic mistake, and the public had been deceived. Not very encouraging words, and they show how dangerous it is even for a great railway magnate to indulge in prophecy.

The one cheering aspect for the directors was that however outsiders might deride and condemn them, the masses of the people remained faithful to the canal. Any quantity of offers were received in the way of small sums, accompanied by most encouraging letters, many working men promising to make weekly payments in aid of the canal, and others to give up luxuries and save the money in order to possess an interest in it.

On 31st August the Second General Meeting of the Shareholders was held in St. James's Hall, and this was looked forward to with much interest, as then would have to be decided the question "What next?" The report of the directors, after recounting the circumstances connected with Messrs. Rothschilds' attempt to issue stock, stated they had conferred with the Mayor of Manchester, Alderman Goldschmidt, and other prominent citizens, and that under their advice they had appointed a Consultative Committee with the Mayor as Chairman to take into consideration the whole scheme of the Ship Canal, including the obtaining of the capital, the strengthening of the position of the company in public estimation, and any other matters affecting its future, and to report to the Board. This Committee was to be entirely independent of the Board, and to consist of leading business men of influence and position, also of landed proprietors whose opinion would command respect in the district. It mattered not if they had previously been opposed to the canal. A report from such a Committee (if favourable) would doubtless have great weight with the public, and render valuable assistance to the directors in securing the necessary funds.

The Chairman, *Mr. Adamson*, regretted that after making satisfactory arrangements with Messrs. Lucas & Aird they had failed to get the capital. They must now try and get the help of the great commercial men of Lancashire, and the large property owners. Also they must show traders that the canal was indispensable to Manchester and Lancashire. The Consultative Committee would sift the Ship Canal scheme to the bottom, and if, as he hoped, the report was favourable it ought to give confidence. There was plenty of money in Lancashire, and if the county did its duty, and raised two and a half to three millions, he had no fear about London providing the remainder.

Sir Joseph Lee had been told by Messrs. Rothschild that it was essential Manchester should show her confidence in the canal if money was to be got in London. So far she had not done so. They had had a great many warm friends, but still in some quarters the scheme had been damned with faint praise. They had placed gentlemen on the Consultative Committee who had been antagonistic to the

To face page 326.

MASONRY FOR ACTON GRANGE VIADUCT.

Birtles, Warrington.

canal, and therefore the report, when it came out, ought to carry force. The ground must be made very sure before they built a new structure. They had been dazzled by the great name of Rothschild; now they must put their own shoulders to the wheel. Adversity had taught them a good lesson, and when the next prospectus was brought forth, he hoped to have the support of the merchants, landowners and the working people of Manchester.

Mr. Houldsworth urged the directors to keep pegging away, and the promoters to stick to their guns. They must not leave a single stone unturned in order to carry the canal.

Mr. Jacob Bright was absent, but wrote: "You know how strong is my conviction of the necessity of this enterprise. It has triumphed over many and great difficulties, and will, I believe, triumph over this last difficulty."

Amongst the earliest to show their renewed confidence in the canal were Mr. Adamson's own workmen. After an address by Mr. Digby Seymour, Q.C., they decided to work two hours per week extra, and that the money received should be invested in the canal to form a kind of trust fund to be managed by a Committee.

The idea of shilling coupons being issued to working men, who could convert them into £1 shares, found much favour with Mr. Adamson, Alderman Bailey and others, whilst it met with strong opposition from other earnest supporters of the scheme. Mr. Ellis Lever asserted "he had no faith in the proposal to raise £5,000,000 by the weekly shillings of working men"

From a Parliamentary return of expenses incurred for Private Bills, it would appear that the promoters of the Ship Canal Bill expended in 1883-84 and 1885 £146,500; to this must be added local expenses, £26,000, making a total cost of £172,500. This did not include the 1886 Bill. It is very probable the numerous opponents spent in the three years as much as the Ship Canal Company. If so, the fight must have cost both sides £345,000.

In the early part of November the heart of Liverpool was stirred by a letter from Mr. Ismay, the Chairman of the White Star Line. He was not satisfied (like Mr. Hornby, Chairman of the Dock Board) with getting over the bar for a limited time twice in twenty-four hours, but maintained it would be quite possible to follow the example of New York, and so improve the bar that at the lowest tide there should never be less than 30 feet of water on it. He advocated also an overhead dock railway, and (greatest change of all) that dock and town dues should be swept away and Liverpool made a free port. But of course these changes were too drastic for the Dock Board to entertain. "Some years back," wrote Mr. Russel Aitken, "I spoke to Captain Graham Hills, marine superintendent, about the scandal of 'the bar' which impedes trade, destroys shipping, and adds another horror to a sea voyage, when he replied to me that it was an utter impossibility to remove the bar, and that I was a madman for proposing such a work."

The Consultative Committee, mentioned by the Chairman as about to be formed at the instance of the directors, consisted of the following gentlemen :—

> P. Goldschmidt, Mayor of Manchester (*Chairman*).
> Charles Moseley, Merchant (*Deputy Chairman*).
> Earl Egerton of Tatton.
> John Alexander Beith, Shipping Merchant.
> John K. Bythell, Shipping Merchant.
> C. P. Scott, Editor, *Manchester Guardian.*
> James F. Hutton, Merchant.
> W. E. Melland, Manufacturer.
> Samuel Ogden, Cloth Agent.
> Robert Bridgford, Estate Agent.
> Alfred Butterworth, Cotton Spinner.
> Hilton Greaves, Cotton Spinner.
> George Robinson, Shipping Merchant.
> Alfred Crewdson, Merchant.
> Henry Theodore Gaddum, Silk Merchant.
> William H. Holland, Cotton Spinner.
> Charles J. Galloway, Machinist.
> Alexander Ireland, Newspaper Proprietor.
> G. F. Fisher, African Merchant.
> Thomas Sowler, Editor, *Manchester Courier.*
> C. S. Carlisle, Merchant.
> James Leigh, Cotton Spinner.
> James Maudsley, Labour Representative.

A more influential and representative body could scarcely have been selected. Several of the members were known to entertain an unfavourable opinion of the canal, and inasmuch as they were to make an exhaustive inquiry and had power to call any witnesses they liked, it was felt their report would either mar or make the undertaking. If they pronounced the scheme sound, it would materially help to restore public confidence, which was absolutely necessary in order to obtain the capital.

Singular to say the Consultative Committee was almost forced on Mr. Adamson

CHARLES MOSELEY, DIRECTOR OF THE MANCHESTER SHIP CANAL
COMPANY.

Van-der-Weyde, London.   *To face page* 328.

—who never liked it—by some of the farseeing directors who felt sure that in order to gain the support of capitalists it would be necessary that the canal scheme in all its bearings should be thoroughly sifted by reliable and impartial men, in whom the public had perfect confidence.

The Consultative Committee sat at intervals for several weeks, and made a most painstaking inquiry into the facts and figures on which the directors had built up their statement that the canal could be made and would pay. They called for evidence on different points, narrowly examined many witnesses, and presented their report on the 26th November. On the 9th December a meeting of the promoters was held in the Mayor's Parlour to hear the unanimous report of the Consultative Committee, which was signed by Alderman Goldschmidt and the rest of his colleagues.

REPORT OF THE CONSULTATIVE COMMITTEE.

1. That the Ship Canal and works are practicable from an engineering point of view, and would permit vessels of the largest class to be safely navigated to and from Manchester.

2. That the canal and works can be completed ready for traffic at a cost within the estimate of £5,750,000, and that the sum of £802,936 set down for the purchase of the necessary land is a safe estimate.

3. That the canal and works should be constructed under a contract fixing a maximum sum, in order to prevent any possibility of the estimate being exceeded.

4. That graving dock accommodation should be provided, and be ready for use on the opening of the canal.

5. That the estimate of £104,200 per annum for the expenses of the working and maintenance is ample. There would probably be a material saving on this item during the first few years after completion.

6. That the acquisition of the Bridgewater undertakings for the sum of £1,710,000, fixed by the Act of Parliament, would be an advantageous purchase for the Ship Canal Company, and that the present average net income of £60,000 would, after the completion of the works, continue to be derived from the Bridgewater Canal and those other portions of the Bridgewater properties not required for the construction of the Ship Canal.

7. That the capital powers of the company, under their Acts of Parliament, amounting to £9,812,000, are sufficient for all the purposes contemplated by their Acts.

8. That the project is a thoroughly sound commercial undertaking, and would speedily become remunerative on the completion of the works. That a large amount of traffic would be at once secured, and that thereafter the increase in traffic and revenue must be steady and continuous.

Having regard to the responsible character of this report, we feel it our duty to state that in our judgment it is necessary, before the issue of any further prospectus, to reconstitute and greatly strengthen the Board of Directors.

Dated this 26th day of November, 1886.

P. GOLDSCHMIDT, *Chairman.*

CHARLES MOSELEY, *Deputy Chairman.*

Etc., etc.

Alderman Curtis, Mayor of Manchester, was in the chair.

Mr. Charles Moseley, before reading the report, said the Consultative Committee had felt an enormous responsibility rested upon them, but he was pleased to say they had come to a unanimous opinion.

Mr. Oliver Heywood and Alderman Harwood thanked the Consultative Committee for their unremitting labours. Knowing the critical and searching inquiry that had been made by gentlemen of the highest standing in the city, Alderman Harwood said he had more confidence in the report than he would have had if it had come from any assembly of gentlemen in London, however highly endowed they might have been. Mr. Charles Moseley paid a high compliment to the work of the old directors, but suggested the necessity of strengthening the Board with men who stood high in the financial world. In the past there had been too much saying, " It is a very good scheme, but I am going to leave some one else to carry it out". Now if it was to be done, that was not the way to do it. They must all lend a helping hand.

Mr. Jacob Bright would not believe the enterprising spirit of Manchester was decaying. After the favourable and unanimous report of a remarkable Committee, he did not think the financial difficulties in the way of the scheme would henceforth be too great. If Manchester—equal in its spirit of enterprise to any part of the country—did not accomplish that work, he should be very much surprised.

The large meeting enthusiastically received the report, and the subscribers left with lighter hearts and with a determination to raise the capital.

To show the earnestness of their convictions the Consultative Committee agreed to put their names down for £68,200, Mr. Hilton Greaves, of Oldham, heading the list with £20,000. The directors had subscribed to the extent of £136,250, of which Mr. John Rylands was responsible for £50,000.

No one took a more intelligent interest in the Ship Canal than did our local

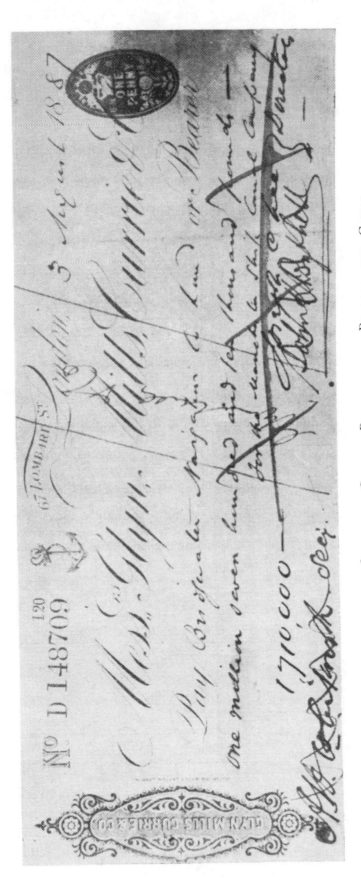

FACSIMILE OF THE £1,710,000 CHEQUE PAID FOR THE BRIDGEWATER CANAL.

To face page 330.

poet, Edwin Waugh.   He was anxious to give it a help, and through the kindness of Sir Leader Williams I am able to give a facsimile of a letter which the
poet wrote when he submitted his verses to Sir Leader for his approval and revision.
It will be noticed that a few minor alterations were made.   The song was afterwards
published in the second series of Waugh's poems, and can be sung to the air of
"Auld Lang Syne".

Church Road,
new Brighton,
Cheshire,
5th October, 1886.

Dear Williams,

How will this do, — for
a start?   Please let me
know, by return   I
think, if it had a popu-
lar tune, the chorus
might
would go.

Yours faithfully,
Edwin Waugh

E Leader Williams, Esqr. C.E.

When the ships come sailing in

By Edwin Waugh

God prosper long the good old town,
  The toilful and the free;
For she has bravely broken down
  The toll-bar of the sea:
And now the victory is won,
  For which we fought so long,
To all the wide world thus shall run
  The burden of my song:—

      Let it float in free, from the open sea;
        Kind brotherhood shall win: town
      And the good old town shall smile again,
        When the ships come sailing in!

The ocean to mankind belongs;
  You cannot take its waves:
'Tis the stormy highway of the strong:—
  With free delight it laves
The shores of earth's far-sundered lands;
  And, on its heaving breast,
With equal pride, from distant strands
  It brings what each yields best

      Let it float in free, from the open sea;
        Kind brotherhood shall win;   town
      And the good old town shall smile again,
        When the ships come sailing in!

The bounteous gifts of nature range,
  Each on its favoured shore;
The wide world is man's great exchange,
  His market and his store!
For kindly harmony designed,
  Earth's varied fruits are sent;
For mutual benefit combined,
  And friendly commerce meant.

      Let it float in free, from the open sea;
        Kind brotherhood shall win;
      And the good old town shall smile again
        When the ships come sailing in!'
                            town

NAVIGATION TO MANCHESTER.

CURIOUS EPITAPH.

Near the west door of St. Elphin's Church, Warrington, is a gravestone inscribed as under :—

This grave is not to be disturbed after the interment of John Leigh.

    The old Quay Flats was my delight;
    I sail'd in them both day and night.
    God bless the Masters, and the Clerks,
    The Packet people, and Flatmen too,
    Horse drivers, and all their crew.
    Our sails are set to Liverpool;
    We must get under way—
    Discharge our cargo, safe and sound,
    In Manchester Bay.
    Now all hands, when you go home,
    Neither fret any, nor mourn;
    Serve the Lord where'er you go,
    Let the wind blow high or low.

Mary Leigh, his sister, died Oct. 6th, 1801, aged 29 years.
Betty, Mother of John and Mary Leigh, died 6th May, 1826, aged 88 years.

    To our God let us pray—
    Keep us from drunkenness and wickedness both night and day.

This stone and grave is free gift of John Yates, Mariner, Captain of the Old Quay Packet.

    God bless all British sailors, Admiral Nelson, and all the English Fleet;
    When we must go, we do not know, sweet Jesus Christ to meet.

THE ABERDEEN UNIVERSITY PRESS LIMITED

Printed in the United States
By Bookmasters